D0501508

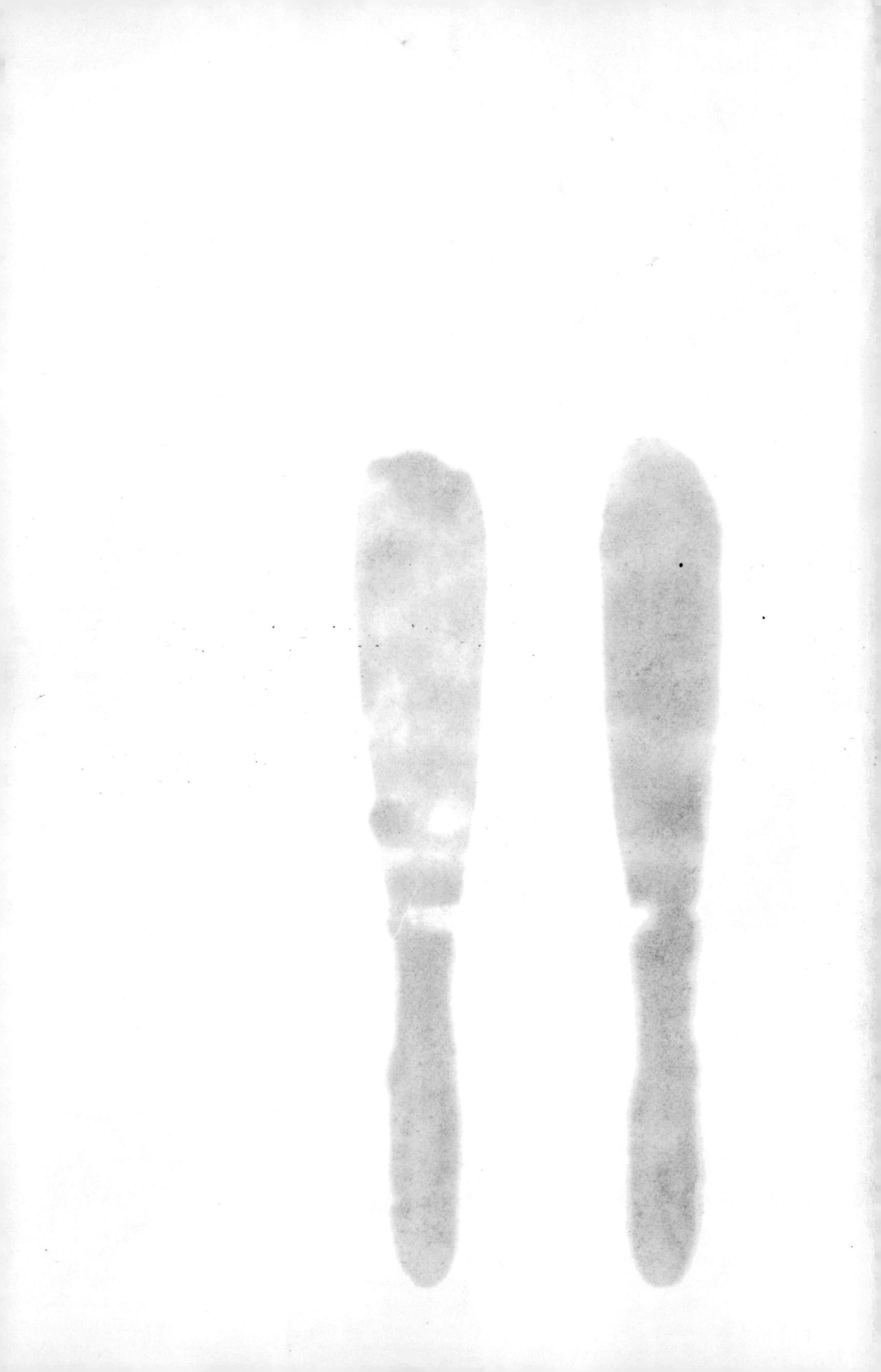

ENERGY SAVING LIGHTING SYSTEMS

ENERGY SAVING LIGHTING SYSTEMS

Prafulla C. Sorcar, P.E.

Butterweck-Sorcar Engineering
Denver, Colorado

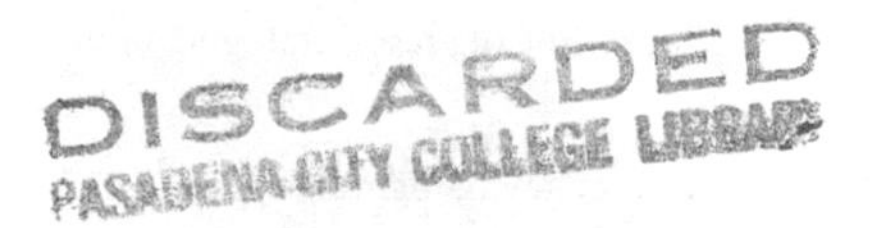

VNR VAN NOSTRAND REINHOLD COMPANY
NEW YORK CINCINNATI TORONTO LONDON MELBOURNE

Library of Congress Catalog Card Number: 81-19773
ISBN: 0-442-26430-5

Manufactured in the United States of America

Published by Van Nostrand Reinhold Company Inc.
135 West 50th Street, New York, N.Y. 10020

Van Nostrand Reinhold Publishing
1410 Birchmount Road
Scarborough, Ontario M1P 2E7, Canada

Van Nostrand Reinhold Australia Pty. Ltd.
17 Queen Street
Mitcham, Victoria 3132, Australia

Van Nostrand Reinhold Company Limited
Molly Millars Lane
Wokingham, Berkshire, England

15 14 13 12 11 10 9 8 7 6 5 4 3 2 1

Library of Congress Cataloging in Publication Data

Sorcar, Prafulla C.
Energy saving lighting systems.

Includes index.
1. Electric lighting. 2. Electric lighting–Energy conservation. I. Title.
TK4188.S59 621.32′2 81-19773
ISBN 0-442-26430-5 AACR2

To my loving daughter Payal

Preface

Ever since the oil embargo of 1973, energy saving has become the most important concern of this nation. Lighting, being a key user of electricity, has been one of the rallying points for mandatory standards to reduce energy consumption.

As of the writing of this preface, according to the U.S. Department of Energy, 93% of the resources used in this country are fossil fuels, i.e., coal, oil, and natural gas. Of these, oil and natural gas are considered the critical fuels since their total world reserve is quickly diminishing and they are nonreplenishable. Of the total resources consumed, approximately 38% is used for generating electricity. Individually, it consists of 64% of total coal, 20% of total oil, 18% of total natural gas, and 100% of hydro and nuclear fuels. From a critical-fuel standpoint, 19% of the total oil and natural gas is used for generating electricity.

Of the total electrical energy, approximately 20% is used to produce artificial light. This means that $0.38 \times 0.20 = 0.076$, or 7.6% of the total energy, and $0.19 \times 0.20 = 0.038$, or 3.8% of the total critical fuel, are being used for producing light today.

Although this percentage of energy usage is small, the conservation of lighting energy is important for two primary reasons: (1) For the currently existing conditions, we must look at all sections of energy use for contribution to energy conservation, and (2) with its efficient use, the operating cost will be minimized. Currently, a survey shows that, in terms of final end user, lighting alone represents 30-50% of the energy cost of a building. As utility rates are skyrocketing, the impact of lighting on operating cost is becoming painfully apparent.

In order to reduce energy consumption and cut operating costs, many different suggestions and techniques have been introduced and practiced in the past few years. One such easy solution suggested is the reduction of the lighting level. Reducing the lighting level certainly saves energy; however, this is not the most effective means. Many early attempts by the government to cut lighting energy by reducing light levels proved counterproductive. The psychological effect on workers and the system reliability are far more important than the marginal energy savings that may be achieved by such means. The key to energy savings thus relies first, on providing adequate visibility with quality illuminance and, second, on achieving this with the minimum consumption of power. It is the purpose of this book to investigate various practical means of attaining this goal.

Special effort has been made to make the book a practical guide, in plain, concise English, for lighting design and decision making. Theoretical information is kept only to the extent where it aids the practical approach. It is written in standard U S. measurement systems because of their popularity. Tables of conversion to metric (SI) systems given at the end of the book will be helpful during transition of one system to another. To aid the beginners, a glossary of terminology and a table of frequently used illuminance levels are also included.

There are nine valuable chapters in the book. The first deals with recognizing the complete range of light sources and their accessories. It extends from incandescents through low-pressure sodium, the extremes of light sources in power consumption. Chapter 2 explains fundamentals and shows how to understand and read photometric reports. Chapter 3 shows the step-by-step procedures of the basic techniques of lighting design, stressing features that have significant impact on energy consumption. Chapter 4 is dedicated to lighting design with fluorescent luminaires. Since more than 80% of today's lighting load is carried by fluorescent lighting, I felt that a thorough discussion was needed of the technical and design aspects of fluorescent lamps, ballasts, and luminaires. It is my belief that this is the first book of its kind that discusses the reasons that a direct application of manufacturer's published "bench-test" data may be invalid in making accurate calculations and explains how the impact of ambient temperature and ballast factor may substantially alter results. Chapter 5 makes an in-depth analysis of industrial lighting systems and design as involved with all types of gaseous discharge sources, including fluorescents, mercury vapor, metal halide, and high- and low-pressure sodium. With numerous graphs, illustrations, and sample problems, step-by-step procedures are shown for a most energy-effective lighting design. Chapters 6 and 7 illustrate the various means of manual and automatic lighting energy-control devices, give an understanding of power factor because of lighting, and show how to handle state-adopted energy conservation standards. Chapter 8 follows the theme of Chapter 4 and deals with the impact of air-conditioning on light output and input power, and vice versa. Chapter 9 is dedicated to exterior lighting. Lighting terminology associated with exterior lighting is explained, and sample problems and illustrations are given for making a lighting design with high-intensity discharge sources and low-pressure sodium.

In all applicable areas, the industry's latest products have been discussed in details to evaluate their merits and the areas where they fit best. Special attempts were made to provide sufficient information so that a designer can feel confident that a lighting design developed within the book's guidelines will be most energy saving and effective.

I sincerely believe this book will be of considerable value and assistance to many technical and nontechnical people involved in energy saving lighting design and making decisions. Students working towards an illuminating, electrical, or architectural degree, consulting engineers, designers, electrical and general contractors, mechanical engineers on HVAC design, architects, interior designers, and maintenance engineers will benefit directly from this book. College and university physical plant administrators, engineers in government agencies, e.g., the Army, Navy, Air Force, General Service Administration, and state and municipal engineering departments, will find this book extremely useful.

I am obliged and grateful to many individuals who provided helpful suggestions and constructive criticism in completing this book. I sincerely thank John E. Kaufman, Technical Director, Illuminating Engineering Society of North America, for reviewing parts of the manuscript and providing many valuable suggestions. My sincere thanks to Terry K. McGowan, Illuminating Engineer, General Electrical Co., for reviewing the first chapter of this book and providing help anytime I needed him. I thank Robert E. Faucett, President, Independent Testing Laboratory, Boulder, Colorado, for reviewing Chapter 2 of this book.

My business partner, Howard W. Butterweck, deserves a special thanks for reviewing the complete manuscript and providing helpful suggestions.

PRAFULLA C. SORCAR

Contents

ENERGY SAVING LIGHTING SYSTEMS

Chapter One
Selection of Light Sources

The different light sources popularly used today are as follows:

Incandescent
Fluorescent
Mercury vapor
Metal halide
High-pressure sodium
Low-pressure sodium

Of these, except for incandescents, all light sources can be termed as gaseous discharge sources. Fluorescent and low-pressure sodium (LPS) light sources operate on low-pressure gaseous discharge, and the mercury vapor, metal halide, and high-pressure sodium (HPS) operate on high-pressure gaseous discharge. Mercury vapor, metal halide, and HPS are commonly known as the high-intensity discharge (HID) light sources.

INCANDESCENT

Principles of Operation and Types

Incandescent lamps produce light by electrically heating high-resistance tungsten filaments to intense brightness. Tungsten is still the most favorable filament since it has high melting point, low evaporation, high strength and ductility, and favorable radiation characteristics. The filaments are of different form and they are designated by a letter as follows: *s* – straight; *c* – coiled; *cc* – coiled coil; *r* – flat or ribbon. Coiling the filament increases the efficacy. For this reason, among the lot, the coiled coil filaments have the best efficacies. The cold resistance of tungsten is lower than its operating resistance. Thus, when the lamp is energized, until the filament builds up its operating temperature, there is a large inrush current at first, for a fraction of a second. See Figure 1-1 for a typical incandescent lamp interior.

There are different types of bulbs for incandescent lamps. The bulb shapes are designed by letters as follows (see Figure 1-2): *A* – standard; *C* – cone; *F* – flame; *CA* – decor; *G* – globular, *GT* – chimney; *P* – pear; *PS* – pear straight; *T* – tubular; *PAR* – parabolic reflector; *R* – reflector. Of the lot, the type A lamps are the most commonly used. The PAR- and R-type lamps are also popular for commercial applications. In general, the bulbs are clear, and when they are clear, they are the best for lighting control when inserted in lighting equipment. Some bulbs are inside-frosted to diffuse the concentrated source. The silica-coated bulbs have a fine coating of silica smoked on the inside surface. This extra coating substantially diffuses the light from the bulb and eliminates the "hot spots" of the inside-frosted lamps. Because of this extra coating, the total light output is reduced by approximately 1%. Over a period of use, the filament evaporates. To prevent this evaporation, 40-watt and higher wattage lamps are gas filled. A filling of inert gas, such as argon and nitrogen, retards evaporation of the filament, thus allowing higher filament temperature and, hence, higher efficacies with less sacrifice of life. Bulbs under 40 watts are evacuated. No inert gas is inserted in these lamps since any gain in efficacy would be offset by loss of energy through conduction of the inert gases.

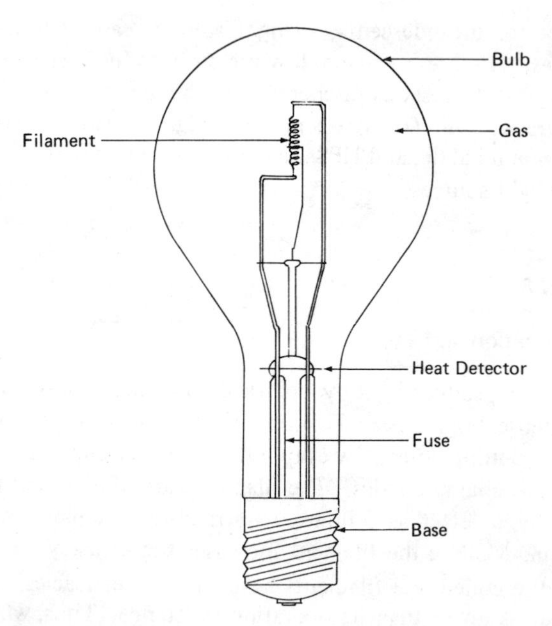

Fig. 1-1. Typical construction of an incandescent lamp.

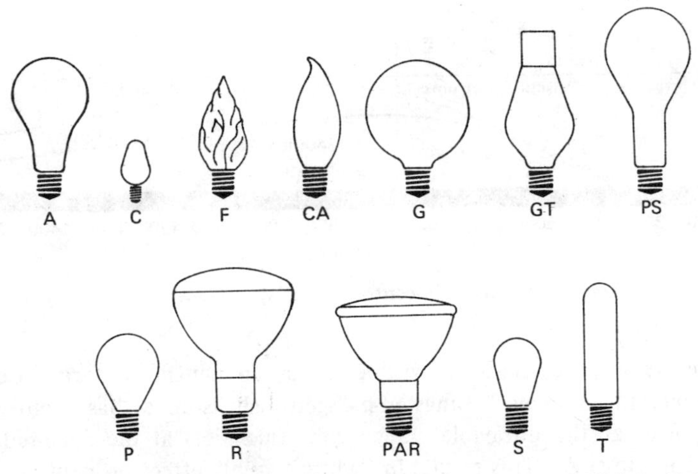

Fig. 1-2. Bulb shapes of incandescent lamps. (Courtesy of GTE Sylvania.)

Another type of incandescent light source is the tungsten-halogen lamp, which is also known as a quartz lamp. These lamps, which contain a halogen gas such as iodine or bromine, operate at high temperature; the outer envelope or the bulb is made of quartz to withstand the heat. The main advantage of these lamps is their lumen-maintenance superiority over the conventional incandescents. At high temperature, the evaporated tungsten associates with a halogen molecule, which prevents it from getting deposited on the bulb wall and instead returns the evaporated tungsten back to the hot filament. To obtain the high temperature, the outer envelope is made smaller.

Color Characteristics

The optical system of a human being responds differently to various wavelengths of radiation. The brain translates these different wavelengths as color. A graphic presentation of the energy emitted by a light source at each wavelength in the spectrum is called a Spectral Power Distribution (SPD) curve. The SPD data are derived through a spectrometer.

Figure 1-3 shows the SPD curves for a typical incandescent source. Incandescent light source, being a thermal emission type, obeys the established physical laws of thermal emission. As can be seen in the figure, the smooth curves originate at near the ultraviolet (short wavelength) range and extend all the way up to the infrared range (long wavelength). Curve A represents a conventional incandescent lamp, which has higher radiant power in the red, orange, and yellow areas than at the ultraviolet range. This phenomenon gives the

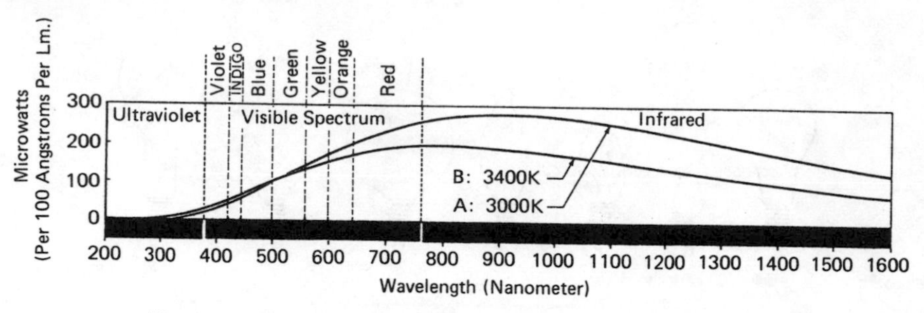

Fig. 1-3. SPD of incandescents. (Courtesy of General Electric.)

incandescents a yellowish-white visible effect on neutral surface. Curve B, which represents a typical tungsten-halogen light source, has slightly more radiant power at the ultraviolet range and much less at the deep-red range, compared to curve A. This results in "white" visible effect on neutral surface. The peak radiant energy actually takes place in the infrared zone. Table 1-1 shows a comparative color performance between a conventional incandescent and that of a tungsten-halogen lamp. An incandescent lamp utilizes about 10% of its total energy for visible radiation. This is shown in Figure 1-4.

Efficacy

In general, incandescent lamps have efficacies varying from 17 to 23 lumens per watt, for up to 500 watts. Higher wattage lamps, e.g., 1000-watt through 10,000-watt lamps, tend to have better efficacies, varying from 23 to 33 lumens per watt. For practical purposes, however, tungsten filament incandescent appears to have a practical efficacy limit of about 40 lumens per watt. The trend, in general, is better efficacy as the wattage goes higher. Table 1-2 shows the initial efficacy of different types of incandescent lamps.

Table 1-1. Incandescent Lamp Selection, by Color.

Lamp Type	Visible Effect on Neutral Surface	Effect on Atmosphere	Color Strengthened	Color Grayed	Effect on Complexion
Incandescent Filament	Yellowish white	Warm	Red, orange, yellow	Blue	Ruddy
Tungsten-Halogen	White	Warm	Red, orange, yellow	Blue	Ruddy

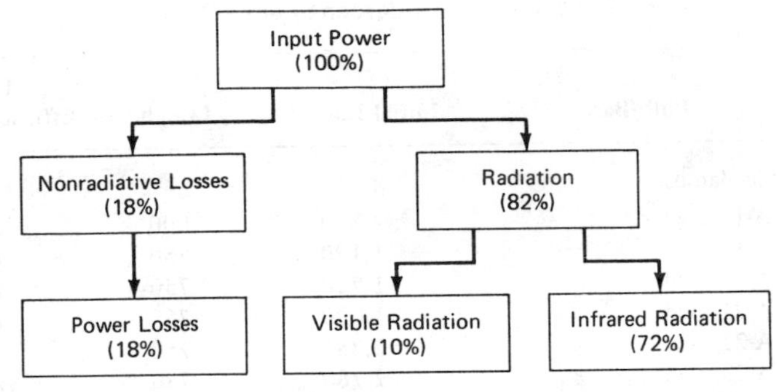

Fig. 1-4. Energy distribution of an incandescent lamp. About 10% of energy is available for visible radiation.

Effect of Burning Position and Depreciation

Burning position of incandescent lamps has no effect on generated lumen output. However, when incandescent lamps are burned over a period of time, the filaments evaporate and become smaller. This evaporation increases the resistance of the filaments, causing reduction in ampere, watts, and lumen output. Figure 1-5 shows the effect of lumen output, the wattage, and the efficacy of different general-purpose lamps for their rated life.

Another significant amount of lumen reduction is caused by absorption of light by deposits of the evaporated tungsten on the bulb. Deposits of evaporated tungsten cause the lamp to blacken inside the bulb. In vacuum lamps (of less than 40 watts) this blackening is uniform over the bulb. In gas-filled lamps the evaporated tungsten particles are carried by convection currents (chimney effect) to the upper part of the bulb. In this respect, the burning position of

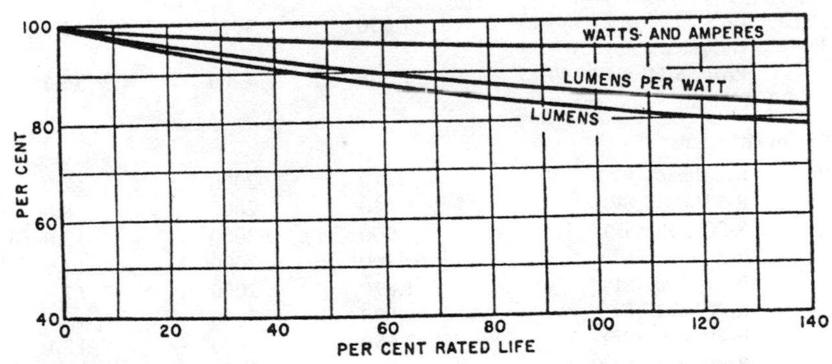

Fig. 1-5. Typical effect on lumen output, wattage, and efficacy for general-purpose incandescent lamps for their rated life.

Table 1-2. Incandescent Lamp Data.

Watts	Bulb/Base	Initial Lumens	Life, h	Initial Efficacy, lm/watt
General-service lamps:				
60	A-19/med.	870	1000	14.5
75	A-19/med.	1,190	750	15.8
100	A-19/med.	1,750	750	17.5
	A-21/med.	1,690	750	16.9
150	A-21/med.	2,880	750	19.2
	A-23/med.	2,780	750	18.5
	PS-25/med.	2,680	750	17.8
200	A-23/med.	4,010	750	20.0
	PS-30/med.	3,710	750	18.5
300	PS-25/med.	6,360	750	21.2
	PS-30/med.	6,110	750	20.3
Projector (PAR) lamps:				
75	PAR-38/med. (SP)	765	2000	10.2
	PAR-38/med. (FL)	765	2000	10.2
150	PAR-38/med. (SP)	1,740	2000	11.6
	PAR-38/med. (FL)	1,740	2000	11.6
200	PAR-46/med. SP (NSP)	2,300	2000	11.5
	PAR-46/med. SP (MFL)	2,300	2000	11.5
250	PAR-38/med. (SP) (tungsten-halogen)	3,180	3000	12.7
	PAR-38/med. (FL) (tungsten-halogen)	3,180	3000	12.7
300	PAR-56/Mog. EP (NSP)	3,840	2000	12.8
	PAR-56/Mog. EP (MFL)	3,840	2000	12.8
	PAR-56/Mog. EP (WFL)	3,840	2000	12.8
500	PAR-56/Mog. EP (NSP) (tungsten-halogen)	7,650	4000	15.3
	PAR-56/Mog. EP (MFL) (tungsten-halogen)	7,650	4000	15.3
	PAR-56/Mog. EP (WFL) (tungsten-halogen)	7,650	4000	15.3
Reflector (R) lamps:				
30	R-20/med. (FL)	210	2000	7.0
50	R-20/med. (SP)	440	2000	8.8
75	R-30/med. (SP)	900	2000	12.0
	R-30/med. (FL)	900	2000	12.0
150	R-40/med. (SP)	1,870	2000	12.4
	R-40/med. (FL)	1,870	2000	12.4
300	R-40/med. (SP)	3,650	2000	12.1
	R-40/med. (FL)	3,650	2000	12.1
500	R-40/Mog. (SP)	6,500	2000	13.0
	R-40/Mog. (FL)	6,500	2000	13.0

Table 1-2. *(Continued)*

Watts	Bulb/Base	Initial Lumens	Life, h	Initial Efficacy, lm/watt
Tungsten-halogen lamps (tubular):				
75	T-3/Min. Scr. (28V)	1,600	2000	21.0
200	T-3/RSC	3,460	1500	17.3
250	T-4/Min. Can.	4,700	2000	18.8
300	T-3/RSC	5,900	2000	19.6
	T-4/RSC	5,650	2000	18.8
325	T-4/Min. Can.	7,800	500	24.0
400	T-4/RSC	7,750	2000	19.3
500	T-3/RSC	10,700	2000	21.4
	T-4/Min. Can.	9,500	2000	19.0

incandescent lamps is important. The base-up position lamps offer more useful lumens over base-down or horizontal position because most of the blackening occurs in the neck area, where part of it is intercepted by the base (see Figure 1-6). This may offer some advantage in light output for luminaires with base-up lamps over horizontal lamps, as shown in Figure 1-7.

Most of this lumen depreciation resulting from bulb-wall blackening is eliminated in the tungsten-halogen incandescent lamps. These lamps produce a relatively whiter color of light than the conventional lamps and have better longevity. As mentioned earlier, they are equipped with a halogen such as iodine or bromine fill gas. The most important feature of these lamps is that at high temperatures the evaporated tungsten, instead of being deposited on the bulb wall, combines with a halogen molcule and is returned to the hot filament, where the halogen is freed and the cycle is repeated. This constant action of cleaning minimizes

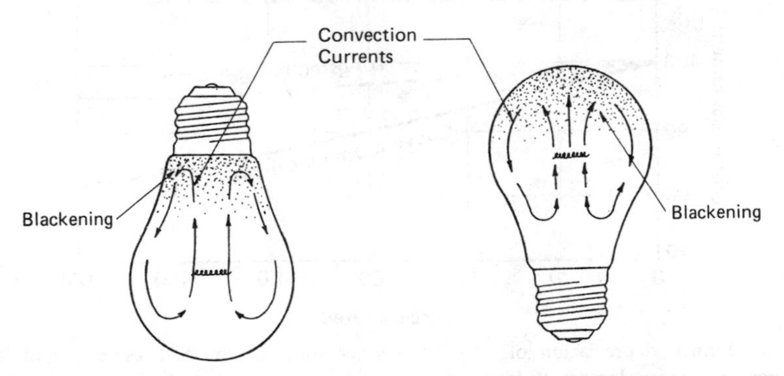

Fig. 1-6. Lamp blackening. Convection currents within a gas-filled bulb, base up (left) and base down (right).

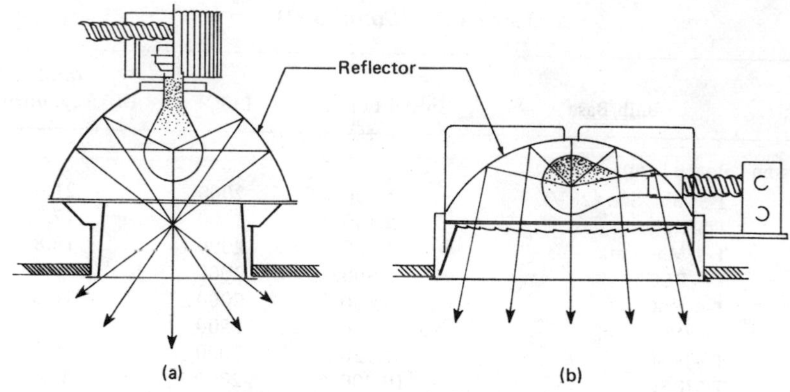

Fig. 1-7. (a) Lamp in base-up position. Blackening occurs in the neck area. Lighting distribution is not critically effected. (b) Lamp in horizontal position. Blackening occurs on the top. This reduces light output from the top and may affect lighting distribution.

bulb-wall blackening and thus maintains initial lumen output. For this reason, while a conventional lamp at the end of its life has 82% of its initial lumens, the tungsten halogen has about 95% (see Figure 1-8).

Like any other incandescent lamp, the tungsten-halogen lamp can be dimmed very easily. When dimmed, operating temperature is reduced by the same proportion with wattage. A low bulb-wall temperature hinders the unifying of tungsten and halogen and may result in blackening on bulb walls. This reduces light output just as in conventional lamps. When the lamp is restored to its full brightness, temperature rises, and the cleaning process begins. At this condition, some of the tungsten deposits are automatically removed and carried back to the source.

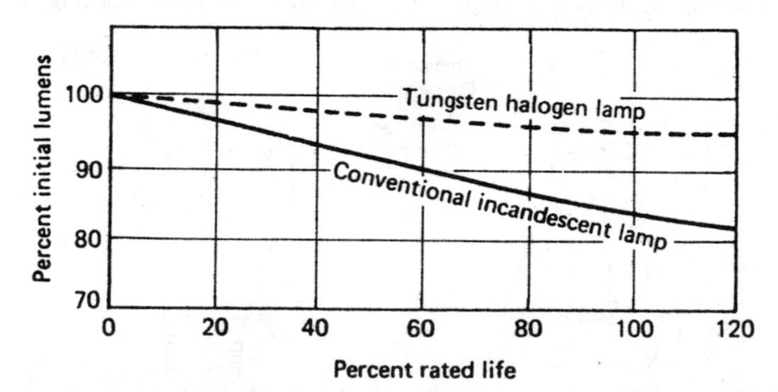

Fig. 1-8. Lumen depreciation of tungsten-halogen and conventional incandescent lamps. In tungsten–halogen lamps, at high temperatures, evaporated tungsten combines with a halogen molecule and is returned to the hot filament. This eliminates lamp blackening and improves lumen maintenance.

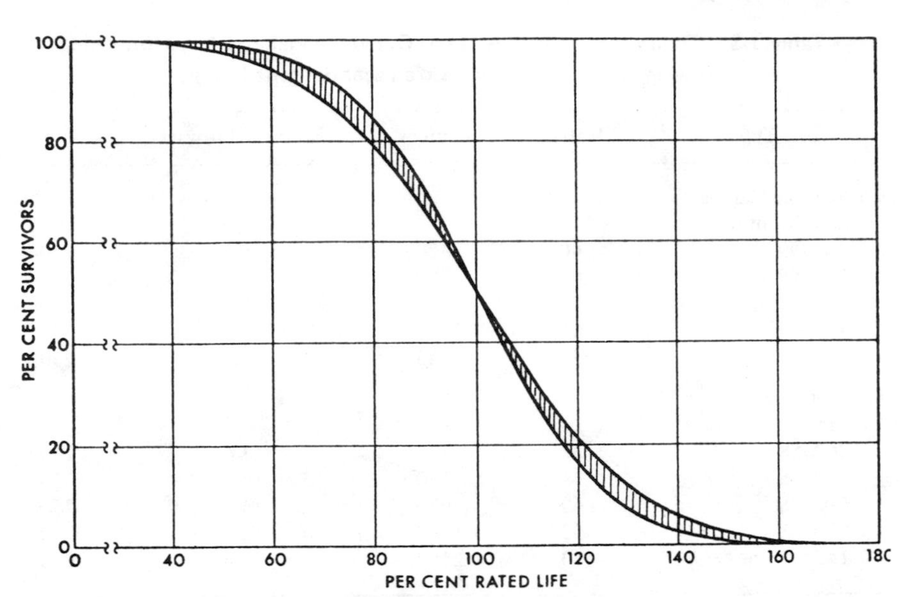

Fig. 1-9. Typical mortality-curve range average for a group of good quality incandescent filament lamps.

Lamp Mortality

While many inherent reasons make it impossible to predict the life of a particular lamp, a large group of lamps fail in a predictable manner. A mortality curve is drawn based on this prediction by each manufacturer for each type of lamp. Mortality curves give two important conclusions. First, with the help of this curve, the life of the lamp is determined; second, it helps in analyzing a group lamp-replacement program. The rated life of a lamp is designated when 50% of a group of lamps burns out. In lighting design, in the determination of the light-loss factor, it is necessary to know when the lamps start burning out and at what stage the lamps are replaced. This is discussed in detail in Chapter 3. A mortality curve helps to determine the number of burnouts before the time of planned replacement is reached. Figure 1-9 shows a typical mortality curve for incandescent lamps. In use, the life of general-service incandescent lamps is seldom as long as predicted by the mortality curve, because of field factors that affect life performance. The most important of these are physical shock and vibration.

Energy-Saving Incandescent Lamps

Because of increased emphasis on energy conservation, the lighting industry has produced several "energy-saving" products. In incandescent lamps there are two

Table 1-3. Data Comparison between Conventional and Krypton-Filled Energy-Saving Extended-Life Incandescent Lamps.

Lamp Type	Watts	Lumens	Efficacy, lm/watt	Life, h
Conventional lamps (extended life)				
40A/99	40	420	10.5	2500
60A/99	60	760	12.7	2500
75A/99	75	970	12.9	2500
100A/99	100	1460	14.6	2500
150A/99	150	2325	15.5	2500
Krypton-filled energy-saving lamps				
36A/SS	36	420	11.6	2500
54A/SS	54	760	14.0	2500
69A/SS	69	970	14.0	2500
93A/SS	93	1460	15.7	2500
143A/SS	143	2380	16.6	2500

Source: Courtesy of GTE Sylvania.

new developments. The first is the krypton-filled lamps; the second is the ellipsoidal reflector lamps.

Krypton-Filled Lamps. These lamps, which make use of krypton instead of the conventionally used argon gas inside the bulb, have been developed primarily to replace the existing 40-watt through 150-watt extended-life incandescent lamps. With this modification, for the same lumen output and rated life, the lamps consume 5 to 10% less wattage. These lamps are ideal for hard-to-reach installations, to reduce frequency of lamp replacements and labor costs. Wattages range from 36 to 143, to replace 40-watt to 150-watt conventional extended-life lamps. The price of these units is higher than that of their equivalent conventional lamps; however, because of the energy saving, the lamps usually pay for themselves within a few months' time. Table 1-3 shows comparison data between the energy-saving lamps and their equivalent extended-life conventional lamps.

Ellipsoidal Reflector Lamps. The ellipsoidal reflector lamps are a new development in the family of reflector lamps; they have the potential for energy saving in some retrofit applications. These lamps are in general for retrofit use, to replace conventional type A and R lamps in an existing cylindrical downlight

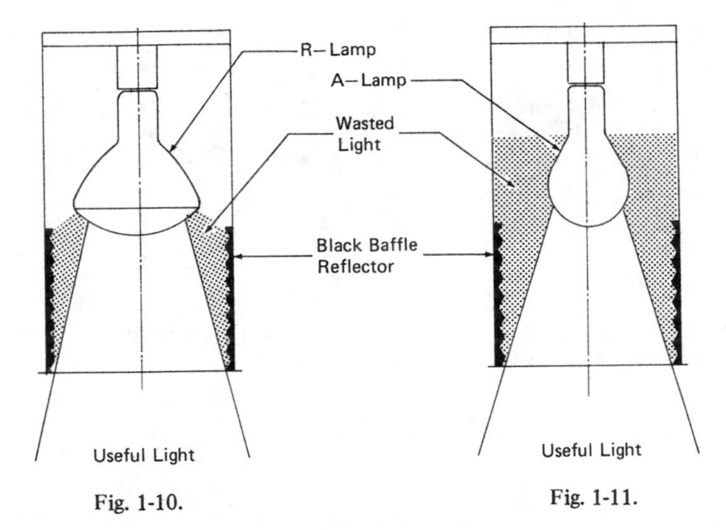

Fig. 1-10.

Fig. 1-11.

Fig. 1-10. Cylindrical downlight with black reflector or baffle, equipped with a type R lamp. A good portion of the light is absorbed by the black surface.

Fig. 1-11. Cylindrical downlight with black reflector or baffle, equipped with a type A lamp. A good portion of the light is absorbed by the black surface.

system with black reflecting surface. On a one-to-one basis, these lamps by themselves do not have better efficacy compared to their conventional equivalents, but when installed inside the luminaire, they emit much more light outside the luminaire than the conventional lamp does, increasing the luminaire efficiency.

In a conventional cylindrical downlight, where a standard type A or R lamp is used, a major portion of the light remains trapped inside the luminaire. If the luminaire has a black baffle or black reflecting surface, there is even further loss, since much of the light is absorbed by these dark surfaces. The main purpose of using a black reflecting material is to minimize the direct glare. Depending on the luminaire housing, shape, size, and the depth of color of the baffle, as low as 40% fewer lumens may be produced by these luminaires (see Figures 1-10 and 1-11).

Much of this loss is eliminated by the use of the ellipsoidal reflector lamps. These lamps are designed on the basic principles of elliptical reflectors. The elliptical reflector has two focal points, as shown in Figure 1-12. The contour of this lamp is so designed that with the lamp filament installed at the first focal point, the light rays emitted from it strike the elliptical surface of the lamp and then emerge through the second focal point. When installed inside a luminaire, the light rays thus avoid contact with the reflecting surface and result in an un-

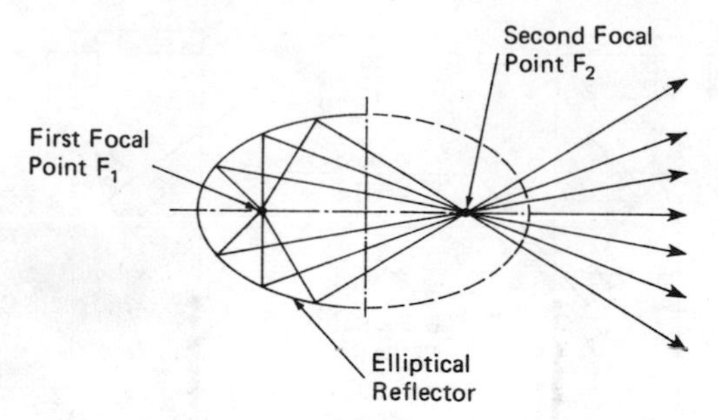

Fig. 1-12. When a point source is placed in the first focal point of an elliptical reflector, all reflected rays merge through the second focal point.

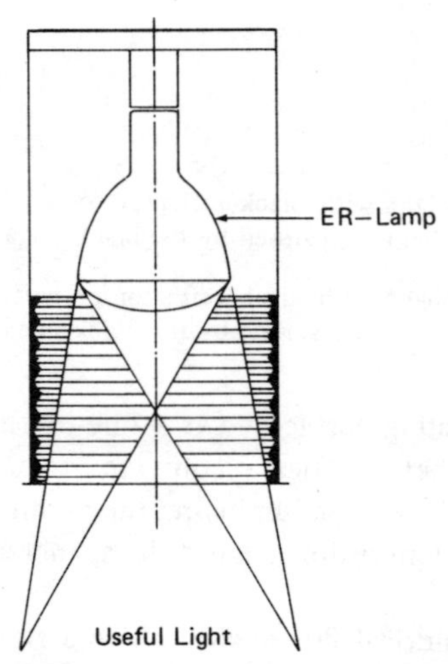

Fig. 1-13. An ER-lamp, working on the principle of elliptical reflector, produces a pool of light from a small opening. This reduces light waste.

interrupted pool of light emitted from a small opening. This is shown in Figure 1-13. With this improvement, the system thus needs relatively lower-wattage ellipsoidal reflector lamps for the same amount of light, resulting in energy conservation. A list of such lamps that are recommended to replace conventional type A or R lamps is shown in Table 1-4.

Before making a decision to use these lamps in an existing system, however, a careful analysis of the photometrics of the luminaires should be made. If the luminaire is specifically designed to work with a special lamp with candle power distribution much dependent on the reflecting surface, the ellipsoidal reflector lamp should be avoided. Use of these lamps in such luminaires will result in lower spacing-to-mounting-height (S/MH) ratio and uneven lighting distribution.

Table 1-4. Suggested Ellipsoidal Reflector Lamps in Place of Existing Type R and A Lamps.

Existing Lamp	Ellipsoidal Reflector Lamp
100A/99	50ER 30
75R30FL.	50ER 30
150A/99	75ER 30
150R FL.	75ER 30
150R FL.	120ER 40
200A/99	120ER 40
300A/99	120ER 40

Source: Courtesy of General Electric.

FLUORESCENT

Principles of Operation and Types

Fluorescent lamps produce light by creating an arc between two electrodes in an atmosphere of very low-pressure mercury vapor and some inert gas in a glass tube. The inside of the glass tube is coated with phosphor. The mercury vapor produces 253.7 nanometer (nm) ultraviolet energy that strikes the crystals of phosphor and excites them to produce light. (See Figure 1-14 for different fluorescent bulb shapes.)

There are different types of fluorescent lamps available. Since fluorescent lamps cannot produce light by direct connections to the power source, they need an auxiliary circuit and devices to get started and remain illuminated. The auxiliary circuit contained inside an enclosure is known as the *ballast*. The different types of lamps are as follows:

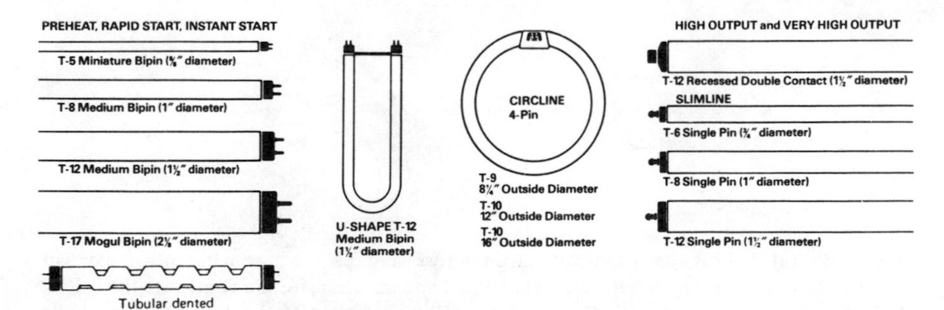

Fig. 1-14. Fluorescent lamp shapes.

1. Preheat: These lamps require a starting circuit for preheating the cathodes to aid in starting (see Figure 1-15a).
2. Instant start and slimline: These lamps mostly use a single pin but do not need the starter and the starting circuit. A ballast provides a high starting

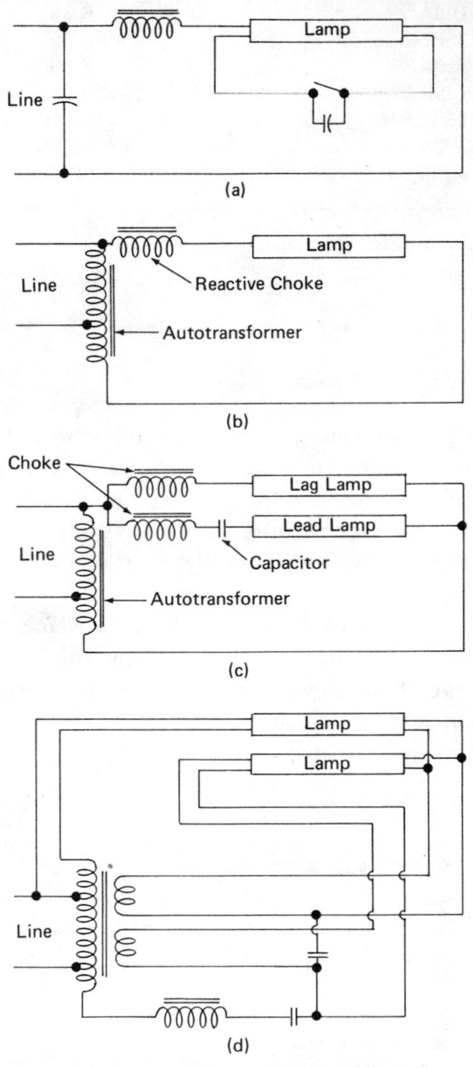

Fig. 1-15. (a) Preheat type of circuit starter across lamp can be automatic or manual. (b) Instant-start type of circuit. Reactive choke coil and autotransformer constitute the ballast components. (c) Two-lamp slimline lead-lag (split-phase) type of circuit. Capacitor, choke coils, and autotransformer constitute ballast components. (d) Two-lamp rapid-start-series type of circuit.

voltage, thus eliminating the starter circuit. Standard lamps operate at 425 milliampere (ma). (See Figure 1-15b for instant start circuit.) The slimline fluorescent lamps use the same starting principles used for instant-start circuits and use single-pin lamps, as shown in Figure 1-15c. Special lampholders which open the primary circuit upon removal of a lamp are used in instant start circuits with slimline lamps to safeguard against the higher circuit voltage.

3. Rapid start: These are the most commonly used lamps for commercial purposes. They make use of the principles of preheat and instant-start circuits but do not require starters. The preheating is supplied by a built-in low-voltage transformer coil in the ballast. These lamps are available in three different types of loading:
   a. Standard loading: These are the kind most used in offices, schools, etc., where 4-foot-long luminaires are used. They operate at 430 ma and use medium bipin bases.
   b. High output: These lamps operate at 800 ma and use recessed double-contact (RDC) bases.
   c. Very high output: These lamps operate at 1500 ma and use recessed double-contact bases.

   See Figure 1-15d for the two-lamp rapid-start-series type of circuit.
4. Trigger start: This is a combination of a preheat lamp and a special ballast having a cathode heating coil, thus eliminating the use of a starter. It has higher preheating voltage for starting than the rapid-start ballasts and is typically used only for lamps below 40 watts.

Color Characteristics

Figure 1-16 shows the SPD curves of different types of fluorescent lamps. The color of a fluorescent lamp strictly depends upon the type of phosphor used on the bulb. The SPD of fluorescents has two components: a continuous smooth curve and a line or bar spectrum. There are two components because there are two types of light sources inside the lamp. The continuous smooth portion is developed because of the excitement of the phosphor, and the bar portion results from the arc inside the lamp.

Fluorescent lamps are commonly manufactured in six standard colors:

Cool white
Deluxe cool white
Warm white
Deluxe warm white
White
Daylight

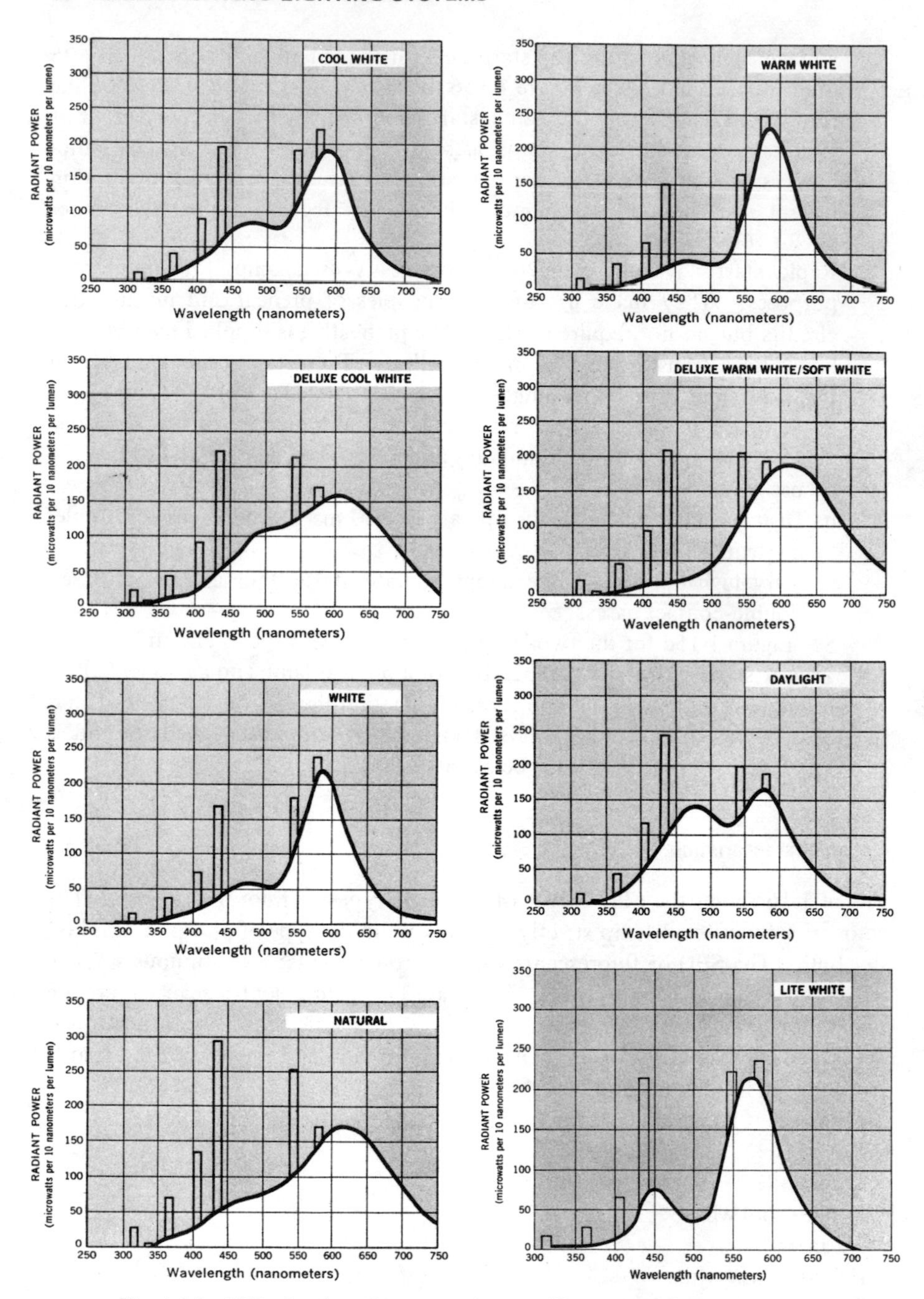

Fig. 1-16. SPD of various fluorescent lamps. (Courtesy of General Electric.)

Table 1-5. Fluorescent Lamp Selection, by Color.

Lamp Type	Visible Effect on Neutral Surface	Effect on Atmosphere	Color Strengthened	Color Grayed	Effect on Complexion	Remark
Cool white	White	Neutral to moderately cool	Orange, yellow, blue	Red	Pale pink	Blends with natural daylight; good color acceptance
Deluxe cool white	White	Neutral to moderately cool	Nearly all colors	None appreciably	Most, natural	Best overall color rendition; simulates neutral daylight
Warm white	Yellowish white	Warm	Orange, yellow	Red, green, blue	Sallow	Blends with incandescent light; poor color acceptance
Deluxe warm white	Yellowish white	Warm	Red, orange, yellow, green	Blue	Ruddy	Good color rendition simulates incandescent light
Daylight	Bluish white	Very cool	Green, blue	Red, orange	Greyed	Usually replaceable with cool white
White	Pale yellowish white	Moderately warm	Orange, yellow	Red, green, blue	Pale	Usually replaceable with cool white or warm white

Source: Courtesy of General Electric.

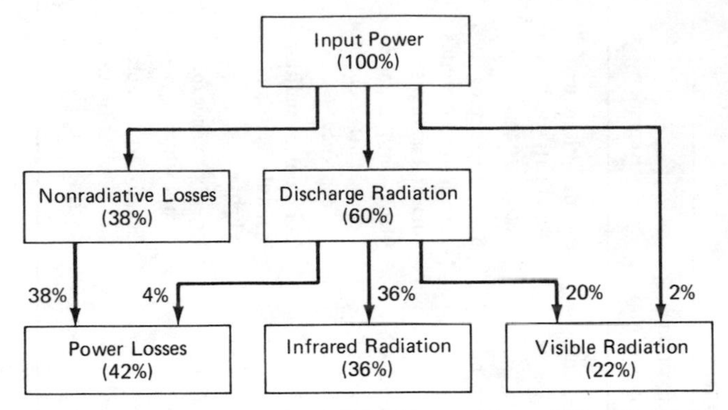

Fig. 1-17. Energy distribution of a fluorescent lamp. About 22% of total energy is available for visible radiation.

Other, less commonly used colors are also available. Cool white, warm white, white, and daylight lamps have better efficacies but are weak in red. The deluxe cool white and deluxe warm white lamps are stronger in red but at the expense of approximately 30% reduced efficacy. Table 1-5 shows color characteristics of different types of fluorescent lamps. Figure 1-17 shows energy distribution of typical fluorescent lamps. About 22% of total energy is available as visible radiation.

Efficacy

The efficacy of most fluorescent lamps is two to three times that of incandescents. The most common fluorescent lamps have efficacies ranging from 50 to 70 lumens per watt, including ballast wattage. The 4-foot-long rapid-start lamps have an average of 60 lumens per watt efficacy. Efficacy improves as the lamps get longer and go towards higher light output.

Fluorescent ballasts are usually designed to operate either one or two lamps. The efficacy of a two-lamp ballast is better than that of a one-lamp ballast. A two-lamp ballast uses about one-fourth the wattage of that of the two lamps it controls, whereas a single-lamp ballast will use about one-third the wattage of the lamp it controls.

Effect of Temperature

Change in ambient temperature has a significant effect on fluorescent lamp performance. A fluorescent lamp is primarily dependent upon mercury vapor pressure for its operation. The mercury vapor pressure in turn is dependent upon the coolest point on the bulb. The operating temperature of the bulb may vary with ambient temperature, lamp wattage, lamp construction, luminaire design, draft condition, etc. A low bulb-wall temperature reduces the mercury vapor pressure inside the lamp, producing less ultraviolet energy and thus reducing the lumen output. Figures 1-18a and 1-18b show typical fluorescent lamp temperature characteristics. The exact shape of curves will depend upon lamp and ballast type; however, all fluorescent lamps have curves of the same general shape, since this depends upon the mercury vapor pressure.

As can be seen in Figure 1-18a, the wattage consumption is also significantly affected by temperature variation. In higher-temperature regions, the lumen output and power consumption depreciate more or less in the same proportion; however, for lower-temperature regions, the lumen output is more significantly affected than the power, resulting in a substantial reduction in lumen efficacy in the lower-temperature region than in the upper.

Light output and luminous efficacy reach optimum values at a bulb-wall temperature of about 100°F. Fluorescent lamps primarily used for indoor-type application are designed so that their coolest bulb-wall temperature will be at 100°F. Luminaires have different heat conditions that affect the operating characteristics of the enclosed lamps. A detailed analysis and thorough discussion of this thermal effect in luminaires follows in Chapter 4.

Change in ambient temperature affects the bulb-wall temperature. In general, fluorescent lamps operate most efficiently at ambient temperatures between 70 and 90°F and probably peak at about 77°F. As a rule of thumb, it can be said that there will be 1% loss in light for every 2°F that the ambient temperature around the lamp exceeds 77°F.

Standard fluorescent lamps, when operated in normal-rated voltage and approved auxiliaries, will produce satisfactory results even at temperatures as low as 50°F. For ambient temperatures lower than 50°F, it is advisable to use low-temperature ballasts (0°F and -20°F), to ensure reliable starting. At low tem-

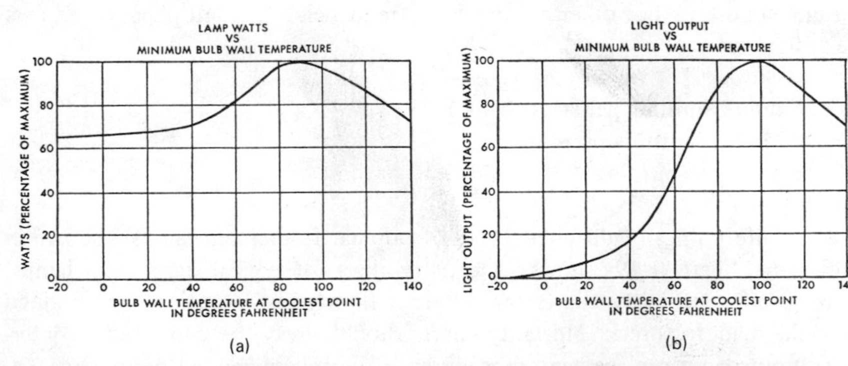

Fig. 1-18. Typical fluorescent lamp temperature characteristics. Light output drops more drastically in the lower temperatures than in the upper. But the power does not drop by same proportion. This results in drastic efficacy reduction at lower temperatures.

peratures, lamps must have some type of thermal protection (enclosed fixture or lamp sleeve) to reach optimum efficacy.

Effect of Humidity

The electrostatic charge built up against the outer side of the glass tube affects the starting voltage as the arc strikes. If moisture and humid air surround the outer side of the glass tube, the lamp requires much higher voltage for starting. Rapid-start lamps are the most sensitive to these effects. This unfavorable effect gets worse significantly as the humidity level reaches 65% and goes higher. The problem is minimized by a coating of silicon on the tube of rapid-start lamps, applied during manufacture.

Mortality and Lumen Depreciation

Each type of fluorescent lamp has a life expectancy that applies when operated under specific conditions. When the ballasts meet Certified Ballast Manufacturers (CBM) standards, lamps meet the American National Standards Institute (ANSI) specifications, and the system is operated in proper temperature and voltage, rated life will be attained. The life of a fluorescent lamp is significantly dependent upon the number of hours it is burned per start. Every time the lamps are started, electron emission material on the electrodes sputters off and reduces the lamp life. The electrodes of a preheat fluorescent lamp take a much more severe shock during starting period than during the actual burning period. Therefore, the longer the lamp is burned at a noninterrupted stretch, the longer its life. The average rated lamp values given in tables by different manufacturers are based on burning cycles of 3 hours per start. The most commonly used lamps, such as the 4-foot rapid start, are rated at 20,000 hours based on 3 hours of burning per start. For other burning hour frequencies the multiplication factors are as follows:

For 6 hours, burning per start	1.25
For 12 hours, burning per start	1.60
For continuous burning	2.50 or more

A list of life ratings of different types of popular fluorescent lamps is shown in Table 1-6. Figure 1-19a shows a mortality chart for typical fluorescent lamps. More specific mortality charts for different fluorescent lamps can be obtained from the manufacturers. Mortality charts should always be consulted to determine group lamp replacement procedures. The recommended period for economical group replacement is at about 60–70% of rated life. Like other types of lamps, fluorescent lamps depreciate in light output as they age. As a lamp ages,

Table 1-6. Fluorescent Lamp Data.

Watts	Bulb	Description	Initial Lumens	Life, h
Rapid-start:				
40	F40 T-12	Cool white	3,150	20,000
	F40 T-12	Warm white	3,150	20,000
	F40 T-12/U	Cool white, U shape (3-$\frac{5}{8}$-in. leg spacing)	2,900	12,000
	F40 T-12/U	Warm white, U shape (3-$\frac{5}{8}$-in. leg spacing)	2,850	12,000
	F40 T-12/U	Cool white, U shape (6-in. leg spacing)	2,900	12,000
	F40 T-12/U	Warm white, U shape (6-in. leg spacing)	2,850	12,000
Slimline:				
75	F96 T-12	Cool white	6,400	12,000
	F96 T-12	Warm white	6,400	12,000
High output:				
60	F48 T-12	Cool white	4,300	12,000
	F48 T-12	Warm white	4,300	12,000
110	F96 T-12	Cool white	9,200	12,000
	F96 T-12	Warm white	9,200	12,000
Very high output:				
110	F48 PG-17	Cool white	7,450	12,000
	F48 PG-17	Warm white	7,000	12,000
215	F96 PG-17	Cool white	16,000	12,000
	F96 PG-17	Warm white	15,000	12,000

general darkening along the entire length of the tube occurs as a normal condition because of mercury streaking. The blackening starts at the ends of the tube by material emitted by the electrodes and indicates that the end of the lamp life is near. Figure 1-19b shows the light-output depreciation of typical fluorescent lamps. Note that all discharge lamps are rated after the first 100 burning hours, since lamps are often somewhat unstable when new and require an initial "burning in" period. The recommended period of 60–70% of rated life for group replacement is based on minimum cost of light, considering lumen output, lamp mortality, energy, and lamp and labor costs.

Ballasts

The principal functions a ballast may perform are the following:

1. Limits the current. A fluorescent lamp is a low-pressure, arc-discharge device. The more current in the arc, the lower the resistance becomes.

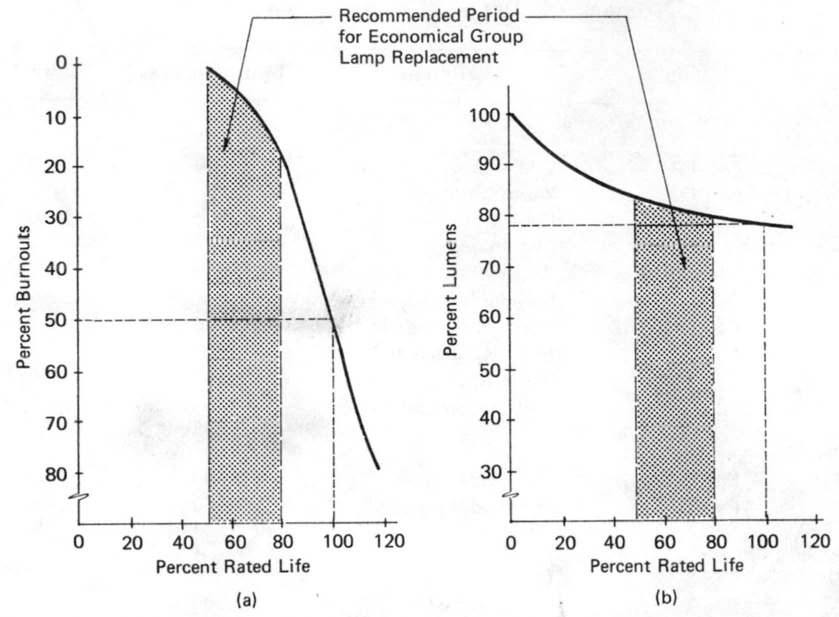

Fig. 1-19. (a) Typical general lighting type fluorescent lamp mortality. (Courtesy of General Electric.) (b) Typical standard white fluorescent lumen depreciation. (Courtesy of General Electric.)

Without a ballast to limit the current, the lamp would draw so much current that it would destroy itself.

2. Provides sufficient voltage to start the lamp.
3. Provides power-factor correction to offset partially the coil's inductive reactance and to help maintain the efficiency of the electrical distribution system.
4. Provides a coil for cathode heating for rapid-start and trigger-start lamps.
5. Provides radio interference suppression.

The light output, life, and starting reliability of a fluorescent lamp depend greatly on the design of the ballast. ANSI specifies the requirements for good lamp performance. The National Electrical Code (NEC) requires "Class P" ballasts in indoor installations; these provide thermal protection of the coil and capacitor with a device to prevent the case temperature from exceeding 230°F because of internal failure or increasing ambient temperatures. Ballasts certified as built to the specifications adopted by the CBM provide light output and power input values that either meet or exceed minimum performance requirements.

No specific voltage is associated with fluorescent lamps since they are not directly connected to the power source. However, the ballasts are rated for specific voltages and must be used with the proper lamp types.

Energy-Saving Fluorescent Lighting Products

In order to reduce energy consumption, cut operating costs, and make the overall system work more efficiently, lighting equipment manufacturers have introduced a variety of new "energy-saving" products in recent years. Some of these are low-power-consuming lamps and ballasts; others are "dummy" tubes that replace standard lamps. A careful analysis and understanding of all these products is necessary before making a decision to use these for any retrofit or new application.

The energy-saving products may be categorized as follows:

Krypton lamps
Lite-white lamps (high-efficacy phosphor)
Low-loss and wave-modified low-loss ballasts
Dummy tubes
Impedance lamps

Krypton Lamps. These special lamps make use of krypton gas instead of the conventional argon as their fill gas. The higher molecular weight of this gas reduces voltage drop across the lamp, thereby reducing the power consumption. The use of these lamps may reduce the power consumption anywhere from 10 to 20%, depending on the type of lamps, at the expense of light reduction of about the same proportion. Table 1-7 compares krypton-filled lamps to

Table 1-7. Comparison of Low-Energy-Consuming and Standard Fluorescent Lamps.

	Standard Lamps			Krypton-Filled Lamps			Lite-White Lamps		
Type	Nominal Watt	Initial Lumens	Lamp Efficacy	Nominal Watt	Initial Lumens	Lamp Efficacy	Nominal Watt	Initial Lumens	Lamp Efficacy
4 ft, Rapid start	40	3,150	78.7	35	2,850	81.4	35	3,050	87.1
8 ft, Slim-line	75	6,300	84	60	5,600	93.3	60	6,000	100
8 ft, High output	110	9,200	83.6	95	8,500	89.4	95	9,100	95.7
8 ft, Power groove	215	16,000	74.4	185	14,000	75.6	185	14,900	80.5

Source: Courtesy of General Electric.

standard lamps. With no reduction of life rating, these lamps normally offer somewhat better efficacy than that of standard lamps. However, because of the relatively lower light output per lamp and higher initial costs, these lamps should be considered for retrofit application only, where the existing system is over-lighted and reduction in light output and power are required. Because there is no change as far as phosphor types are concerned, these lamps are available in all standard colors, such as cool white, warm white, deluxe cool white, deluxe warm white, daylight, and white.

Lite-White Lamps. The lite-white lamps are either standard or krypton filled, but use a special phosphor to improve light output. A combination of low wattage and high lumen output results in the best luminous efficacy of the whole family of fluorescent lamps. Depending on the type, these save between 14 and 20% of power consumption at the expense of only 3-12% of light output when compared with their conventional counterpart. Table 1-7 shows such a comparison. For 4-foot rapid-start lamps, which are the most popular, the loss in light output is only 3%. Because of their special phosphor coating, these lamps are presently only available in the lite-white color. However, this color has been designed to be comparable to standard cool white lamps. The color is well balanced in all wavelengths except red, where it is somewhat lower.

Low-Loss and Wave-Modified Low-Loss Ballasts. The use of low-loss ballasts and the newer wave-modified low-loss ballasts further reduce the total power consumption of a luminaire. The wave-modified low-loss ballast consists of an electromagnetic circuit that increases light output more than low-loss ballasts, especially when equipped with lite-white lamps. When these ballasts are equipped with regular or energy-saving lamps, a wide range of light output is achieved. It can be categorized as follows:

1. Reduces light output, consuming less power.
2. Maintains almost same light output with less power.
3. Increases light output, with less power.

Obviously, while the first two characteristics are suitable for retrofits, item 3 encourages the use of such products for a new application.

There should be careful analysis before a decision is made to use any combination of these products. The energy-saving lamps operating at a slightly higher current cause a proportionate increase in ballast capacitor voltage. If these lamps are used in existing standard ballasts, they impose undesirable higher voltage across the existing capacitors and may cause ballast failure; this is especially the case for older ballasts. Typically, at normal room ambient temperature of 70°F or higher, or in enclosed luminaires that have higher luminaire ambient temperature, the voltage rise is minimal (about 2-7%). At lower temperature, however, the capacitor voltage may rise to 15%. Where these lamps are equipped with suitable ballasts, the lower overall heat generation because of lower wattage would increase the ballast life by as much as a factor of two. The capacitor voltage rise in use with high-output and very high-output reduced-wattage lamps is negligible and has insignificant effect on ballast failure.

The low-power-consuming lamps are not recommended to be used with low-power-factor or dimming ballasts, and they should be used at room ambient temperatures of 60°F or higher.

A combination of these energy-saving lamps and ballasts, producing light with lower wattage, generates less heat than their conventional counterparts. When installed inside an enclosed luminaire, this loss of heat has significant effect in luminaire efficiency and electrical characteristics. This is discussed in detail in Chapter 4.

Dummy Tubes. A fluorescent luminaire containing two or a multiple of two lamps is generally equipped with two-lamp ballasts; these operate a pair of lamps connected in series. When an area is overlighted and a reduction in light output is necessary for energy savings, removing some of the lamps is the simplest solution to this problem. However, because each pair of lamps is usually connected in series with a common ballast, removal of one lamp would cause the other one to go out. Disconnecting some of the luminaires would be a costly affair, and it would also result in uneven lighting.

The problem can be minimized with the use of the dummy tubes. By replacing a lamp with a dummy tube, the series circuit of the lamps is completed, enabling the remaining lamp to operate properly without its pair. The dummy tube, by itself, does not produce any light when paired with a lamp, the system light output reduces to about 33% of the two-lamp situation. It is constructed of a glass tube (some are plastic) similar to that of a conventional fluorescent lamp, with a tuned capacitor inside to complete the circuit. These tubes are available for the replacement of all types of conventional fluorescent lamps, including F20T12, F30T12, F40T12, F40U/6, F96T12HO and F96T12VHO with rapid-start ballasts, and F48T12 and F96T12 lamps with instant-start slimline ballasts. While these tubes can be inserted on either side of a series ballast, the slimline models must be connected to the blue leads of the ballast. A note of importance is that these tubes are not to be used with lead-lag, preheat, dimming, or emergency ballasts.

When used with a dummy tube, the remaining lamp actually produces 66% of its own light output, causing an overall light reduction to 33% in the luminaire. The power consumption for this amount of light is also of the same proportion, resulting in an 67% overall reduction in energy. The effect on light output, input power, and power factor for different types of fluorescent lamps in pair with dummy tubes is shown in Table 1-8.

When some ballasts are operated without lamps, they draw a certain amount of power because of the magnetizing current. This is also shown in Table 1-8. Magnetizing current, being inductive, causes a low, lag-type power factor. Thus,

Table 1-8. Effect on Electrical Characteristics with Dummy Tube.

Lamps	1 Ballast + 2 Standard Lamps			Ballast Only			1 Ballast + 1 Dummy Tube and 1 Standard Lamp		
	Watts	Current	P.F.	Watts	Current	P.F.	Watts	Current	P.F.
1a. F20 normal p.f.	39	0.53	0.61	4	0.19	0.17	19	0.19	0.83
1b. F20 high p.f.	50	0.44	0.94	4	0.21	0.15	20	0.21	0.79
2. F30	80.1	0.66	1.0	7	0.34	0.17	30.4	0.27	0.93
3. F48 (slimline)	106	0.88	1.0	n/a	n/a	n/a	51	0.66	0.64
4. F40	100	0.85	0.98	10	0.38	0.21	30	0.28	0.89
5. F96 (slimline)	172	1.56	0.91	n/a	n/a	n/a	72	0.70	0.85
6. F96/HO	267	2.25	0.98	13	1.22	0.08	88	0.84	0.87
7. F96/VHO	437	3.85	0.94	21	1.45	0.12	114.4	1.16	0.82
8. F40 (energy-saving lamp)	85	0.76	0.93	10	0.38	0.21	28	0.25	0.93
9. F96 (slimline, energy-saving lamp)	142	1.22	0.96	n/a	n/a	n/a	59	0.57	0.86
10. F96/HO (energy-saving lamp)	218	1.85	0.98	13	1.22	0.08	70	0.62	0.94

Note: 1. Operating voltage is 120 volt
2. Slimline ballasts automatically disconnected when lamps are removed.

in a situation where lamps are removed from every two of three luminaires for a 33% overall light level, the system would not only produce uneven lighting; it also would reduce the power factor. Use of dummy tubes would minimize this problem, offering a better overall power factor with even lighting.

The main disadvantage of a dummy tube is the unsymmetrical light-source pattern that is observed when the luminaire is looked at directly. The candle-power distribution is also affected to a certain extent because there is only one lamp burning on one side of the luminaire. However, from the standpoint of its use in retrofits for areas such as lobbies, corridors and hallways, etc., these disadvantages may not be very critical.

Impedance Lamps. The impedance lamps are retrofit fluorescent tubes that add additional degrees of flexibility in the management of the energy of lighting. Like the dummy tubes, these are used to replace one of the two fluorescent tubes in a standard two-lamp ballast circuit, but with a significant difference. Unlike the dummy tubes, these lamps glow by themselves and are restricted to F40 rapid-start lamp applications only. Like other retrofit lamps, these should also be used when reduction of energy is important with a lower illuminance.

The impedance lamps are similar to regular 4-foot fluorescent lamps except that a section of approximately $3\frac{1}{2}$ inches at one end contains the special circuit (primarily a capacitor). The net result of this circuit is to increase the impedance in the series circuit containing the two lamps, thereby reducing current and wattage and glowing both the lamps evenly. For this reason, the ballast that operates this lamp with a standard F40 will operate cooler and, in addition, the voltage across its capacitor will decrease when compared to operating two standard F40 lamps. These factors will improve the life of the ballast.

The impedance lamps are available to reduce the energy consumption to two levels, 33% and 50%. The light output is almost proportional to energy consumption when used in an enclosed luminaire. For others, such as open strips or vented industrials, etc., the light output suffers slightly greater loss than the wattage because of lower lamp operating temperatures.

These lamps are recommended for use only with two-lamp, series type, rapid-start ballasts and at 60°F or higher ambient temperature. The use of these lamps with any other combination may result in overheating of the ballast and over-stressing components of the lamp. They should not be used with any energy-saving lamps or low-loss and dimming ballasts.

HIGH-INTENSITY DISCHARGE SOURCES

Principles of Operation

Mercury vapor, metal halide, and high-pressure sodium (HPS) are the high-intensity discharge (HID) sources. Like fluorescent lamps they produce light by estab-

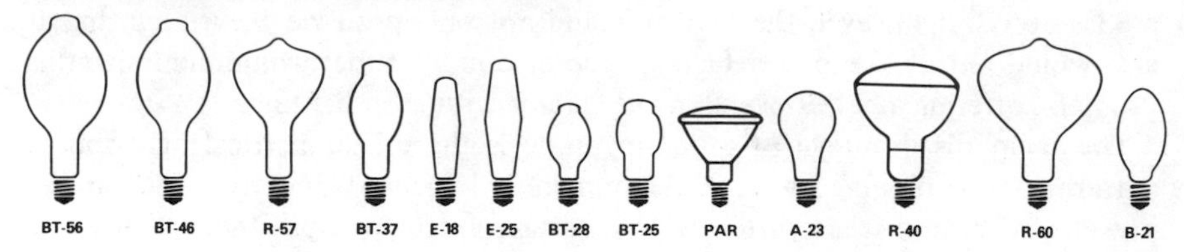

Fig. 1-20. Bulb shapes of HID lamps. (Courtesy of GTE Sylvania.)

lishing an arc between two electrodes; however, in HID lamps the electrodes are only a few inches away from each other, enclosed in opposite ends of a small, sealed, translucent or transparent arc tube. The arc tube is then closed in an outer bulb that is filled with nitrogen and an inert gas in mercury vapor lamps; there is either a vacuum or nitrogen in metal halide lamps; and there is a vacuum or xenon in HPS lamps. The outer glass is made of borosilicate glass, which can withstand high temperature (750°F maximum) and is resistant to thermal shocks, thus reducing the possibility of shattering when the hot lamp comes in contact with rain or snow. An arc of electricity spanning the gap between the electrodes inside the inner tube generates heat and pressure much higher than those in fluorescent lamps. This vaporizes the atoms of various metallic elements contained within the arc tube. This vaporization causes the atoms to emit a high amount of electromagnetic energy in the visible range, producing light. The metallic element in mercury lamp arc tubes is a precise amount of high-purity mercury. In metal halide, a small quantity of scandium and sodium iodides are added to the basic mercury. In HPS it is a combination of sodium and a small amount of mercury. Various bulb shapes of HID lamps are shown in Figure 1-20.

Purpose of Ballasts

All HID lamps require a ballast to function properly. The basic reasons for the ballast for HID lamps are primarily similar to those for fluorescent lamps. They are categorized as follows:

1. Limits the current. All arc-discharge lamps have negative resistance characteristics. If these lamps are directly connected to a nonregulated voltage supply, the arc discharge will draw an unlimited amount of current almost immediately and the lamp will destroy itself. The main function of the ballast is to limit this current flow through the arc tube.
2. Provides a voltage kick to start the arc discharge by transforming the available voltage to that required by the lamp.

3. Provides power-factor correction partially to offset the coil inductive reactance and help maintain the efficiency of the electrical distribution system.
4. Provides the correct voltage to allow the arc discharge to stabilize.
5. Prevents any voltage or current surge caused by the arc discharge from reflecting into the line circuit.
6. Adjusts lamp volts and current to the requirements of the lamp as it ages.

What to Look for in Ballasts

The following are the operating characteristics inherent in HID ballasts. These features should be observed closely in selecting any type of ballast for any HID lamp. Each of them plays an important role in the optimum performance of the lamp and the total system selected.

Efficiency. All HID lamps need ballasts to perform properly. The ballasts consume power. The user pays the power bill for the combined energy consumption of the lamps and the ballasts. Therefore, the efficiency of the ballast selected is very important from an energy-saving standpoint. Ninety percent efficiency means that 90% of the power is used towards producing light and the remaining 10% is absorbed by the ballast.

Power Factor. A ballast can be classified as high power factor or low power factor (sometimes called *normal power factor*). A low-power-factor ballast draws more line current than a high-power-factor ballast, thus needing larger conductors and circuit protection devices. In a large application where hundreds of low-power-factor ballasts are used, the combined effect may reduce the total building power factor substantially. In many states, there is a penalty for less than the specified power factor. Other states, which have enacted energy conservation laws, require that a specified minimum power factor be maintained. If the overall power factor is less than specified, it must be improved. A discussion on this subject is done in Chapter 6. "High power factor" of a ballast indicates the value is at least 90%. Anything less than 90% is considered as "low power factor" or "normal power factor."

Line Current. On some ballast types, the line current at start up is less than operating current at steady state condition. In this situation, the circuit protective device and the conductor sizes that have been designed based on the operating current are enough for starting as well. For other ballasts, the starting current can be as much as 150% more than the operating current. Obviously, this system will require much larger conductor sizes as well as protective devices.

Line Voltage Regulation. Line voltage regulation controls the change in lamp watts due to variation in input line voltage. Most electrical distribution systems have a tolerance of ±5% of nominal line voltage. Each type of ballast has a specific design tolerance that has to be observed carefully. Deviation from these figures may result in poor lamp and ballast performance. In general, if the line voltage will vary more than ±5%, a regulating-type ballast should be used.

Crest Factor. In a sinusoidal current or voltage waveform, the ratio of the peak value to the root mean square (rms) value is known as the *crest factor*. In an ideal sinusoidal curve, this factor is 1.41. The crest factor of a given ballast is determined by the magnetizing circuit design dealing with the core and coil of the ballast. It is a design criterion of the ballast itself and is totally independent

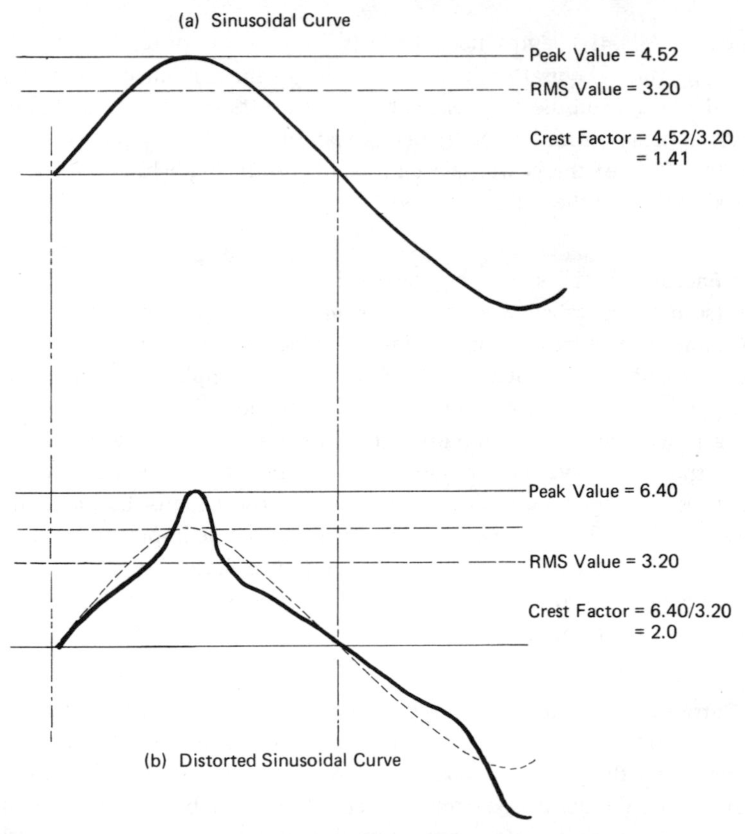

Fig. 1-21. In an ideal sinusoidal curve, the crest factor is 1.41, as shown in (a). A distorted sinusoidal curve may have a crest factor higher than 1.41, such as that shown in (b); this may hasten lamp lumen depreciation and shorten the lamp's useful life.

of lamps and fixtures used. However, it plays an important role in maintaining the lumen output of the lamps. A high lamp-current crest factor can hasten lamp lumen depreciation and shorten a lamp's useful life.

Figure 1-21a shows a pure sinusoidal current waveform whose peak value = 4.52 amp.; the rms value = 3.20 amperes. The current crest factor in this case is 4.52/3.20 = 1.41. A distortion in the waveform can create a condition where the crest factor will be higher than 1.41. Figure 1-21b shows an example of that. In this case, the current crest factor is 6.40/3.20 = 2.0. ANSI standards permit a maximum ballast current crest factor of 2.0 for mercury vapor lamps and 1.8 for metal halide lamps and HPS lamps. The high peak current associated with a ballast of high crest factor causes electrode deterioration, resulting in reduced lumen output.

Line Voltage Variation. All HID lamp ballasts carry a label for their designed operating voltage and frequency. They should be operated on the specific voltage and frequency ratings. Some ballasts are tapped to accommodate more than one line voltage, such as 120/240 volt, and some have taps for line voltages that differ from nominal values, such as 110/120 volts. The lamp performance will be significantly altered when the input voltage to the ballast differs from the design value.

MERCURY VAPOR

Principle of Operation and Types

Figure 1-22 shows a typical mercury vapor lamp. The outer bulb is filled with nitrogen and encloses the inner arc tube, which contains high-purity mercury and argon gas. As the circuit is energized, the starting voltage is provided across the starting electrode and the adjacent main electrode to create an argon arc. This increases heat and vaporizes the mercury. The ionized mercury atoms decrease the resistance across the main electrodes and cause the main arc to strike.

Mercury vapor lamps for general applications are available from 40 to 1000 watts. They can be either clear or phosphor-coated, but all bulb shapes are limited to the following (see Figure 1-20): *A* – arbitrary or artistic; *BT* – bulged-tubular; *E* – elliptical; *T* – tubular; *R* – reflector; *PS* – pear-shaped; *PAR* – parabolic aluminized reflector.

Color Characteristics

There are four different colors of mercury vapor standard lamps that are manufactured today:

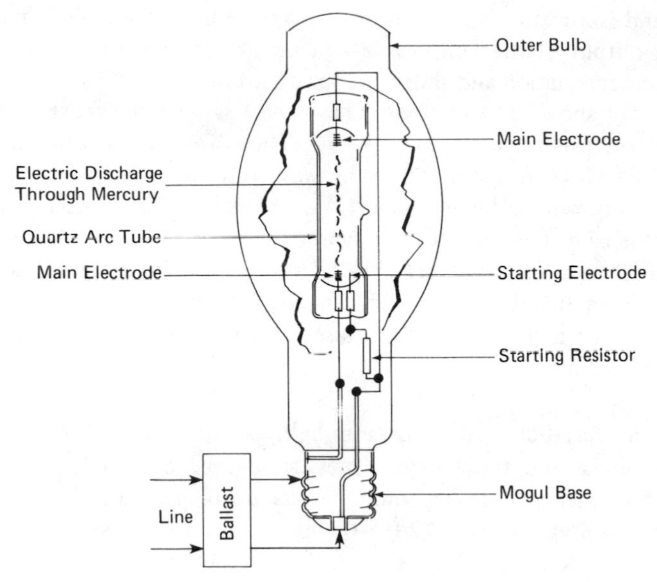

Fig. 1-22. Typical design circuit and construction of a mercury vapor lamp.

Clear mercury
Color improved
Deluxe white
Warm deluxe white

Figure 1-23 shows the SPD curves of mercury vapor lamps. The clear mercury lamps have predominantly blue-green color characteristics, mainly because of the arc. This is shown with the line spectrum. The color-improved lamps make use of a coating of phosphor, which improves the color but still retains a poor rendition of red. The deluxe white lamp has a coating of phosphor that produces more red and improves the overall color. The warm white deluxe lamps use thicker coating of the same type of phosphor used in the deluxe lamps, providing excellent color but with a reduction in lumen output. Table 1-15 shows a color comparison of all these lamps. The energy distribution of a 400 W mercury vapor lamp is shown in Figure 1-24.

Warm-Up and Restriking Time

Mercury vapor lamps take about 5 to 7 minutes to warm up, depending on ambient temperature. When a momentary interruption or outage occurs, because of the circuit's being turned off manually or because of a power outage, or when lamp operating voltage drops below that required to sustain the arc, the mercury atoms in the arc tube are deionized and radiation stops. The lamp is unable to

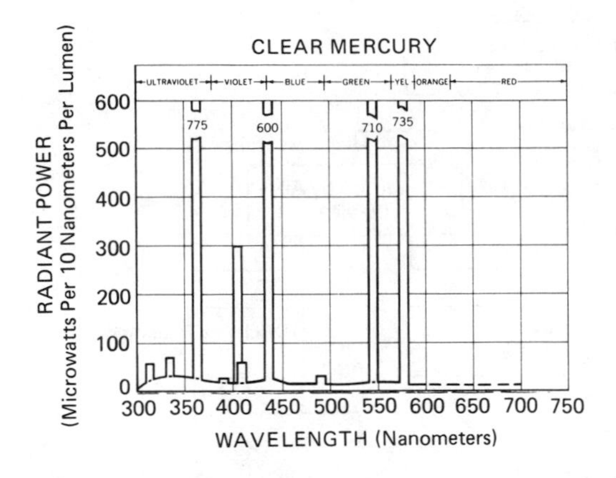

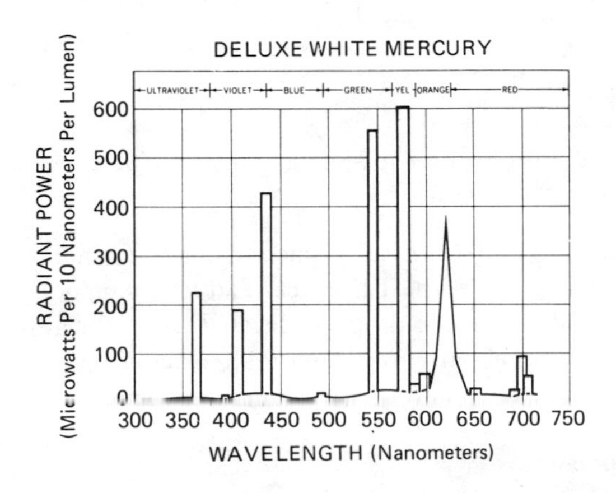

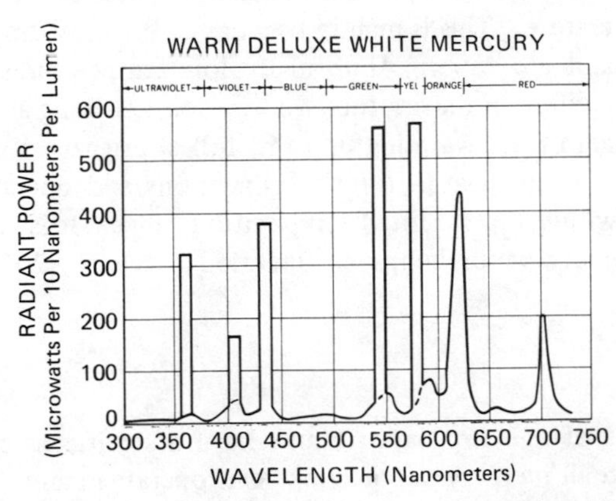

Fig. 1-23. SPD of mercury vapor lamps. (Courtesy of General Electric.)

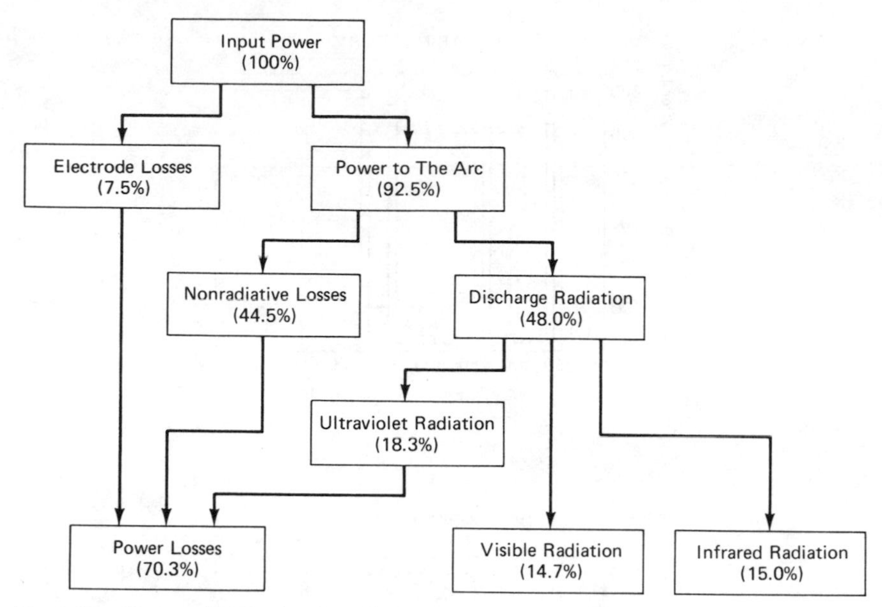

Fig. 1-24. Energy distribution for a 400-watt mercury vapor lamp. About 14.7% of energy is available for visible radiation.

reionize the argon and mercury gases, since in the hot condition, it needs much higher starting voltage than the ballast can actually supply. This requires the arc tube to cool and pressure to drop to a level at which the available voltage can restrike. This period is about 3 to 6 minutes, depending on the wattage, temperature characteristics, and operating conditions of the lamp.

Effects of Temperature

Unlike fluorescents, mercury vapor lamps are not critically affected by changes in ambient temperature. This is mainly because of the insulating effect between the outer bulb and arc tube. Abnormally low temperatures may develop a vapor-pressure condition in the arc tube that prevents striking at the rated voltage or prevents the lamp from warming up to its fullest brightness. The user should always follow the lamp manufacturer's instructions and operate the lamp for optimum usage within the specified temperature range. It is recommended, in general, that mercury vapor lamps be operated between -20° and 104°F for best performance.

Burning Position

Burning positions of mercury vapor lamps have a significant effect on performance. Although all mercury vapor lamps can operate in any burning position, lumen output reduces when they are burned in a horizontal position. Depending on the wattage of the lamps, this reduction could range anywhere from 3 to 6% of the lumens produced by vertically operated lamps. When the mercury vapor lamp is operated horizontally, the arc tends to bow upwards because of convection current effects. The effect, obviously, becomes more severe as the arc length gets longer for different wattage lamps. See Table 1-9 for data on vertically operated and horizontally operated mercury vapor lamps.

Efficacy

Mercury vapor light has been one of the most popular light sources in the whole family of lighting. However, its luminous efficacy is very poor; it is only better than that of the incandescents. Table 1-9 shows lumen output of different mer-

Table 1-9. Mercury Vapor Lamp Data.

			Initial Lumens		
Watts	Bulb	Description	Vert.	Hor.	Life, h
50	E-17	DX	1,575	1,490	16,000
75	E-17	Clear	2,700	2,550	16,000
	E-17	DX	2,800	2,650	16,000
100	A-23	Clear	3,700	–	18,000
	A-23	DX	4,000	–	18,000
	E-23½	Clear	3,850	3,650	24,000+
	E-23½	DX	4,200	4,000	24,000+
	R-40	Flood	2,850	–	24,000+
	R-40	DX (WFL)	2,850	–	24,000+
175	E-28/BT-28	Clear	7,950	7,500	24,000+
	E-28/BT-28	DX	8,600	8,100	24,000+
	E-28	Color-improved	7,250	6,900	24,000+
	R-40	Flood	5,700	5,400	24,000+
	R-40	DX (WFL)	5,700	5,400	24,000+
250	E-28/BT-28	Clear	11,200	10,500	24,000+
	E-28/BT-28	DX	12,100	11,400	24,000+
400	E-37/BT-37	Clear	21,000	20,000	24,000+
	E-37/BT-37	DX	22,500	21,500	24,000+
	E-37	Color-improved	20,500	19,000	24,000+
1000	BT-56	Clear	57,000	55,000	24,000+
	BT-56	DX	63,000	60,000	24,000+
	BT-56	Color-improved	55,000	53,000	24,000+

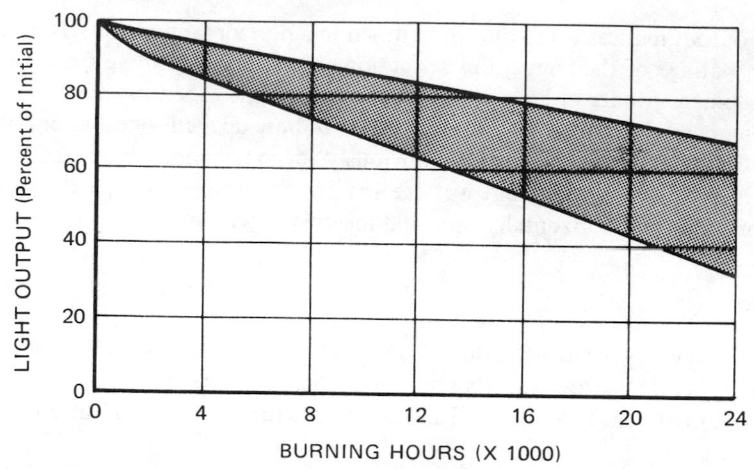

Fig. 1-25. Lumen depreciation of mercury vapor lamps. The shaded area indicates operation on various ballasts. (Courtesy of General Electric.)

cury vapor lamps operated vertically and horizontally and their corresponding efficacies. The efficacies increase as the wattage of lamps increases.

Lumen Maintenance and Depreciation

Like any other light source, the mercury vapor lamps radiate fewer lumens as they age. This depreciation is usually the result of the combined effects of physical changes inside the arc tube and the deposition of light-absorbing particles of electrodes or their coatings on the tube that are sputtered off from the prolonged impact of arc particles. Figure 1-25 shows lumen depreciation of different mercury vapor lamps as they age with time. Like fluorescents, the mercury vapor lamps are seasoned for the first 100 hours before readings are taken.

Some mercury vapor lamps are phosphor-coated, mainly to convert the ultraviolet energy, which is generally absorbed by the outer glass in a clear lamp, to visible radiation. This improves the color and also increases the total lumen output for certain lamps. For a clear mercury lamp the maintenance characteristics are determined only by the arc tube. For a phosphor-coated lamp, however, the maintenance characteristics are determined by both the arc and the phosphor. Figure 1-26a shows the lumen maintenance characteristics of clear, color-improved, and white lamps. The color-improved mercury lamps have always been better in maintenance characteristics; the white lamps have been good in initial lumens. Today's deluxe white lamps combine both these virtues. The phosphor for warm deluxe lamps is the same as that for the deluxe lamps; however, because the coating is much heavier, the warm deluxe lamps produce

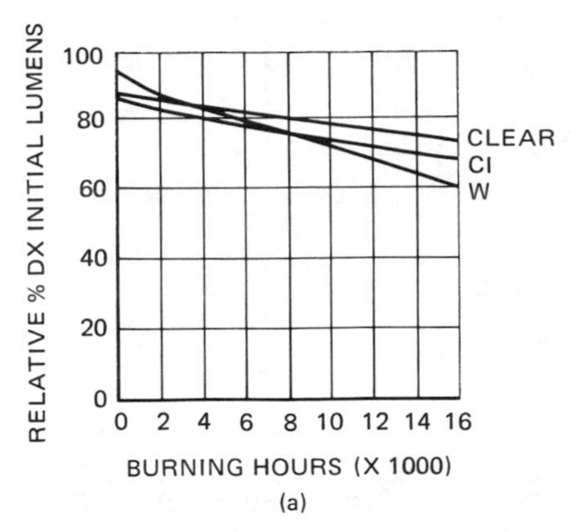

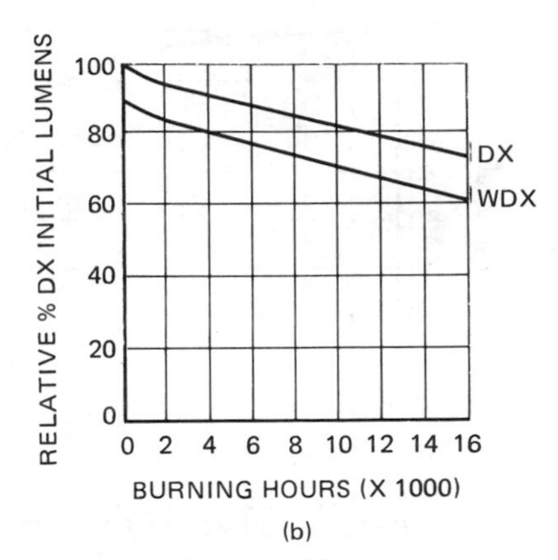

Fig. 1-26. (a) Lumen depreciation of clear, color-improved, and warm white mercury vapor lamps. (Courtesy of General Electric.) (b) Lumen depreciation of deluxe and warm deluxe mercury vapor lamps. (Courtesy of General Electric.)

fewer lumens than the deluxe lamps. Figure 1-26b shows the lumen-maintenance characteristics of typical mercury vapor deluxe and warm deluxe lamps.

Mortality

Perhaps the main reason for mercury vapor lamps' popularity is their outstanding life. The general service lamps of 100 to 1000 watts have average rated life of 24,000+ hours; others are in the range of 18,000 hours. All ratings are based on ten or more burning hours per start for the lamps with 24,000 hours of life and five or more burning hours per start for the lamps with 18,000 hours of life. As in fluorescents, the life of mercury vapor lamps is also dependent on the number of hours the lamp is burned per start. Every time the lamp is started, the electron emission material on electrodes sputters off and reduces the lamp life.

Normally, for any lighting source the average rated life of a lamp is designated when 50% of the lamps in a group still operate. For mercury vapor lamps, however, at the end of their rated life of 24,000 hours, statistically about 67% in a group still remain active (see Figure 1-27). Although a 50% mortality occurs much further than 24,000 hours, this was accepted as the rated life, since mercury vapor lamps are poor in lumen efficacy. The + sign indicates that its 50% burn-out occurs in excess of 24,000 hours.

Mercury Lamp Ballasts

Since there are no specific standards for HID ballasts and manufacturers are not held to specific regulations, there are large variations in lamp performance. Bal-

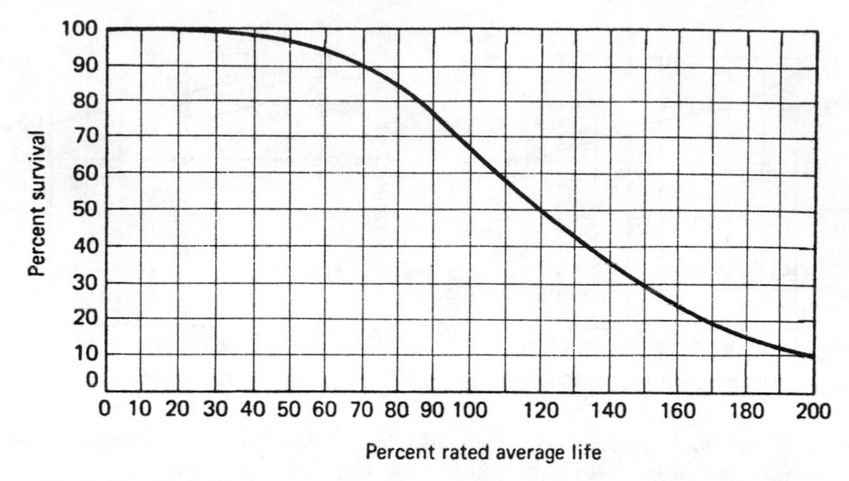

Fig. 1-27. Typical mercury vapor lamp mortality. (Courtesy of General Electric.)

last terminology, as a result, has been confusing since much overlapping has occurred by naming ballasts by their electrical characteristics. A brief summary of different types of mercury ballasts includes the following:

1. Inductive ballast:
   - High-power-factor reactor ballast
   - Low-power-factor reactor ballast
   - Nonstabilized ballast
   - Reactor ballast
2. Lag ballast:
   - Autoreactor ballast
   - Autotransformer ballast
   - High-power-factor autotransformer ballast
   - Low-power-factor autotransformer ballast
   - Nonstabilized ballast
3. Regulator ballast:
   - Constant-wattage (CW) ballast
   - Lead ballast
   - Premium CW ballast
   - Saturated inductive and capacitive ballast
   - Stabilized ballast
4. Autoregulator ballast:
   - Autostabilized ballast
   - Autotransformer and regulating ballast
   - Combined inductive and capacitive ballast
   - Constant-wattage autotransformer (CWA) ballast

Inductive Ballast (Figure 1-28a). This is the simplest type of ballast in the family. It is simply a wire coil wound on an iron core connected in series with the lamp, which serves to limit the current. Unless an across-the-line capacitor is connected, this highly inductive circuit produces very low power factor, as low as 50%.

Operational Characteristics:

Ballast losses	Low.
Power factor	Pure inductive circuit, 50%. With capacitor across-the-line, over 90%.
Starting current	$2\frac{1}{2}$ times the normal operating current.
Voltage regulation	Nonregulating; use is limited to lines where ±5% line

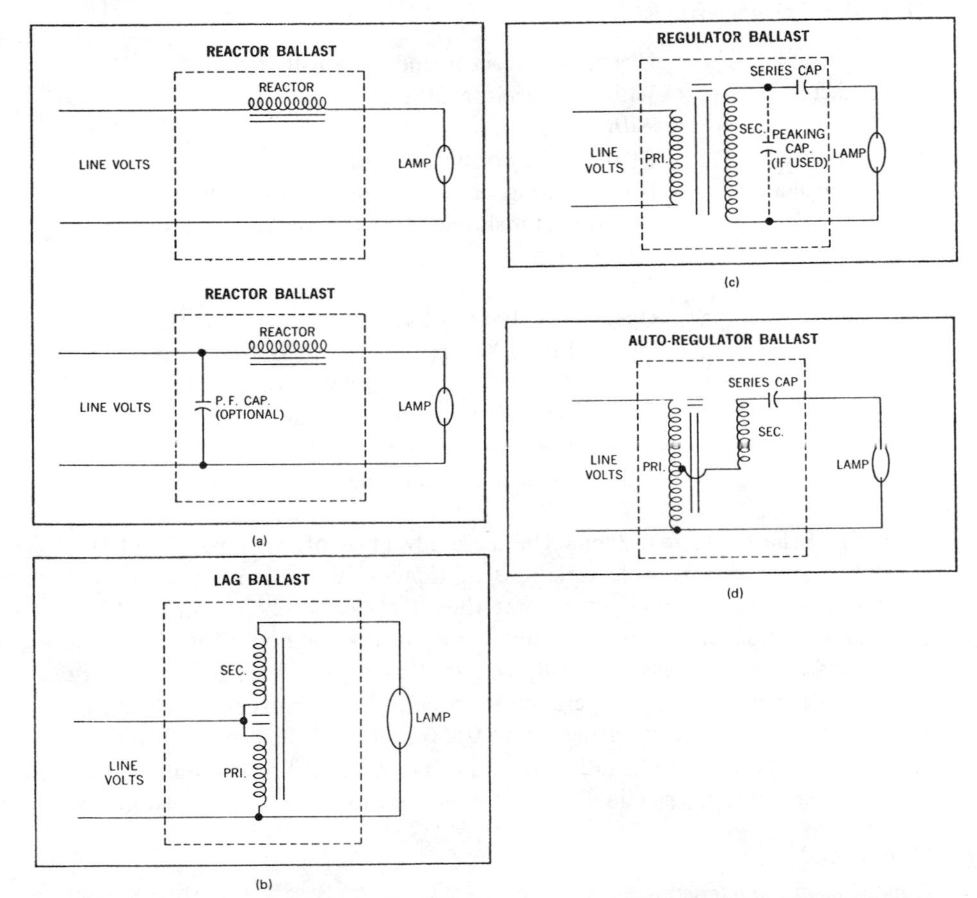

Fig. 1-28. Mercury vapor ballasts. (Courtesy of General Electric.) (a) Inductive ballast with or without a capacitor. (b) Lag ballast. (c) Regulator ballast. (d) Autoregulator ballast.

	voltage is maintained. Lamp watts vary ±10% with a ±5% change in line volts.
Current crest factor	1.4–1.5.
Line voltage variation	Minimum line voltage 240 and 277 volts for all wattage lamps except for 750, 1000, and 1500 watt lamps, which operate on 480 volts.
Others	Least expensive, simple, light and quiet.

Lag Ballasts (Figure 1-28b). This type of ballast includes autotransformers and reactors combined on single structures. Unless capacitors are used, their inherent power factor is very low, as in the reactor type.

Operational characteristics:

Ballast losses	Higher than reactor (inductive ballasts).
Power factor	Without capacitor, 50%. With capacitor, over 90%.
Starting current	$2\frac{1}{2}$ times the normal operating current.
Voltage regulation	Nonregulating, use is limited to lines where ±5% line voltage is maintained. Lamp watts vary ±10% with a ±5% change in line volts.
Current crest factor	1.4–1.5.
Line voltage variation	Operating voltage is 240 and 277 for all wattage lamps except for 1000- and 1500-watt lamps, which operate on 480 volts. If the line voltage is different from rated operating voltage, the autotransformer steps it up to the required voltage.
Others	Higher cost than reactor type.

Regulator Ballast (Figure 1-28c). The main advantage of this type of ballast is its ability to operate at wide variations of input voltage. It has primary and secondary windings wound on a core; they are electrically separated. The secondary portion of the core operates in magnetic saturation. The current in the secondary side remains essentially constant over a wide variation of primary voltage. The current limiting operation is provided by a series-connected capacitor that shares the limiting current at a 50:50 ratio with the ballast. The series capacitor makes it a "lead" rather than a "lag" circuit. For low temperature starting, sometimes a "peaking" capacitor is used to increase the open circuit voltage.

Operational Characteristics:

Ballast losses	Ballast losses are highest.
Power factor	95%.

Starting current	Lower than operating current.
Voltage regulation	Lamp watts vary only ±2 or 3% with line voltage changes of ±13%.
Current crest factor	1.7–2.0.
Line voltage variation	Any voltage for all wattage lamps.
Others	The most expensive ballast. Physical size is larger than other ballasts.

Autoregulator Ballast (Figure 1-28d). This type of ballast combines the auto-transformer and the regulator circuit. The degree of regulation depends on the amount of primary voltage coupled into secondary. Being a trade-off between the regulator and the lag ballasts, this type includes the most popular ballasts in new installations.

Operational Characteristics:

Ballast losses	Ballast losses are slightly lower than regulator ballasts.
Power factor	90%.
Starting current	Lower than operating current.
Voltage regulation	Lamp watts vary about 5% with line voltage changes of 10%.
Current crest factor	1.7–2.0.
Line voltage variation	The system will work with any line voltage, for any wattage lamp.
Others	These ballasts are smaller, lighter, and less expensive than the regulator ballasts.

Two-Lamp Ballasts

Two-lamp ballasts are in general of two different types: two-lamp series ballasts and two-lamp parallel ballasts. The main advantage in a two-lamp ballast is the reduced cost of installation and relatively lower per unit cost, compared to having two individual units for the same size of lamps. The other advantage is the somewhat reduced power consumption, compared to two individual ballasts of the same lamp sizes. For example, the input wattage of two individual 400-watt mercury vapor lamps operating on two individual ballasts is 454 + 454 = 908 watts. Power consumption for the same two lamps operating on a two-lamp, similar type of ballast is 880 watts.

Of the two, the two-lamp series ballast circuit is by far the more popular, mainly because of the lower cost, minimum size, and lesser weight. The connection of the wires is also simple, being confined to only two leads to accept the two lamps in series (see Figure 1-29a). The main disadvantage of these types of ballasts is that when one lamp is burned out, the other one goes out and will

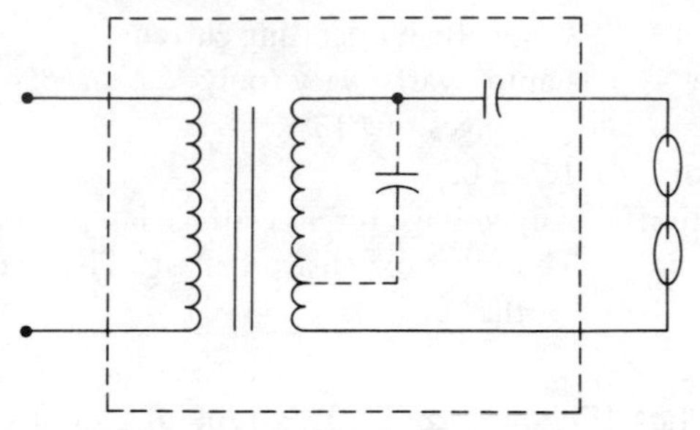

Typical Two-Lamp Series Ballast – Regulator Circuit

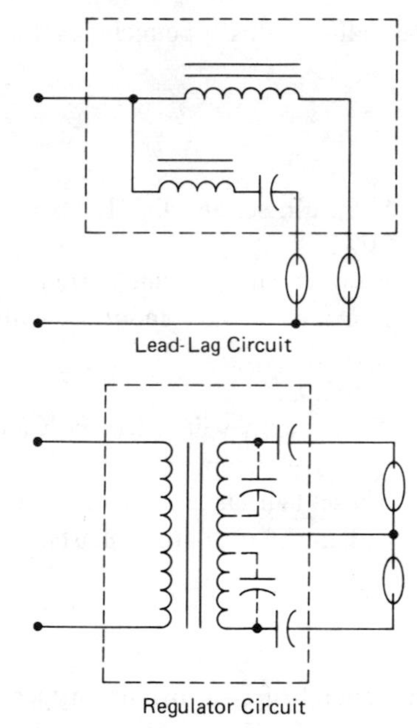

Typical Two-Lamp Parallel Ballasts

Fig. 1-29. Two-lamp ballasts for mercury vapor lamps. (Courtesy of General Electric.) (a) Two-lamp series ballast regulating circuit. One disadvantage with this system is that if one lamp burns out, the other one goes out, too, and will not glow until the first one is replaced. (b) Two-lamp parallel ballast circuit. These basically incorporate two separate circuits for the two lamps. Burnout of one lamp does not affect the other one.

not glow until the first one is replaced. The two-lamp parallel ballast consists of an independent circuit for each of the two lamps and thus is more expensive than the series design (see Figure 1-29b). These have three supply leads to be connected to thc two lamps. Failure of one lamp does not compel the other one to go out.

Self-Ballasted Mercury for Incandescent Retrofits

As their name implies, these special lamps do not have separate ballasts, but the ballast circuits are an integral part of the filaments themselves. These use medium-skirted mogul bases and screw-in sockets just like incandescents. In general, self-ballasted mercury lamps are 50% less efficient than standard mercury vapor lamps but 50% more efficient than incandescent lamps. Depending on the type of lamp and wattage, the life ratings vary as much as 8 to 20 times longer than that of their incandescent counterpart. The combined quality of easy installation, higher lumen efficacy, and longer life make these ideally suitable for replacement of incandescents located in high-ceiling areas. These are also sometimes suitable for areas where a high amount of light is required, but installation of separately mounted ballasts is impractical.

There are two types of self-ballasted mercury lamps: (1) incandescent filament lamps and (2) solid-state starting lamps. The incandescent filament self-ballasted mercury lamps make use of a bimetal switch, a filament heater coil, and a filament ballast. Heat from the filament ballast opens the bimetal switch and the filament heater coil, which helps in the ionization of the gas as the starter electrode starts an arc to one of the main electrodes. As the arc stabilizes, the filament ballast limits the current. These lamps are recommended to operate in ambient temperatures between a minimum of 5°F and a maximum of 122°F (see Figure 1-30a).

The solid-state starting self-ballasted mercury lamps do not have any bimetal switch. The input 120-volt AC is rectified to full-wave dc voltage. An approxi-

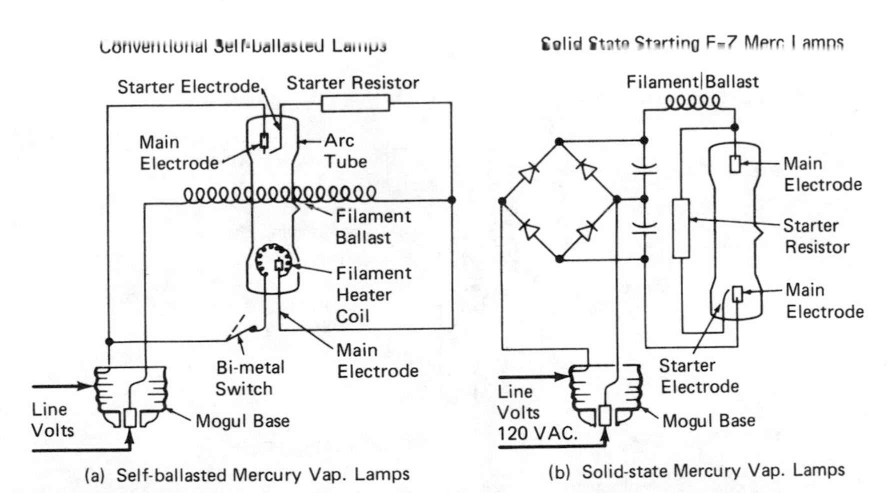

Fig. 1-30. Conventional and solid-state, self-ballasted mercury vapor lamps. (Courtesy of General Electric.) (a) Design circuit of an incandescent filament self-ballasted mercury vapor lamp. (b) Design circuit of a solid-state–starting self-ballasted mercury vapor lamp.

mate 300 peak voltage is provided by the capacitors for the starting pressure, which strikes the arc between the starter electrode and the cathode. The starter electrode circuit becomes highly resistive as the arc tube voltage builds up (see Figure 1-30b). These lamps are recommended to operate in ambient temperatures between 50°F and 122°F. The average rated life of these lamps is about 14,000 hrs. They provide approximately 50% higher efficacy than the incandescent filament self-ballasted lamps: The starting warm-up and restriking time is $1\frac{1}{2}$ minutes and 6 minutes, respectively. Because of high heat build-up, these lamps should not be used in enclosed luminaires.

METAL HALIDE

Principles of Operation and Types

The metal halide lamps are similar to mercury vapor lamps in operating principles. They are also very similar in physical appearance to conventional mercury vapor lamps (see Figure 1-31). The main difference in the two is that the metal halide lamps contain some important metallic additives in addition to argon and mercury in the arc tube, which partially vaporize and produce different color rendering in the overall light output. Two typical combinations of metal additives are generally used in metal halide lamps: (1) sodium iodide and scandium iodide and (2) sodium iodide, thallium iodide, and indium iodide. Of the different metal

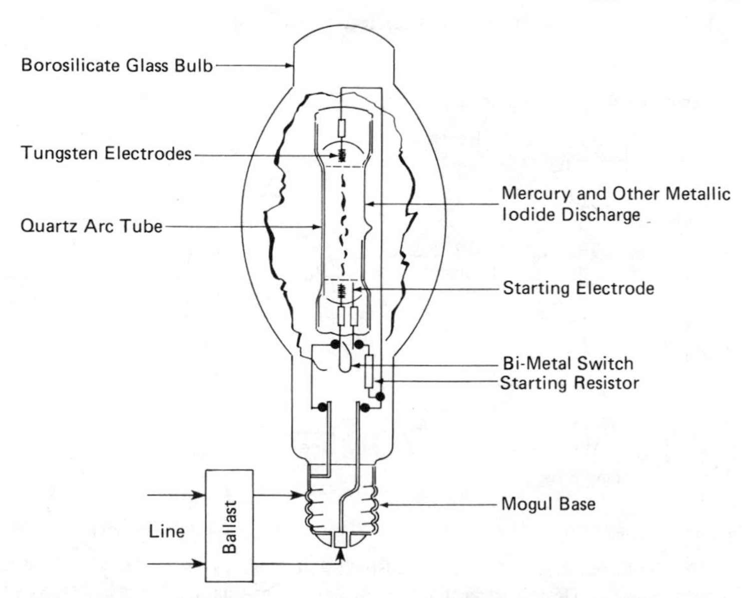

Fig. 1-31. Typical design circuit and construction of a metal halide lamp.

additives, the sodium, thallium, and indium are the principal producers of dominant color spectra, while the others produce multiline spectra across the full visible region. The other main difference is that the arc tubes in metal halide lamps are usually smaller than those in the mercury lamps with equivalent wattage. Some lamps include a bimetal switch to short-circuit the starter electrodes (i.e., starter electrode and the main electrode) after the arc has been struck; this eliminates the possible build-up of a small voltage between the electrodes that could cause electrolytic failure of the pinch seal. Metal halide lamps that make use of such bimetal switches require two types of lamps for each wattage, chiefly because the bimetal switch has to be properly positioned to receive the maximum heat inside the lamp. For the base-down vertical position, the location of the bimetal switch is at the top, away from the base; in the base-up position, its location has to be near the base, at the top. As mentioned earlier, because of the various properties of the metals added, radiations occur in various colors. Among the lot, the sodium vaporizes at the very end and produces orange, yellow, and a little of red (which happen to be the colors to which human eyes are most sensitive). Any reduction in operating wattage causes less ionization of the sodium iodide, which results in reduction of these warm tones in the total light output of the metal halide lamps. Conversely, when wattage is increased, more sodium is ionized, resulting in a shift of color towards the pink.

In general, metal halide lamps having a greater number of metal additives (as in the second common combination mentioned above) are more sensitive to color variations. For this reason, for an area where the source of light is easily visible and color rendering and consistency are of prime importance, it is better to use lamps with fewer metal additives, as in the first typical combination. Manufacturers should be consulted for such a decision.

Metal halide lamps are available from 175-watt to 1500-watt. They are either type E (elliptical) or BT (bulged-tubular) (see Figure 1-20). All wattage lamps, except for the 1500-watt, are available in base-up and base-down design. The lamps are either clear or phosphor coated. A phosphor coating improves color rendition slightly. In both the vertical operations (base-up or -down), the lamps can be operated successfully up to ±15 degrees from the vertical axis.

Color Characteristics

Figure 1-32 shows the SPD curves for metal halide lamps. The metal halide lamps provide energy at all wavelengths across the visible spectrum, creating a well-balanced color rendering. A phosphor coating is not a must for color improvement since the excellent color rendering is directly obtained from the arc itself. A phosphor coating helps to diffuse the source and should be considered for low-ceiling applications. Color consistency from one lamp to another depends on the type of metal additives in lamps, the supply voltage, ballasts, and

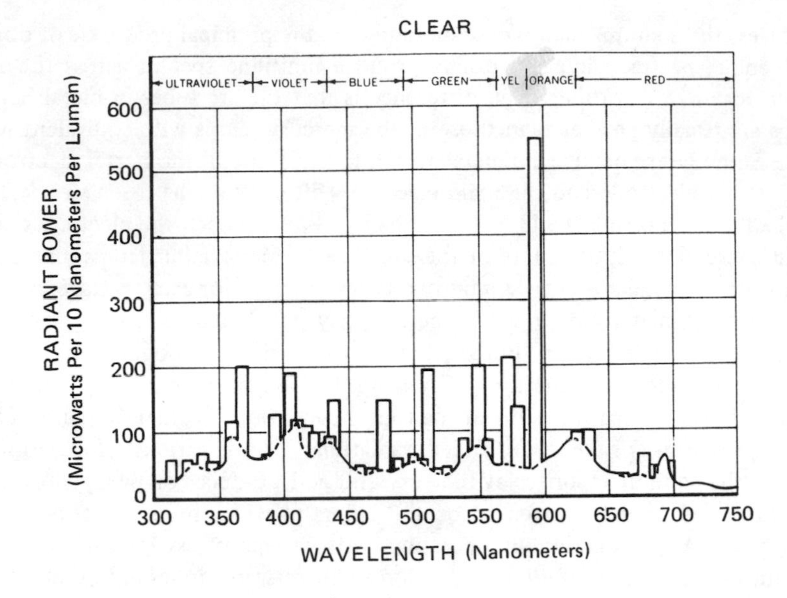

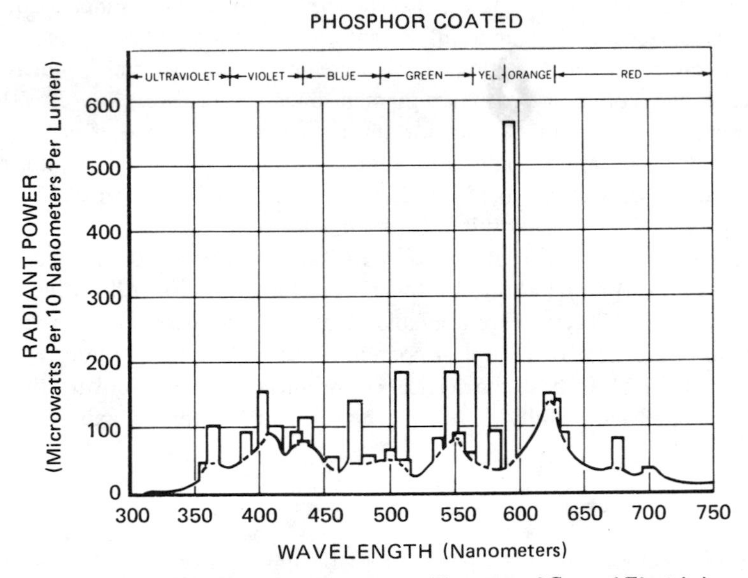

Fig. 1-32. SPD of metal halide lamps. (Courtesy of General Electric.)

the lamp age. When a color consistency is critical among a group of lamps, lamps should be replaced in group rather than on a spot burn-out basis. Table 1-15 shows color comparison of metal halide and other lamps. Energy distribution of a 400-watt metal halide is shown in Figure 1-33.

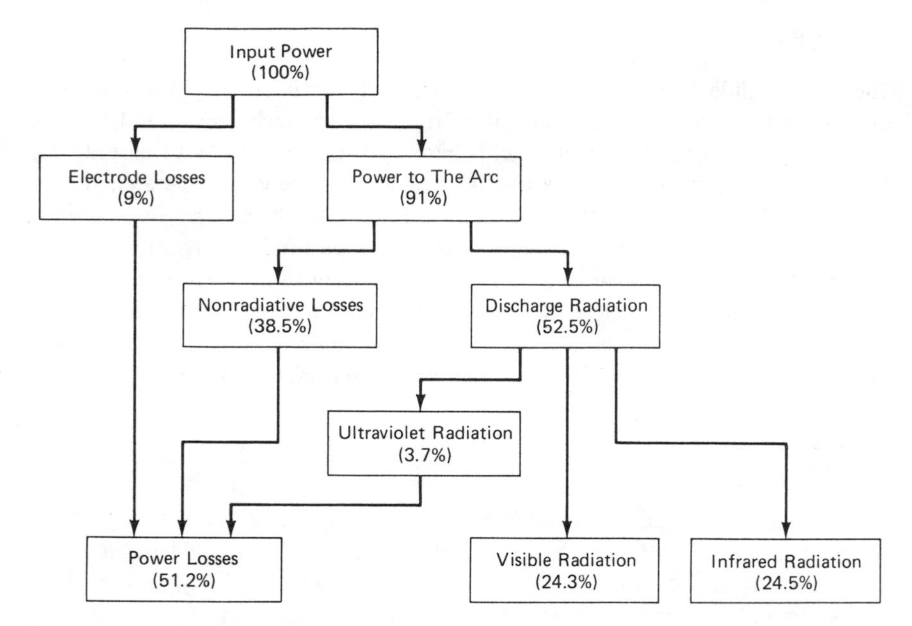

Fig. 1-33. Energy distribution of a 400-watt metal halide lamp. About 24.3% of total energy is available for visible radiation.

Warm-Up and Restriking Time

The basic principles of starting metal halide lamps are the same as for mercury vapor lamps. However, because of the presence of different metallic additives, the starting voltage is normally higher than that for the mercury vapor lamps. As the lamp warms up, it produces different color changes with the vaporizations of the metal additives, and several minutes are required before a full brightness is established. The metal halide operates at a relatively higher temperature than a mercury lamp. As a result, its total time to cool down and warm back to full equilibrium brightness is about 10 to 15 minutes. The starting warm-up time is about 5 minutes.

Effects of Temperature

As in mercury vapor lamps, metal halide is not critically affected by change in ambient temperature. The recommended minimum starting temperature is -20°F for single lamp ballasts and 0°F for the twin lamp ballasts. Maximum operating ambient temperature is 130°F. Abnormally low temperature may develop a condition of vapor pressure in the arc tube, preventing it from striking at rated voltage or preventing the lamp from warming up to its fullest brightness.

Burning Position

The metal halide lamps are extremely sensitive to their burning position. For the proper functioning of the bimetal switch, each vertically operated lamp has two separate designs, one to burn with base up and the other with base down. The lumen output for either is identical as long as they are operated in the vertical position, with a tolerance of maximum ±15° from the vertical axis. However, when operated horizontally, these lamps produce significantly low lumen output and also change in color. When the burning position of a metal halide is changed, as much as 6 hours may be required before the lamp color, lumen output, and other electrical characteristics stabilize. See Table 1-10 for data comparing the vertically operated lamps to the horizontally operated.

Efficacy

Perhaps the biggest advantage of metal halide lamps over the mercury vapor lamps of the same wattage is their substantially higher efficacy. When the metal halide lamp is operated in vertical position, its efficacy range may vary from 66 to 100 lumens per watt, compared to 22-60 lumens per watt for mercury, ballast losses included. The efficacy of the 1500-watt lamp is as high as 100 lumens per watt, ballast losses included. This substantially high efficacy along with better color rendering properties makes these sources ideally suitable for high-ceiling areas where a lot of light with good color is required. See Table 1-10 for the performance of metal halide lamps.

Table 1-10. Metal Halide Lamp Data.

			Initial Lumens		
Watts	Bulb	Description	Vert.	Hor.	Life, h
175	E-28/BT-28	Clear	14,000	12,000	7,500
	E-28/BT-28	Coated	14,000	12,000	7,500
250	E-28/BT-28	Clear	20,500	19,500	7,500
	E-28/BT-28	Coated	20,500	19,500	7,500
400	E-37/BT-37	Clear	34,000	32,000	15,000
	E-37/BT-37	Coated	34,000	32,000	15,000
1000	BT-56	Clear	110,000	107,800	10,000
	BT-56	Coated	105,000	100,000	10,000
1500	BT-56	Clear	155,000	150,000	3,000

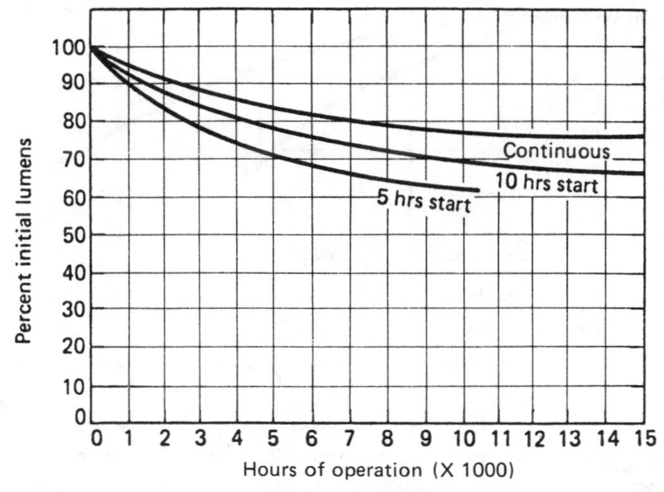

Fig. 1-34. Metal halide lumen depreciation.

Lumen Maintenance and Depreciation

In the family of HID lamps, the metal halide lamps depreciate the most and fastest. A 400-watt lamp that initially produces 34,000 lumens goes down to as low as 70% of its initial lumen after only 10,000 hours of burning, compared to 80% for mercury lamps of similar wattage. See Figure 1-34 for lumen depreciation of typical metal halide lamps. Like mercury vapor lamps, the depreciation is mainly the result of the deposition of light-absorbing particles of electrodes or their coatings on the tube that are sputtered off from the prolonged impact of arc particles.

Mortality

The main disadvantage of the metal halide lamps is perhaps their relatively shorter life. The 175- and 250-watt lamps are only rated 7500 hours; the 400-watt lamps are rated 15,000 hours (some are now 20,000), compared to a consistent 24,000+ hours for most mercury vapor vapor lamps (all are rated for a minimum of 10 hours of burning per start). The mortality curves are shown on Figure 1-35 for different wattage lamps. As can be seen, except for some specially made 400-watt metal halide lamps, all lamps are down to about 50% of their rated lumen output within 10,000 hours of burning. The horizontally operated lamps depreciate to 50% of rated lumen within 6000 hours. The 1500-watt lamps, which have the maximum efficacy, have the least life, as low as 3000 hours (10 hours per start), which is only comparable to some incandescent lamps. The substantially high lumen output and severely low life make these suitable

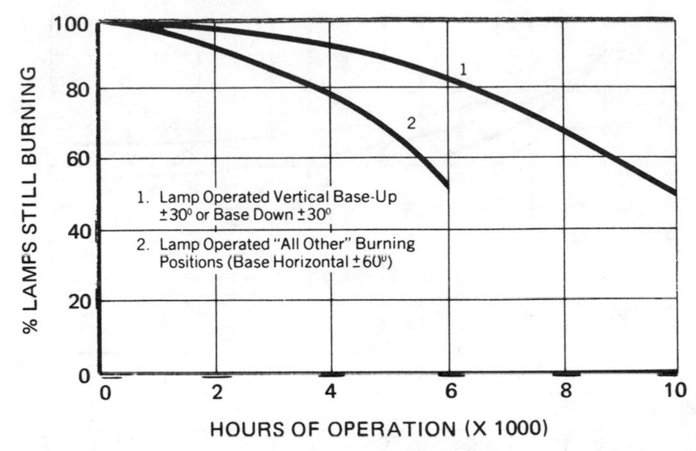

Fig. 1-35. Typical mortality curves of 400- and 1000-watt standard metal halide lamps. When operated vertically, base up or base down, lumen maintenance is better than in any other operating position. This is shown with curves 1 and 2. (Courtesy of General Electric.)

only where high amounts of lighting of good color are required for relatively shorter periods, such as in sports arenas and stadiums.

Ballasts

Although metal halide and mercury vapor lamps are quite similar in operational characteristics, all metal halide lamps cannot be used with mercury vapor ballasts. Certain metal halide lamps may be used with certain lag-type reactor mercury ballasts; however, the manufacturer should be consulted before such a move is made. Manufacturers of HID ballasts normally will provide a list of such interchangeability.

Because of their ionization phenomenon, the metallic additives in a metal halide lamp require a higher ballast open-circuit voltage and a higher reignition voltage. The voltage normally available from a mercury ballast is much less than this requirement. As a result, when an unsuitable mercury ballast is used with a metal halide lamp, the arc will be extinguished, and the lamp will cool, restart, and continue to repeat the cycle. At times the ballast may operate the lamp satisfactorily when it is new, but as the lamp ages, trouble will start. Some manufacturers have specially designed some metal halide lamps that can operate with mercury vapor ballasts. This is discussed later in this section.

A special ballast, known as the "lead-peaked" ballast, overcomes this problem, making it suitable for all metal halide lamps. These ballasts basically adopt the same circuitry as the autoregulator mercury vapor ballast but provide the highly peaked wave shape of open-circuit voltage by having one or more large slots in a portion of the core under secondary winding (see Figure 1-36).

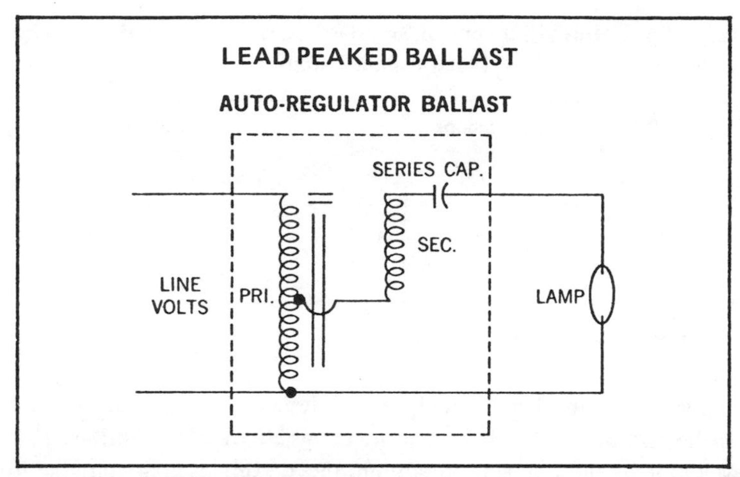

Fig. 1-36. Lead-peaked ballast of a metal halide lamp. These adopt the same circuitry as the autoregulator mercury vapor ballast but provide the highly peaked wave shape of open-circuit voltage by one or more large slots in a portion of the core under secondary winding. (Courtesy of General Electric.)

Operational Characteristics:

Ballast losses	Equivalent to mercury regulator.
Power factor	90%.
Starting current	Lower than operating current.
Voltage regulation	For a ±10% line voltage change, the lamp watts will vary about ±2%.
Current crest factor	1.6 to 1.8.
Line voltage variation	System will work at any voltage.

Two-Lamp Ballasts

Similar to the mercury vapors, the metal halide two-lamp ballasts also offer the advantage of reduced cost of installation and relatively lower per unit cost. These also provide somewhat better energy saving compared to the situation in which there are two individual ballasts. For example, the input watts of two individual ballasts is 460 + 460 = 920 watts. Power consumption for the same two lamps operating on a two-lamp ballast is 880 watts.

Latest Development in Metal Halide Lamps

Recent developments in metal halide lamps have brought a wider choice in lamp selection for new installations as well as retrofits. They include (1) Curved arc tube and (2) lamps for mercury vapor retrofits.

Table 1-11. Initial Lumens of Standard Lamps vs. Curved Arc Tube Metal Halide Lamps.

	Standard Lamps			Curved Arc Tube Lamps		
Watts	Vertical	Horizontal	Life	Vertical	Horizontal	Life
175	14,000	12,000	7,500	n/a	15,000	12,000
400	34,000	32,000	20,000	n/a	40,000	20,000

Source: Courtesy of GTE Sylvania.

The curved arc tube lamps are specially designed to have an arc tube that matches the contour of the arc when the lamps are operated horizontally. With this development, the temperature around the arc tube is maintained uniformly, producing much higher lumen output than that of their conventional, vertically operated counterparts. These lamps are available in 175-watt and 400-watt. The data for comparison are shown in Table 1-11. The 175-watt lamp produces 25% better efficacy with 33% longer life, and the 400 watt lamp produces 25% better efficacy for the same life rating as compared to the horizontally operated standard lamps. These lamps are to be used in suitable enclosed luminaires, with the curved arc tube bowing upwards. They are available either clear or phosphor coated; in either case, the operating characteristics remain the same. The lamps are to be operated horizontally within ±15° tolerance.

Metal halide lamps specially designed to replace existing mercury vapor lamps with approved ballasts are available only in 325, 400, and 1000 watts. With the use of these lamps, the lumen output increases substantially, as is shown in Table 1-12. The life rating, however, decreases to as low as 50%. Manufacturers' instructions must be followed to determine the type of ballasts the existing lamps must have in order to operate these energy-saving lamps successfully.

Table 1-12. Metal Halide Lamps for Mercury Vapor Retrofits.

	Mercury Vapor		Metal Halide	
Watts	Life	Lumens	Life	Lumens
400	24,000+	21,000	15,000	34,000
1000	24,000+	58,000	10,000	103,000

Source: Courtesy of General Electric.

HIGH-PRESSURE SODIUM

Principles of Operation and Types

The physical appearance, construction, operation, and radiation is somewhat different for high-pressure sodium (HPS) lamps than for the HID lamps discussed above. The arc tube is made of a ceramic, translucent aluminum oxide, that can withstand a temperature as high as 2372°F. The outer envelope is made of a borosilicate glass that can withstand a temperature as high as 750°F. The arc tube diameter is much smaller than that for mercury vapor or metal halide tubes, which do not permit having the starting electrode inside the arc tube. The ballast contains a special starting circuit to ionize the xenon gas across the main electrode gap by means of a low-energy, high-voltage pulse on each cycle or half cycle. The arc tube contains xenon as the starting gas, and a small quantity of sodium-mercury amalgum that gets partially vaporized at operating temperature. The main function of the mercury gas is to raise the gas pressure and operating voltage of the lamp to a practical limit. See Figure 1-37 for a typical interior of a high-pressure sodium lamp.

In general, there are two bulb shapes available. The bulbs ranging from 50-watt through 400-watt are type E (elliptical) and the 1000-watt bulbs are type T (tubular) (see Figure 1-20).

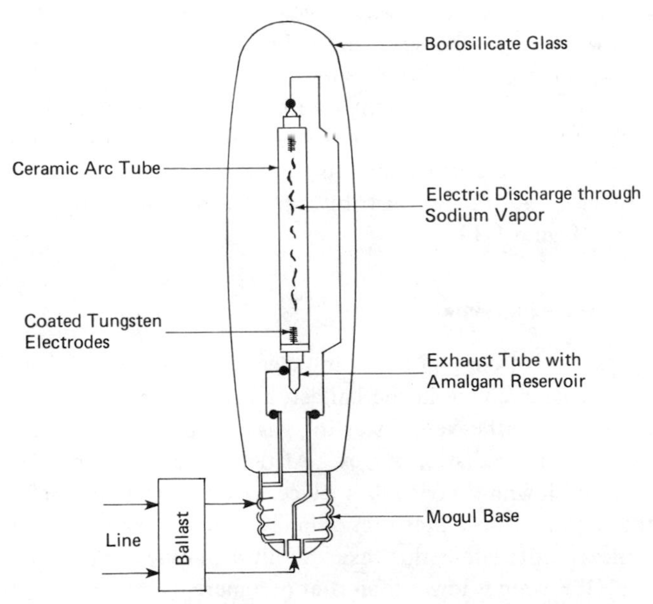

Fig. 1-37. Typical design circuit and construction of HPS lamps.

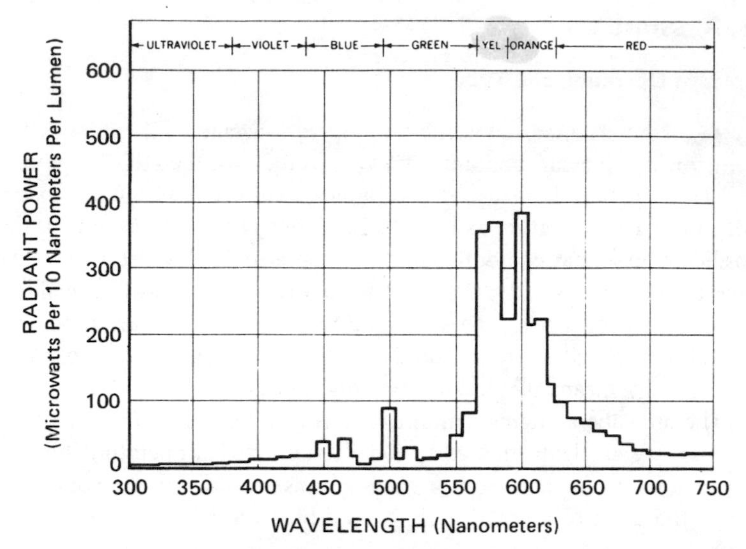

Fig. 1-38. SPD of HPS lamps. (Courtesy of General Electric.)

Color Characteristics

Figure 1-38 shows the SPD curves of HPS lamps. These lamps produce energy at all wavelengths; however, the major portion of the energy is concentrated in the yellow-orange part of the spectrum. Because of this color characteristic, red objects look orange and blue or green objects appear gray. Color consistency from lamp to lamp is better than with metal halide lamps, although color shifts can occur because of different ballast characteristics and input voltage variation. A color characteristic comparison of different types of HID lamps can be seen in Table 1-15. Energy distribution of a 400 W high-pressure sodium lamp is shown in Figure 1-39.

Warm-Up and Restriking Time

The HPS lamps do not contain a starting electrode. The starting is done with the help of a special circuitry in the ballast; a high voltage with high-frequency pulse is used to ionize the xenon starting gas. As the lamp warms up, it goes through different color radiation stages. At first, when the xenon and mercury are ionized, a bluish-white glow takes place. Next comes the monochromatic yellow of the sodium at low pressure. Finally, a broad spectrum appears to its fullest brightness. Altogether this takes about 4 minutes. Since the operating pressure of an HPS lamp is lower than that of a mercury vapor lamp, the restriking time is much shorter: about $\frac{1}{2}$ to $1\frac{1}{2}$ minutes.

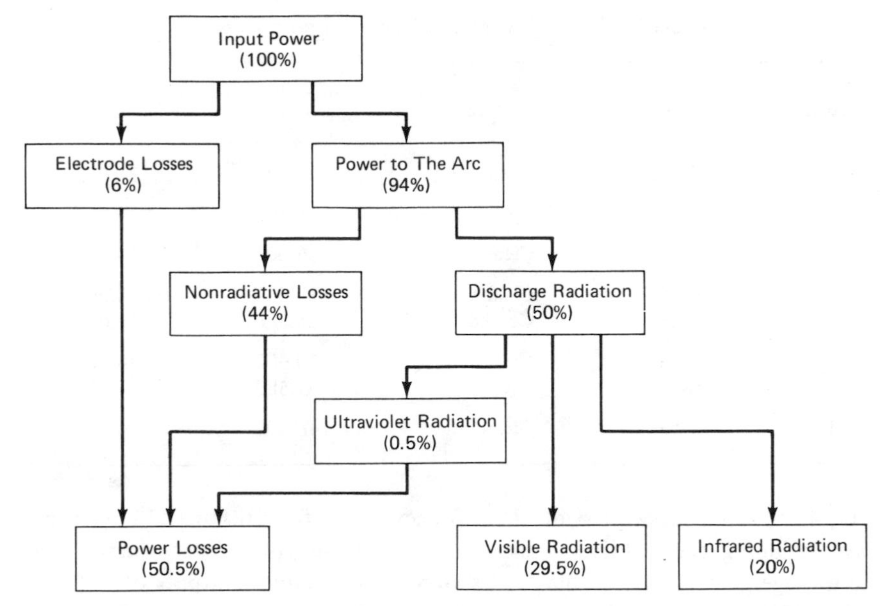

Fig. 1-39. Energy distribution of a 400-watt HPS lamp. About 29.5% of energy is available for visible radiation.

Effects of Temperature

HPS lamps, like other HID lamps are not critically affected by a change in ambient temperature. Since the arc is contained in an arc tube that is enclosed inside the outer tube, a good thermal insulation is maintained between these two tubes. Abnormally low temperatures may develop a vapor-pressure condition in the arc tube that prevents striking at rated voltage or prevents the lamp from warming up to its fullest extent.

Burning Position

When the HID lamps are burned horizontally, the arc tends to bow upwards because of convection currents. In HPS lamps, the arc tube diameter is so small that the convection current effect is insignificant. This results in successful operation of all HPS lamps to their rated output in any position.

Efficacy

Perhaps the main advantage of HPS lamps is their outstanding efficacy. As of this publication, HPS lamps are available in 50-, 70-, 100-, 150-, 200-, 215-, 250-,

Table 1-13. High-Pressure Sodium Lamp Data.

Watts	Bulb	Description	Initial Lumens (any position)	Life, h
70	E-23½	Clear	5,800	24,000
100	E-23½	Clear	9,500	24,000
150	E-23½	Clear	16,000	24,000
250	E-18	Clear	27,500	24,000
	E-28	Coated	26,000	24,000
400	E-18	Clear	50,000	24,000
	E-28	Coated	47,500	24,000
1000	T-18	Clear	140,000	24,000

310-, 400-, and 1000-watt sizes. The efficacy ranges from 60 to 127 lumens per watt, including ballast losses. Like other HID lamps, the efficacy increases as the wattage goes higher. Table 1-13 shows a list of lumen outputs of different types of HPS lamps. With moderate color rendition and outstanding lumen efficacy and life rating, these lamps are the best energy savers in the family of high-intensity discharge lamps.

Lumen Maintenance and Depreciation

Figure 1-40 shows lumen maintenance curves of all HPS lamps. The mean lumens for all lamps occur at approximately 91% of initial outputs and terminate at about 67% of initial outputs at the end of rated life (24,000 hours).

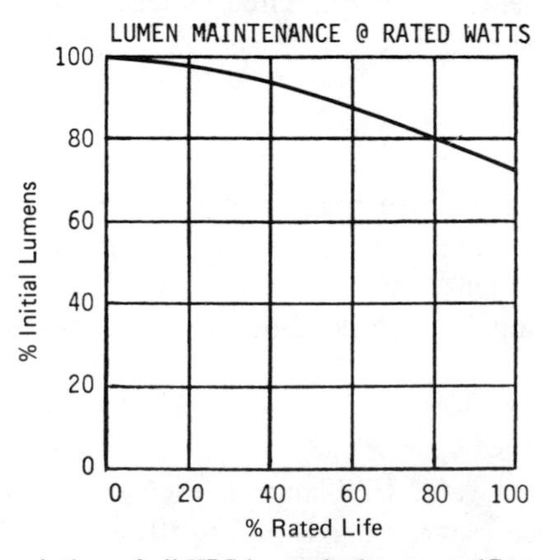

Fig. 1-40. Lumen depreciation of all HPS lamps is the same. (Courtesy of General Electric.)

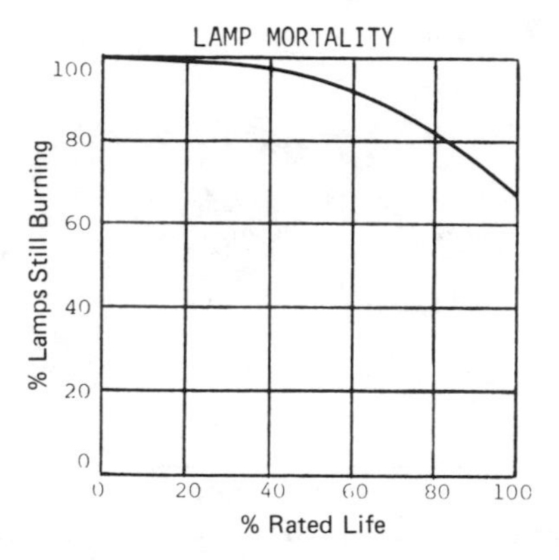

Fig. 1-41. Typical mortality curve of all HPS lamps. (Courtesy of General Electric.)

Mortality

Based on a minimum of 10 hours of operation per start, HPS lamps have a life of 24,000 hours which is an added attraction of these lamps. The life of an HPS lamp is directly related to the rate of lamp voltage rise. The lamp volts rise as the lamp gets older and keep increasing until the maximum of ballast voltage is available. After this, the normal life of the HPS lamp ends. Figure 1-41 shows the mortality curve of all HPS lamps.

Ballasts

The design criteria involved in a HPS ballast are considerably different from those of the other HID ballasts noted earlier. Two of the major differences are as follows: (1) It must be able to provide a much higher starting voltage pulse, and (2) it must control the lamp watt variation within a narrow range. The starting voltage of all lamps, 70- through 400-watt, requires 2500 volts, and that for 1000-watt is 3000 volts. The amalgam reservoir releases consistent amounts of sodium to be vaporized, resulting in a lower voltage rise and longer lamp life. As the lamp ages, the stabilizing voltage requirement increases. If the line voltage is altered to operate the lamp wattage other than as rated, the lamp voltage also changes because of the corresponding change in the amalgam vaporized. A change in lamp wattage due to change in lamp voltage can be plotted as shown in Figure 1-42. The ballast curve represents the operating characteristics of the ballast associated with the lamp, the exact shape of which will vary with the type of design. The intersecting point of the two denotes the operating point of the specific lamp and ballast combination at a particular time of the life of the lamp. Note that as the lamp voltage increases, the wattage proportionately increases to a certain point and then gradually decreases.

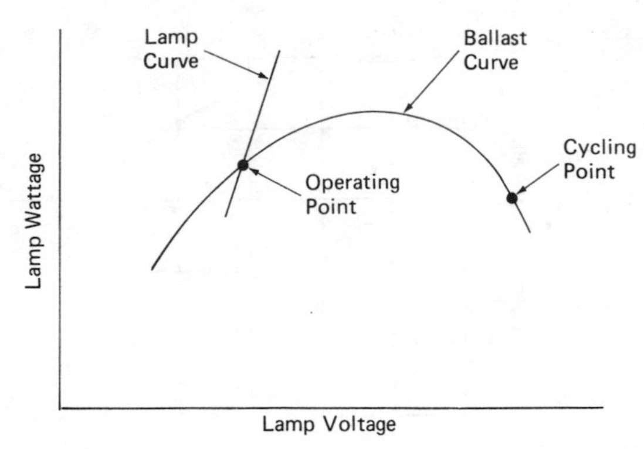

Fig. 1-42. Typical operating curves of an HPS lamp and ballast. The ballast curve represents the operating characteristics of the ballast associated with the lamp, the exact shape of which will vary with the type of ballast design.

Towards the end of a lamp's life, the higher operating temperatures require a higher stabilizing voltage beyond what the ballast can actually supply. This causes the light to be extinguished. As the lamp cools down, its stabilized voltage requirement being somewhat lower than that at high-temperature operating condition restrikes the arc and sustains the light until the temperature rises to extinguish it again. The process repeats itself, indicating that the end of life is near. This is known as the *recycling point* of the curve. This is shown in Figure 1-42. A relamping period must be selected before the recycling point is reached.

An increment in lamp watts will indicate a rise in lamp lumens. As the lamp lumen depreciates with age, the increase in lamp wattage makes up for the loss to a certain point, after which the lumen values drop consistently. The satisfactory operating range of a lamp-ballast combination is expressed by a "trapezoid" or "operating window," as shown in Figure 1-43. In selecting an HPS

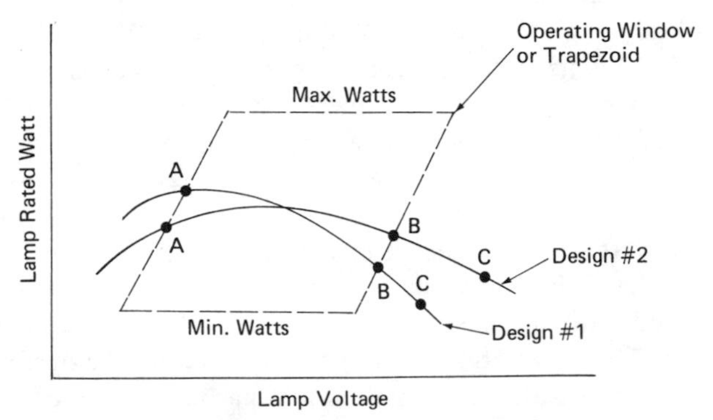

Fig. 1-43. Two types of HPS ballast design operating within a trapezoid or window. Point A represents the minimum voltage of initial operation; B is the relamping point; and C is the cycling point for each ballast. Choice is obviously design 2.

lamp-ballast combination, these data should be studied. The perimeter of the trapezoid indicates the ballast operating limits in terms of the maximum and minimum voltage and wattage it can provide for the lamp's satisfactory operation of the ballast design. Two curves, designed by #1 and #2, are shown in Figure 1-43, each representing a different type of ballast design, both to operate the same lamp. In both curves, point A represents the minimum voltage of initial operation, B the point of relamping, and C the point of recycling. The choice is obviously curve #2, since the light output will start below nominal value, but as the lamp voltage increases along the ballast characteristic curve, the system light output will increase with wattage–and stay fairly constant until relamping period. Curve #1, on the other hand, requires higher wattage and voltage in initial operation and then steadily decreases in light output with shorter life and relamping period.

In general, there are four different types of ballasts that are used for HPS lamps.

Reactor Ballast (Figure 1-44a). These ballasts can be successfully used where the system voltage is proper for operating the lamp. However, sometimes an autotransformer is used to compensate for any input voltage difference.

Operational Characteristics:

Ballast losses	Equivalent to mercury regulators.
Power factor	90%.
Starting current	Higher than operating current.
Voltage regulation	With ±5% input voltage variation the lamp watts vary about ±11%.
Current crest factor	1.4–1.5.
Line voltage variation	The system will work at any voltage when operated with an autotransformer.

Magnetic Regulator Ballasts (Figure 1-44b). The magnetic regulator ballast for HPS lamps has the circuitry to compensate for variation in lamp voltage and input-line voltage. A conventional mercury regulating ballast cannot be used for HPS lamps since it has constant current output characteristics that increase the lamp wattage in direct proportion with any variation in the line voltage.

Operational Characteristics:

Ballast losses	Highest ballast losses.
Power factor	95%.
Starting current	Lower than operating current.
Voltage regulation	With ±10% input voltage variation, the lamp wattage varies about ±4%.

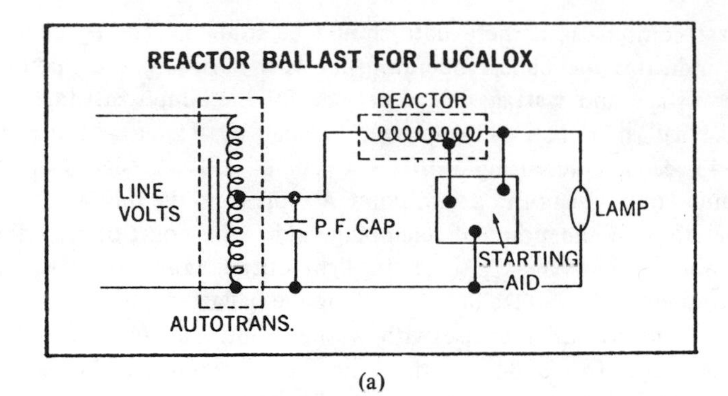

(a)

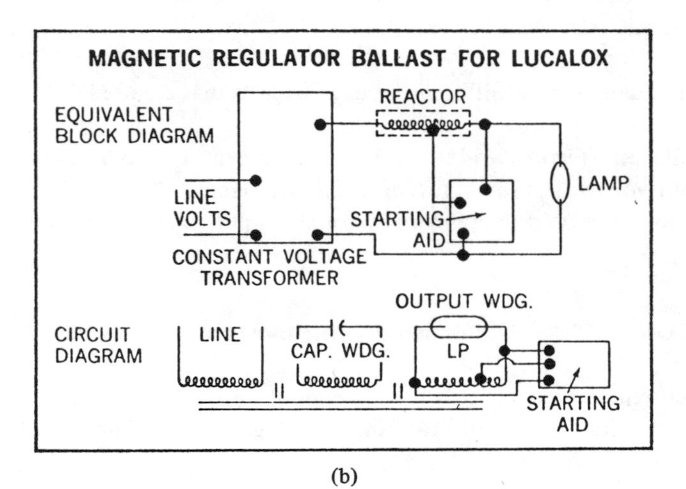

(b)

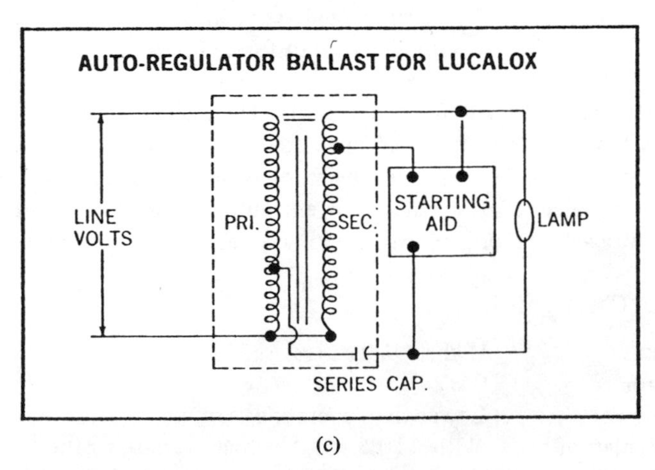

(c)

Fig. 1-44. HPS ballasts. (Courtesy of General Electric.) (a) Reactor ballast. (b) Magnetic regulator ballast. (c) Autoregulator ballast. (d) Electronic regulator ballast.

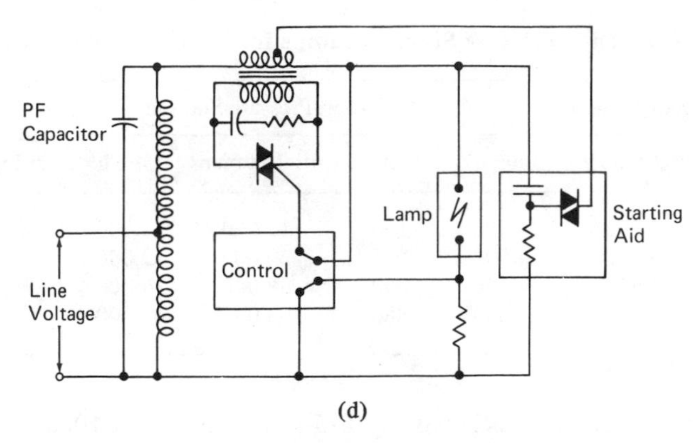

(d)

Fig. 1-44. *(Continued)*

Current crest factor	1.7.
Line voltage variation	The system will work at any voltage.

Autoregulator Ballasts (Figure 1-44c). These are similar to the regular mercury autoregulator ballasts, but with the help of special core slots and leakage reactance these meet the special requirements of HPS characteristics.

Operational Characteristics:

Ballast losses	Equivalent to the mercury vapor reactor.
Power factor	95%.
Starting current	Lower than operating current.
Voltage regulation	±10%.
Current crest factor	1.6–1.8.
Line voltage variation	The system will operate at any voltage.

Electronic Regulator Ballasts (Figure 1-44).

Operational Characteristics:

Ballast losses	Lower than magnetic or autoregulator ballasts.
Power factor	85%–90%.
Starting current	Higher than operating current.
Voltage regulation	±10%.
Current crest factor	1.6.
Line voltage variation	System will work at any voltage.

HPS Lamps for Mercury Vapor Retrofits

In recent years some manufacturers have developed special HPS lamps to replace mercury vapor lamps, using existing ballasts. These are available in 150-, 215-,

Table 1-14. High-Pressure Sodium Lamps for Mercury Vapor Retrofits.

Mercury Vapor			High-Pressure Sodium			% Improvement
Watts	Initial Lumens	Life, h	Watts	Initial Lumens	Life, h	in Light Output
175	8,500	24,000	150	13,000	12,000	53
250	13,000	24,000	215	20,000	12,000	53
400	23,500	24,000	360	38,000	16,000	62
1000	63,000	24,000	880	102,000	12,000	62

360-, and 880-watt to replace existing 175-, 250-, 400-, and 1000-watt mercury vapor lamps. Manufacturers' specifications should be carefully consulted to see which type of mercury vapor ballasts these can operate on. The greatest advantage of these lamps is an improvement in lumen output of more than 60%, at a sacrifice of 50% of life rating. Most of these lamps are rated for 12,000 hours of life, based on 10 or more burning hours per start. Table 1-14 shows the comparative data of these lamps with mercury vapor lamps.

The design of these special lamps is similar to the regular HPS lamps except that, in place of xenon, a neon-argon starting gas is used. The design also makes use of a bimetal switch inside the lamp, which eliminates the regular high-voltage starting pulse and converts the starting system to operate with mercury vapor ballasts.

These lamps are to be used only for mercury vapor retrofits. For a new installation, these should not be considered, since the regular HPS lamps of replaceable wattage produce much more light with much greater life.

LOW-PRESSURE SODIUM

Principles of Operation and Types

The light-producing element in a low-pressure sodium (LPS) is a U-shaped arc tube, which is doubled back on itself with its limbs very close together. The electrodes are sealed-in at the pinches of the arc tube mounted inside an outer vacuum jacket. The outer jacket is made of borate glass. The starting gas is neon with a small amount of xenon, argon, or helium. In some lamps the arc tube is dimpled to maintain a uniform distribution of sodium throughout. The arc is carried through the vaporized sodium under a pressure of 5×10^{-3} mm. of mercury producing intense light of yellow-orange color. The operating temperature of the arc tube is approximately 500°F.

The lamps are available from 18-watt to 180-watt, and the shape of all bulbs is tubular. The base is a bayonet base (BAY-Bl) that will maintain the U-shaped arc tube in a horizontal position (see Figure 1-45).

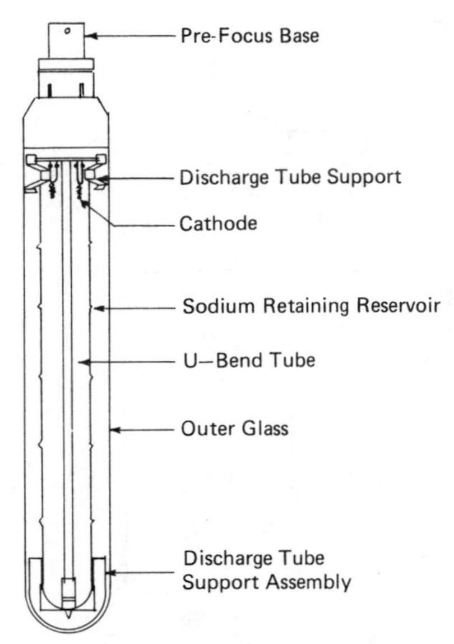

Fig. 1-45. A low-pressure sodium lamp.

Color Characteristics

Figure 1-46 shows the SPD curves of a typical LPS lamp. The light produced is monochromatic, consisting of a double line in the yellow region of the spec-

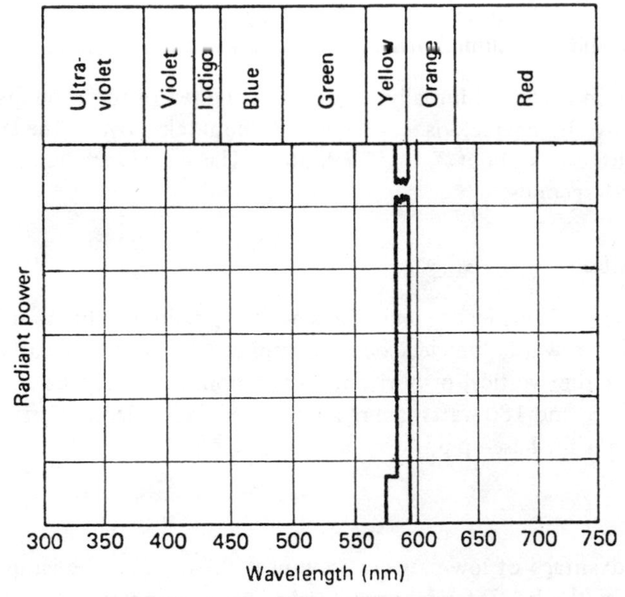

Fig. 1-46. SPD of LPS lamps.

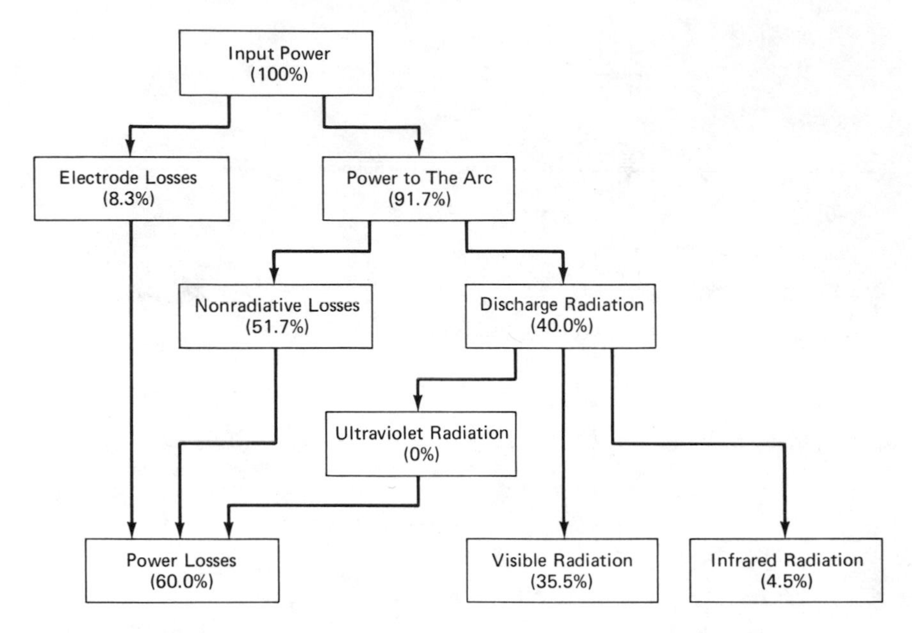

Fig. 1-47. Energy distribution of LPS lamps. About 35.5% of energy is available for visible radiation.

trum at 589 and 589.6 nm as shown. The energy distribution of a low-pressure sodium lamp is shown in Figure 1-47.

Warming-Up and Restriking Time

When the lamp is started initially, it produces the red of the neon discharge and then gradually the characteristic yellow of sodium takes over. The starting time to full brightness is about 7 to 15 minutes. The restriking time may vary between 1 to 30 seconds.

Burning Position

Burning position is critical to lamp life since lamp failure is due to the migration of the sodium towards the electrodes. Lamps ranging up to 55 watts are recommended for either vertical or horizontal operation; however, the higher-wattage lamps (90, 135, and 180 watts) are recommended for horizontal use only. Vertical operation is for base-up position only.

Efficacy

The great advantage of low-pressure sodium is the outstanding lamp lumen efficacy, which is the best of any type of light source commercially available. The

Table 1-15. High-Intensity Discharge Lamp Selection, by Color.

	Clear Mercury	White Mercury	Deluxe White Mercury	Metal Halide	High-Pressure Sodium
Efficacy	Low	Low	Low	Medium	High
Effect on neutral surface	Greenish blue-white	Greenish white	Purplish white	Greenish white	Yellowish
Effect on atmosphere	Very cool, greenish	Moderately cool, greenish	Warm, purplish	Moderately cool, greenish	Warm, yellowish
Colors strengthened	Yellow, green, blue	Yellow, green, blue	Red, yellow, blue	Yellow, green, blue	Yellow, orange, green
Colors grayed	Red, orange	Red, orange	Green	Red	Red, blue
Effects on complexions	Greenish	Very pale	Ruddy	Greyed	Yellowish
Remarks	Very poor color rendering	Moderate color rendering	Color acceptance similar to C.W. fluor.	Color acceptance similar to C.W. fluor.	Color acceptance similar to W.W. fluor.

lamp efficacy starts from 137 lumens per watt for the 35-watt lamps and ranges up to 183 lumens per watt for the 180-watt lamps. Table 1-16 shows the lumen output and life of different LPS lamps. As the lamp ages, its light output remains almost constant; but consumes much more power, thereby reducing the system efficacy to as low as 77% of the original value. This has been shown in Table 1-17.

Table 1-16. Low-Pressure Sodium Lamp Data.

Watts	Bulb	Description	Initial Lumens	Life, h
35	T-17	Monochromatic yellow	4,650	18,000
55	T-17	Monochromatic yellow	7,700	18,000
90	T-21	Monochromatic yellow	12,500	18,000
135	T-21	Monochromatic yellow	21,500	18,000
180	T-21	Monochromatic yellow	33,000	18,000

Table 1-17. Efficacy of Low-Pressure Sodium Lamps.

Lamp Nominal Watts	Lumens	Ballast Losses (Watts)	Lamp Wattage			System Efficacy (Lamp & Ballast)		
			Initial (100 h)	Mean (9000 h)	End (18,000 h)	Initial (100 h)	Mean (9000 h)	End (18,000 h)
35	4,650	25	36	41	44	76.2	70.4	67.3
55	7,700	26	55	60	64	95	89.5	85.5
90	12,500	35	90	116	122	100	82.7	79.6
135	21,500	43	130	173	178	124.2	99.5	97.2
180	33,000	40	176	220	241	152.7	126.9	117.4

Lumen Maintenance and Depreciation

Low-pressure sodium is the only one in the complete range of commercially available light sources whose lumen output actually slightly increases over the life of the lamp. The lumen output is said to be constant over an operating temperature range of 14°-104°F. This feature, along with the absence of ultraviolet rendition, does not attract insects, resulting in an increase in overall "Light Loss Factor" of this type of light source.

Mortality

Low-pressure sodium lamps have long, reliable life. All lamps are rated for 18,000 hours, and 90% are found to survive at 10,000 hours.

Ballasts

The standard low-pressure sodium ballast is an autoleakage transformer maintaining nearly constant current. However, because of the constant increment in lamp operating voltage, the total power consumption increases with age. Open-circuit voltage for ignition of the arc is from 390 to 575 volts.

REFERENCES

Collins, B. R., and S. A. Mule. "HPS Lamp Design—A Diversification in Design Wattage and Optical Control." *Lighting Design & Application,* March 1977, pp. 26–31.

Collins, B. R., W. E. Smyser, and J. T. Suter. "Low-wattage, Low-voltage HPS Lamp Design and Application." *Lighting Design & Application*, April 1975, pp. 30–36.

Fromm, O. C., J. Seehawer, and Wagner, W. J. "A Metal Halide High-pressure Discharge Lamp with Warm White Colour and High Efficacy." *Lighting Research and Technology*, 11, 1979.

General Electric Company. *Ballasts for High Intensity Discharge Lamps, Rating and Data*, GEA-8928C, December 1977.

General Electric Company. *Light and Color*, TP-119, January 1974.

Helms, Ronald N. "Energy and Lighting Design–Part Two." *Electrical Construction and Maintenance*, December 1979, pp. 53–61.

IES. "Light Sources." *IES Lighting Handbook*, 5th ed., New York: Illuminating Engineering Society, 1972, pp. 8-1 to 8-44.

Koedam, M., R. DeVann, and T. G. Verbeek. "Further Improvements of the LPS Lamp." *Lighting Design & Application*, September 1975, pp. 39–45.

Lowry, E. F., and E. L. Magner. "Some Factors Affecting the Life and Lumen Maintenance of Fluorescent Lamps." *Illuminating Engineering*, 44, February, 1949, p. 98.

McGowan, Terry K. "What about Low-pressure Sodium?" *Electrical Contractor*, May 1977, pp. 34–36.

McGowan, Terry K. "High-pressure Sodium–Ten Years Later." *Lighting Design & Application,* January 1976, pp. 45–51.

Sorcar, Prafulla C. *Rapid Lighting Design and Cost Estimating*. New York: McGraw-Hill, 1979.

Chapter Two
Photometric Analysis

A good lighting design is dependent upon the proper selection of luminaires to produce light of required quality and quantity at the right location. In selecting the luminaires, it is thus important to recognize their operational characteristics. Photometrics is the state of art that analyzes the luminaires from their operational standpoint and aids the lighting designer in making a selection. In general, each type of luminaire manufactured by a company is designed to distribute light in a particular manner that may be totally different from another type of luminaire using the same lamp. In fact, the lighting distribution characteristics of the same luminaire will differ substantially from one type of lamp to another, from wattage to wattage. Thus, it is of extreme importance to recognize the operational behavior of each type of luminaire that the designer intends to use, with the right type of lamp. The photometrics of each type of luminaire are available from the manufacturer, and are prepared in the format recommended by the Illuminating Engineering Society of North America (IES). The selection of the parameters and the techniques of tests are guided by the IES; however, the actual tests are done either by the manufacturer at its own facility or by an independently owned testing laboratory hired by the manufacturer.

A photometric report of an indoor-type luminaire includes the following:

A. Luminaire identification and general description
B. Candlepower distribution curve
C. Candlepower summary: average candlepower and zonal lumens/flux
D. Zonal lumens and percentage, or The lumen summation
E. Luminaire efficiency
F. Maximum and average brightness, or The luminance summary
G. CIE–IES luminaire classification
H. Shielding angle
I. S/MH ratio and SC
J. CU

For most indoor-type applications, the above information may be sufficient to select a luminaire; however, additional information, such as the visual comfort probability, equivalent sphere illumination, etc., may be required for certain applications where considerations of direct glare of luminaire and measurement of contrast on the task are critical. These are explained in Chapter 3. In this chapter we will analyze each of the items listed above, discuss how their values are determined, and recognize their role in energy-saving lighting design.

As it can be imagined, photometric reports of different luminaires will be different, and there is no typical report that can hold true for all luminaires. However, for the purpose of explaining, we will consider a commonly used luminaire that may cover most of the questions that arise concerning almost all luminaires. Figure 2-1 shows a typical photometric report of a 2 X 4-foot fluorescent troffer containing four lamps and a plastic lens. For convenience, each item to be discussed in the photometric report has been enumerated alphabetically.

Luminaire Identification and General Description

As the name indicates, this part of the photometric report identifies the luminaire with its manufacturer's name and product number (catalog number); illustrates with a cross-sectional picture its physical dimensions and general description; and narrates the number and type of lamps used, with the rated number of lumens produced per lamp. In this example, the luminaire selected has 0.125 inches thick acrylic prismatic lens with four 4-foot-long, cool white, 3200-lumen, rapid-start lamps and two premium ballasts.

Additional information, such as the number of planes of photometric, is also included in this section. A photometric report can be based on one-, three-, or five- or more plane analysis (see below, in this chapter). In this example, the photometric has been based on five planes, and hence is called a *5-plane photometry*.

Candlepower Distribution Curve

The candlepower distribution curve is drawn on a polar coordinate system to illustrate the luminous intensity of the luminaire. This is also known as the luminous intensity distribution curve. To determine the curve, the luminaire is placed at the center of an imaginary sphere having a radius equal to the test distance. The mathematics involved in the determination of the candlepower readings make use of the inverse square law method, which requires the luminaire to be a point source. Luminaires holding visible, bare clear bulbs with filaments, or an arc tube, such as incandescent or HID sources, will qualify as

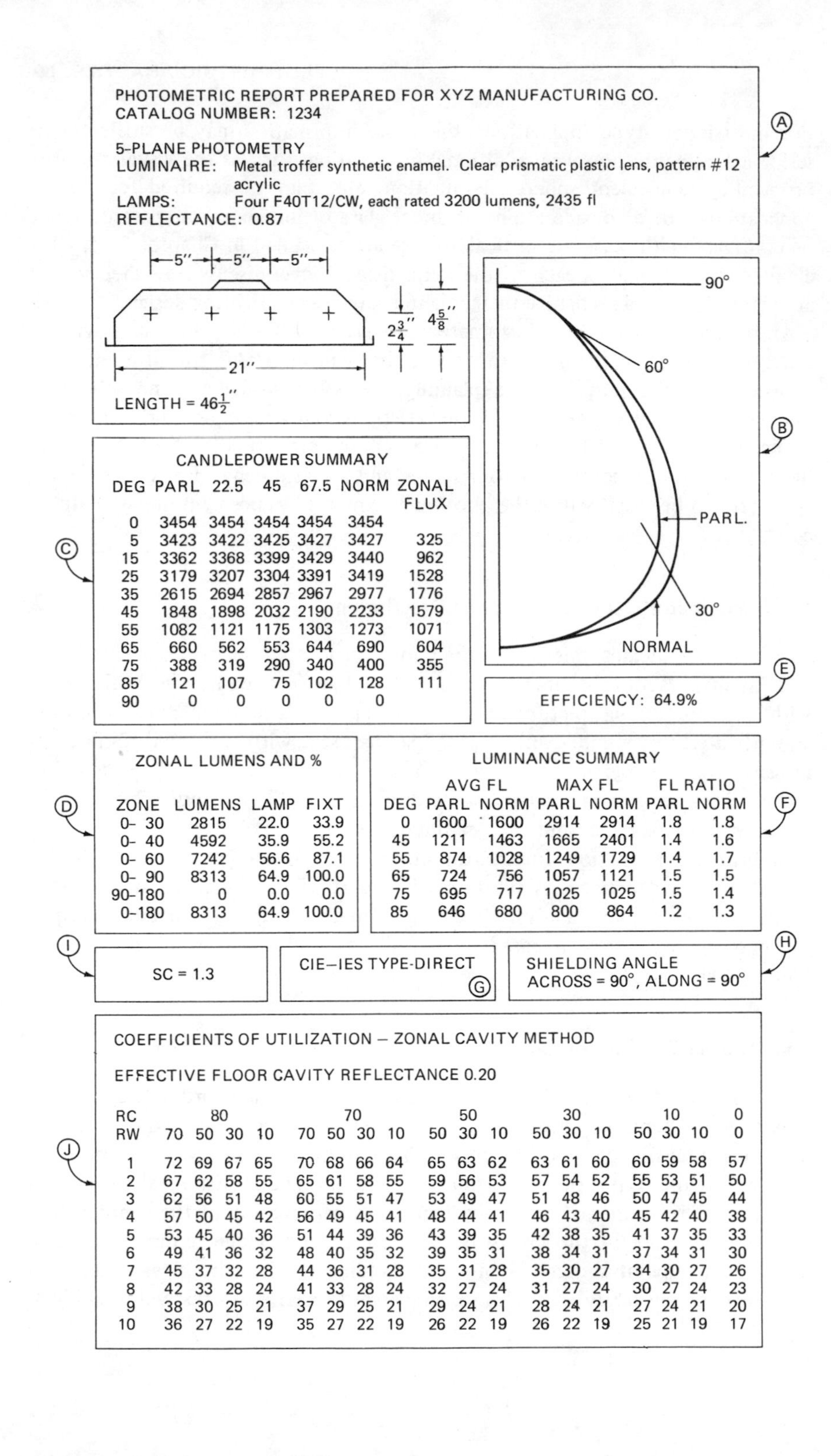

PHOTOMETRIC REPORT PREPARED FOR XYZ MANUFACTURING CO.
CATALOG NUMBER: 1234

5-PLANE PHOTOMETRY
LUMINAIRE: Metal troffer synthetic enamel. Clear prismatic plastic lens, pattern #12 acrylic
LAMPS: Four F40T12/CW, each rated 3200 lumens, 2435 fl
REFLECTANCE: 0.87

CANDLEPOWER SUMMARY

DEG	PARL	22.5	45	67.5	NORM	ZONAL FLUX
0	3454	3454	3454	3454	3454	
5	3423	3422	3425	3427	3427	325
15	3362	3368	3399	3429	3440	962
25	3179	3207	3304	3391	3419	1528
35	2615	2694	2857	2967	2977	1776
45	1848	1898	2032	2190	2233	1579
55	1082	1121	1175	1303	1273	1071
65	660	562	553	644	690	604
75	388	319	290	340	400	355
85	121	107	75	102	128	111
90	0	0	0	0	0	

EFFICIENCY: 64.9%

ZONAL LUMENS AND %

ZONE	LUMENS	LAMP	FIXT
0– 30	2815	22.0	33.9
0– 40	4592	35.9	55.2
0– 60	7242	56.6	87.1
0– 90	8313	64.9	100.0
90–180	0	0.0	0.0
0–180	8313	64.9	100.0

LUMINANCE SUMMARY

	AVG FL		MAX FL		FL RATIO	
DEG	PARL	NORM	PARL	NORM	PARL	NORM
0	1600	1600	2914	2914	1.8	1.8
45	1211	1463	1665	2401	1.4	1.6
55	874	1028	1249	1729	1.4	1.7
65	724	756	1057	1121	1.5	1.5
75	695	717	1025	1025	1.5	1.4
85	646	680	800	864	1.2	1.3

SC = 1.3

CIE–IES TYPE-DIRECT

SHIELDING ANGLE
ACROSS = 90°, ALONG = 90°

COEFFICIENTS OF UTILIZATION – ZONAL CAVITY METHOD

EFFECTIVE FLOOR CAVITY REFLECTANCE 0.20

RC	80				70				50			30			10			0
RW	70	50	30	10	70	50	30	10	50	30	10	50	30	10	50	30	10	0
1	72	69	67	65	70	68	66	64	65	63	62	63	61	60	60	59	58	57
2	67	62	58	55	65	61	58	55	59	56	53	57	54	52	55	53	51	50
3	62	56	51	48	60	55	51	47	53	49	47	51	48	46	50	47	45	44
4	57	50	45	42	56	49	45	41	48	44	41	46	43	40	45	42	40	38
5	53	45	40	36	51	44	39	36	43	39	35	42	38	35	41	37	35	33
6	49	41	36	32	48	40	35	32	39	35	31	38	34	31	37	34	31	30
7	45	37	32	28	44	36	31	28	35	31	28	35	30	27	34	30	27	26
8	42	33	28	24	41	33	28	24	32	27	24	31	27	24	30	27	24	23
9	38	30	25	21	37	29	25	21	29	24	21	28	24	21	27	24	21	20
10	36	27	22	19	35	27	22	19	26	22	19	26	22	19	25	21	19	17

point sources. However, area sources, such as the fluorescent troffers, or the linear sources, such as the fluorescent strip lights, would cause inaccurate results since they are not point sources. For this matter, when luminaires are relatively larger, the test distance should be five or more times the maximum diameter or dimension of the luminaire, so that the test luminaires can be considered as a point source. If the maximum source dimension is no more than one-fifth the distance to the point of calculation, the error will be less than 1%. The light intensity measured is expressed in candelas and they are measured in every direction from the luminaire. Symmetrical luminaires, with circular reflectors around vertically mounted lamps, are sometimes rotated on the lamp axis during test. In plotting the candlepower distribution, the luminaire is assumed to be placed at the center of the imaginary sphere in its normal burning position that corresponds to the luminaire position drawn at the center of the polar coordinate. Luminous intensity is then plotted at each angle on the polar coordinate; this represents the candlepower curve. Figure 2-2 shows a typical illustration of the candlepower curve drawn on a polar coordinate. The polar coordinate, as can be seen in the figure, has the reference axis at 0°, which is at the bottom (nadir), and 180° at the top. The readings are normally taken at intervals of 5°. If the distribution is symmetrical and the curve is smooth, reading intervals of 10°-15° are not uncommon. For a nonuniform, unpredictable distribution, however, readings may have to be taken at smaller increments. The curve connecting all such candlepower readings at various angles is known as the candlepower distribution curve.

As can be imagined, the candlepower curves for a symmetrical luminaire will always be identical at any vertical plane. This is shown in Figure 2-3. For asymmetrical luminaires, however, this is not true. In an ideal situation, for an asymmetrical luminaire, an infinite number of candlepower distribution curves have to be plotted for different vertical planes and angles, the result of which will represent a three-dimensional distribution solid. This is impractical, of course, so for an asymmetrical luminaire, distribution curves are plotted for a minimum of three vertical planes (0°, 45°, and 90°). If further accuracy is required, curves are plotted at 22.5° and 67.5° as well. Conventionally, the three-plane analysis is known as the *3-plane photometry*, and the five-plane analysis is known as the *5-plane photometry*. Note that for a symmetrical luminaire, "1-plane photometry" is sufficient, since the distribution curves for all vertical planes will be identical. Actual measurements, however are usually made through several planes, and the average is plotted.

In our example, a 5-plane photometry has been shown, since the luminaire is asymmetrical and an accuracy in the distribution study was required. This is shown in the section B of Figure 2-1.

Fig. 2-1. Typical photometric report of a 2 × 4-foot fluorescent troffer, equipped with four lamps and acrylic prismatic lens.

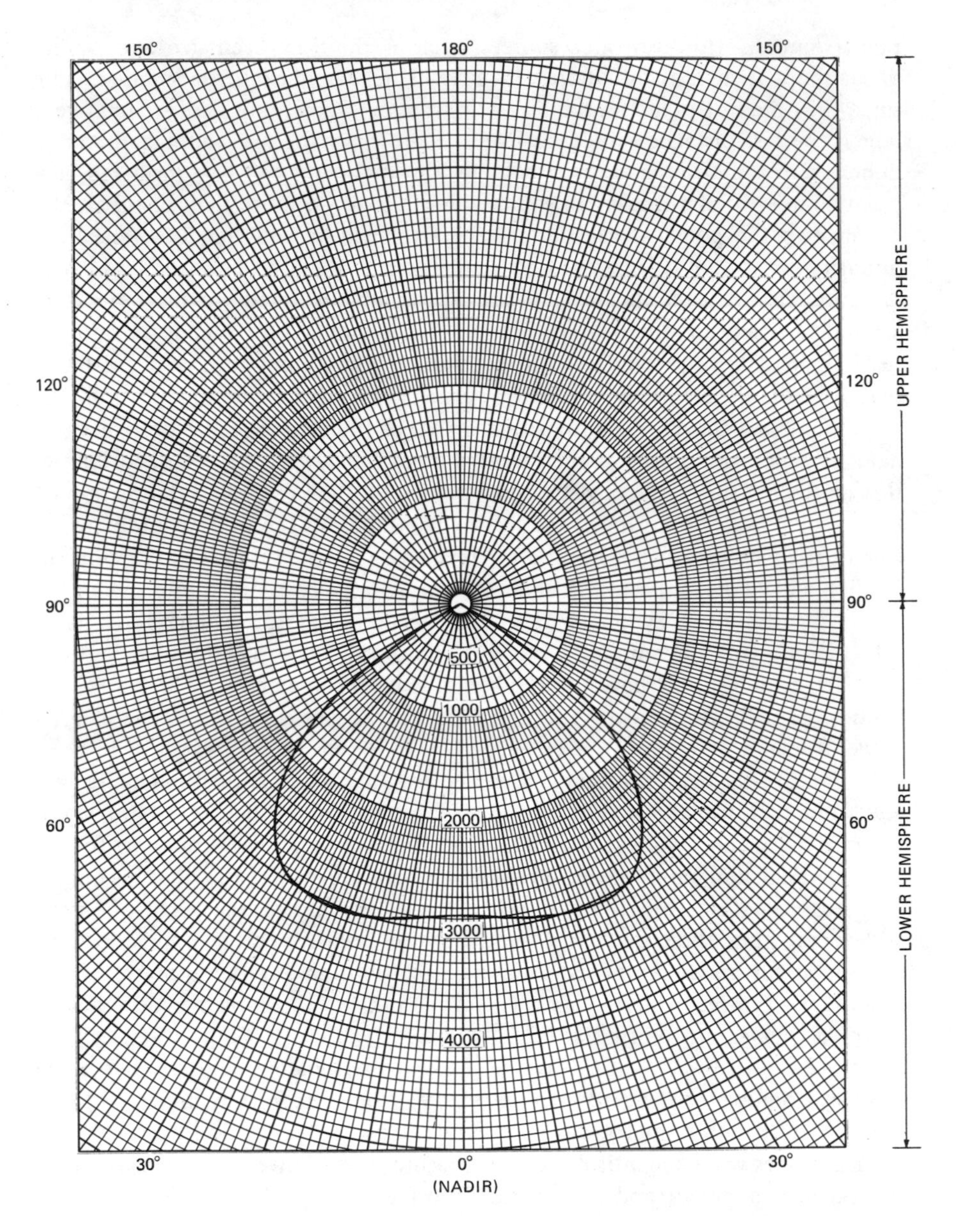

Fig. 2-2. Candlepower distribution curve drawn on a polar coordinate. Vertical angles are measured from the nadir, which is at the bottom of the luminaire. The vertical angles are measured up to 180°.

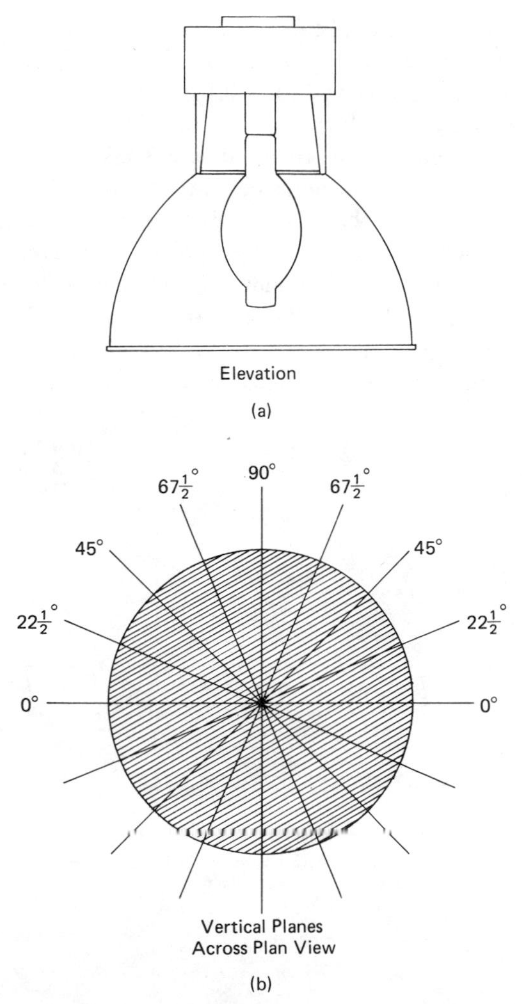

Fig. 2-3. (a) A typical elevation of symmetrical luminaires, with a round reflector and a vertically mounted lamp. (b) At all vertical planes (perpendicular to this page) the cross-sectional view of the luminaire is identical; hence, the candlepower distribution is always the same. For asymmetrical luminaires, this is not true.

Candlepower Summary: Average Candlepower and Zonal Lumens/Flux

The candlepower summary part of the photometric report basically serves two purposes. First, it tabulates all the candlepower readings for each angle measured at vertical planes; second, with the help of these figures, the zonal lumens or the zonal flux is determined. Part C of Figure 2-1 shows the candlepower summary of the example. The first column represents the actual angles at which the candlepower readings were taken. In this example, the readings were taken at an interval of 10 degrees. The second, third, fourth, fifth, and sixth columns represent the candlepower values in candelas for 0, 22.5, 45, 67, 67.5, and 90 degree vertical planes, respectively.

In determining the zonal lumens, the luminaire is assumed to be placed in the center of an imaginary sphere of radius equal to the test distance. Let us suppose, this radius is R. The sphere is now divided into horizontal slices corresponding to the locations of each angle in the polar coordinate. This is shown in Figure 2-4a. Thus, to find a zone bounded by two angles, θ_1 and θ_2, we draw horizontal lines from the polar coordinate at these angles. The horizontal slice obtained in the sphere (see Figure 2-4b) would represent the zone. To determine the total number of lumens produced at this zone, it is necessary to know

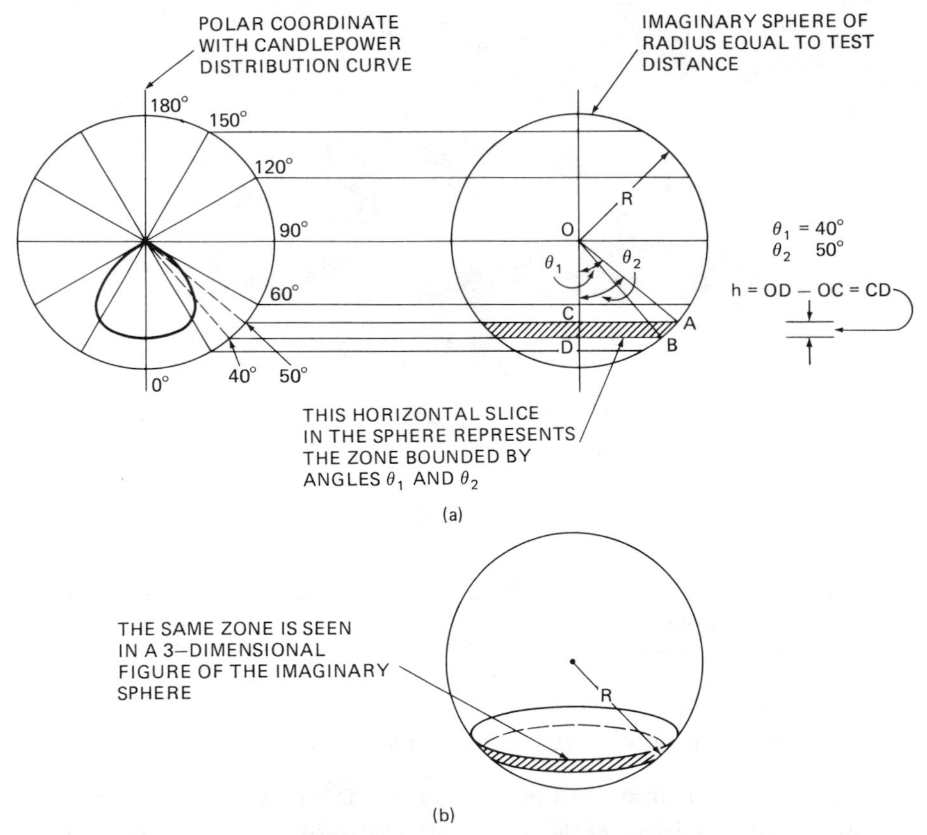

Fig. 2-4. (a) In determining the zonal lumens, the luminaire is assumed to be placed in an imaginary sphere of radius R, which is equal to test distance. The sphere is divided into horizontal slices corresponding to zonal angles on a polar coordinate. The shaded area represents the zone bounded by $\theta_1 = 40°$ and $\theta_2 = 50°$. (b) The same zone is seen in a three-dimensional figure of the imaginary sphere.

the average or mean candlepower of the midzone and a zonal constant given by $2\pi (\cos \theta_1 - \cos \theta_2)$. The zonal lumens is given by the product of the two. The formula is actually derived from the basic relation of candlepower, area, and the test distance, which is given by:

$$L = \frac{CP \times A}{R^2},$$

where

L = lumen
CP = candlepower
R = test distance (Radius of the sphere).
A = area

As shown in Figure 2-4a, the surface area of the zone bounded by θ_1 and θ_2 is given by:

$$A = 2\pi R \times h$$
$$= 2\pi R \times CD$$

where

$$CD = OD - OC$$

Now

$$OD = OB \cos \theta_1 = R \cos \theta_1$$

and

$$OC - OA \cos \theta_2 = R \cos \theta_2$$

So

$$CD = R (\cos \theta_1 - \cos \theta_2).$$

Thus the total zonal area $= 2\pi R \times R (\cos \theta_1 - \cos \theta_2) = 2\pi R^2 (\cos \theta_1 - \cos \theta_2)$. Applying this value of A into the original formula,

$$\text{Zonal lumens} = \frac{\text{Candlepower} \times 2\pi R^2 (\cos \theta_1 - \cos \theta_2)}{R^2}$$
$$= \text{Candlepower} \times 2\pi (\cos \theta_1 - \cos \theta_2).$$

The next step is to find the candlepower value at the zone under consideration. Since the upper and lower limits of the zone is bounded by two angles, θ_1 and θ_2, an average candlepower value must be determined from the midzonal total candlepower and the number of planes. This is done by adding all the candlepower values obtained at the midzonal angle at all planes and dividing by the number of vertical planes. Note that, because there is only one 0° and one 90° vertical plane and two each of all other planes, the total number of planes for a 3-plane photometry (0°, 45°, and 90°) will be four and that for a 5-plane (0°, 22.5°, 45°, 67.5°, and 90°) will be eight. This can be calculated from Figure 2-3.

Let us suppose we want to determine the zonal lumens of an area bounded by 40° and 50°. The midzonal (45°) average candlepower can be found as follows:

$$CP \text{ avg. } 45^\circ = \frac{CP_0 + 2CP_{67.5} + 2CP_{45} + 2CP_{22.5} + CP_{90}}{8}$$

$$= \frac{1848 + 2(2190) + 2(2032) + 2(1898) + 2233}{8}$$

$$= 2040.$$

$$\begin{aligned}\text{The zonal constant} &= 2\pi\,(\cos\theta_1 - \cos\theta_2)\\ &= 2\pi\,(\cos 40^\circ - \cos 50^\circ)\\ &= 2\pi(0.7660 - 0.6428)\\ &= 0.774,\end{aligned}$$

so the zonal lumens = 2040 $\times$ 0.774 = 1579.

Zonal Lumens and Percentage, or The Lumen Summation

This part of the photometric report basically shows the total lumens in larger zones representing various key angles; it also shows their percentage in comparison with the total bare-lamp lumens and total luminaire lumen output. There are no set rules what particular zones have to be shown. However, the zones representing 0-40, 0-90, 90-180, and 0-180 degrees are in general always shown since they represent the critical areas that are important in any lighting design. Light emitted at the 0-45° zone is usually called the *direct light zone* since it falls directly onto the task without causing any direct glare to the viewer. The 45-90° zone is known as the *direct glare zone* since the degree of brightness caused by the luminaire is usually dependent on the candela at these angles. Total lumens at the 90-180° zone represents the light at upper hemisphere, and that at 0-90° represents the lower hemisphere. Finally, the total of lumens at the 0-180° zone represents the total output of the luminaire. In section D of Figure 2-1 the first column represents the different key zones; the second column shows the total lumens at these zones; the third and fourth columns show the lumen percentage compared with the total output by bare lamps and the total output from the luminaire. Let us take the example at the 0-30° zone:

Zone	Lumens	% Lamp	% Luminaire
0-30°	2815	$\frac{2815}{4 \times 3200} = 22$	$\frac{2815}{8313} = 33.9$

Note that the total number of lumens per lamp and the number of lamps used to determine the third column was obtained from section A of the photometric report. The total luminaire lumen output to determine the fourth column was obtained from the total lumens at the 0-180° zone.

Efficiency

The efficiency of a luminaire is the ratio of the total number of lumens emitted by a luminaire to the total number of lumens produced by the bare lamp. Thus, the third column of section D of the photometric, showing the percentage of lumens at the 0-180° zone compared to the bare lamp lumens, is actually the efficiency of the luminaire. In the example, the efficiency of the luminaire is 64.9%.

It is of extreme importance to note that high efficiency of a luminaire does not necessarily mean a more energy-effective luminaire. A luminaire may have much of its light produced in nonuseful areas and yet have high efficiency. The efficiency of a luminaire only shows the percentage of total lumens produced by the luminaire as compared to the total of the bare-lamp lumens. It does not take into account whether the lumens are falling on required areas.

A true measure of the system energy effectiveness can be more reliably judged by the coefficient of utilization (CU) values of the luminaire. The CU of a luminaire represents that portion of the total lumens of the bare lamps that either directly or indirectly falls on the task location. This is discussed in more detail later in this chapter.

Maximum and Average Brightness, or The Luminance Summary

One of the important items in the photometric report is the luminance or the brightness of the luminaire. This section shows the average and maximum brightness of a luminaire at the glare zone as seen across and along the luminaire. It also shows the ratio of the maximum to the average brightness for each of the glare-zone angles. With these values, a preliminary idea can be achieved in making a luminaire selection in regard to the amount of discomfort the luminaire can provide when looked at directly. The unit of luminance is a footlambert. The mathematical relation between luminance (footlambert) and the luminous intensity or the candlepower (candela) is given by

$$\text{Avg. luminance at angle } \theta = \frac{\text{Candlepower at } \theta \times 144\pi}{\text{Projected area at } \theta \text{ in square in.}};$$

alternatively, it can be said that for a specific viewing angle, for a projected area of one square inch, the luminance of the luminaire is 144π, or 452.16 times the candlepower. In this example the luminance summary is shown in item F of Figure 2-1. As can be seen, the average, the maximum and the ratio of maximum to average brightness are tabulated for each of the angles that fall into the glare zone. The tables are separately made for luminance as viewed in the direction of parallel and perpendicular to the lamp axis.

The average luminance of a luminaire is found with the help of the formula shown earlier. In determining the area in the formula, it is important that the projected area, as viewed from the specific angle, is used for calculation. Obviously, representing only the apparent surface of the luminaire, the projected area will differ from angle to angle. In addition, some luminaires may have certain areas in the projected area that may not be bright enough to be considered. In the past it was accepted that the projected area was to represent only that portion of the luminaire, which could produce luminance of 150 footlambert or more. This included any part of the luminaire including sides and pendant hanger. This limitation was removed by the IES in 1976, since in some lighting systems, such as in coffers or indirect lighting systems, the immediate surrounding reflecting surfaces are also considered as a part of the luminous area. For calculation purposes it is now necessary that all luminous areas, which may or may not be an integral part of the luminaire, be considered. Figure 2-5 shows the accepted projected area for different types of luminaires, as currently recommended by the Illuminating Engineering Society of North America.

Let us now see how the values 1463 and 1211 footlamberts were determined for average luminance at 45° as viewed across and along the lamps respectively.

The luminaire in our example is recessed. The only part producing light is the bottom of the luminaire, of dimensions 21 × 46.5 inches. This is the actual area as viewed perpendicular to its surface. Referring to Figure 2-6, when the luminaire is viewed at an angle of 45°, the projected area will be smaller than the actual area. Looking crosswise, the length of the luminaire will remain the same; however, the effective width or the apparent width will be much smaller than the actual width. Thus, crosswise (Figure 2-6a),

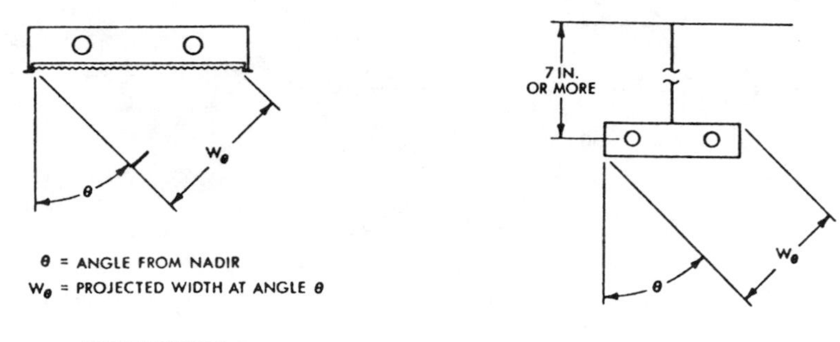

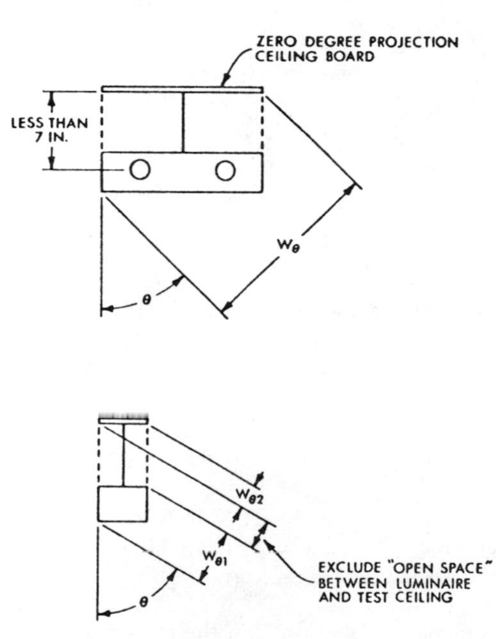

Fig. 2-5. Accepted projected areas of some luminaires, as recommended by the IES.

$$
\begin{aligned}
\text{Projected area at angle } \theta &= L \times W_\theta \\
&= L \times W \sin(90 - \theta) \\
&= L \times W \cos\theta; \\
\text{if } \theta = 45^\circ \text{, then} \quad &= 46.5 \times 21 \times \cos 45 \\
&= 46.5 \times 21 \times 0.707 \\
&= 690 \text{ square inches.}
\end{aligned}
$$

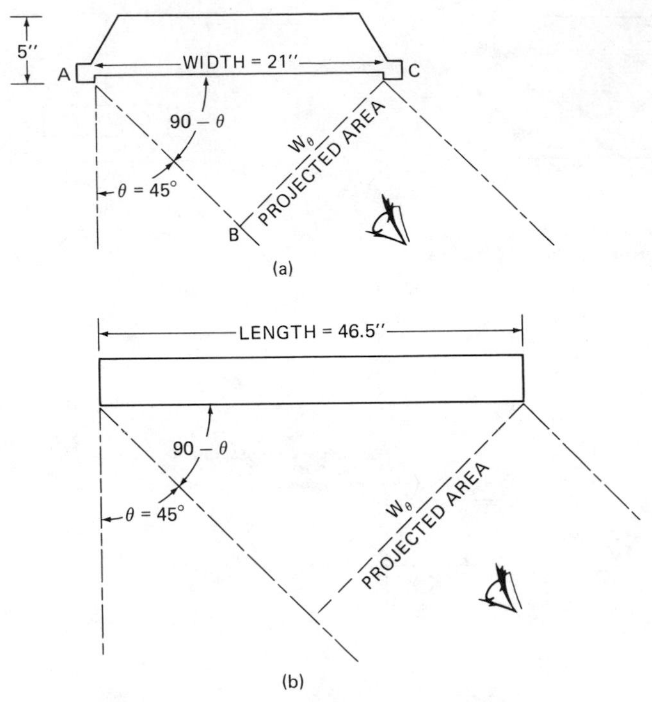

Fig. 2-6. (a) Projected area, looking lengthwide at angle θ. (b) Projected area, looking widthwise at angle θ.

Similarly, lengthwise (Figure 2-6b),

$$\begin{aligned}
\text{Projected area at angle } \theta &= W \times W_\theta \\
&= W \times L \sin(90 - \theta) \\
&= 21 \times 46.5 \cos\theta; \\
\text{now, at } \theta = 45^\circ, \quad &= 21 \times 46.5 \times 0.707 \\
&= 690 \text{ square inches.}
\end{aligned}$$

Now from candlepower summary of section C of the photometrics,

$$CP_{45^\circ} \text{ (across)} = 2233 \text{ candela, and}$$
$$CP_{45^\circ} \text{ (along)} = 1848 \text{ candela.}$$

Applying the formula,

$$\begin{aligned}
\text{Avg. luminance (fl), looking across} &= \frac{CP \text{ at } \theta \times 144\pi}{\text{Projected area at } \theta} \\
&= \frac{2233 \times 144 \times 3.14}{690} \\
&= 1463 \text{ footlamberts (across)}
\end{aligned}$$

and looking along the lamps,

$$= \frac{1848 \times 144\pi}{669}$$

$$= 1211 \text{ footlamberts (along).}$$

The maximum luminance of a luminaire, such as shown in the fourth and fifth column of section F of Figure 2-1, is a measured quantity. In its determination, a photometer with a circular aperture of one square inch is used to search the surface of the lens or louvers for the maximum luminance at the glare angles. These values are different for different vertical angles and tabulated as shown in our example. Average luminance as determined by the formula is not sufficient to determine the visual comfort of a luminaire. Peak brightness, or the maximum luminance, is more of a problem where glare is concerned. The ratio of the two indicates a level of visual comfort. The Direct Glare Committee of RQQ Committee of the IES states that

> . . . direct glare may not be a problem in lighting installations if all three of the following conditions are satisfied:
>
> 1. The Visual Comfort Probability (VCP) is 70 or more.
> 2. The ratio of maximum to average luminance does not exceed 5 to 1 (preferably 3 to 1) at 45, 55, 65, 75 and 85 degrees from nadir crosswise and lengthwise.
> 3. Maximum luminaire luminance crosswise and lengthwise do not exceed the following values:

Angle	Maximum Luminance (fl)
45	2250
55	1605
65	1125
75	750
85	495

In selecting the luminaire, an initial judgment can be made by looking at the average and maximum luminance at the glare zone. The VCP then would confirm its visual comfort level. The VCP, the method currently accepted by the IES for evaluating direct glare in a room from a group of luminaires, represents a percentage of people who probably would find the glare produced in the space not objectionable. A discussion of this method is found in Chapter 3.

CIE-IES Luminaire Classification

The CIE-IES* classification system describes a luminaire by the percentage of light in the upper hemisphere and lower hemisphere of the candlepower dis-

*CIE = International Commission on Illumination.

tribution curve. There are six categories of the CIE–IES classification: (1) direct, (2) semi-direct, (3) direct-indirect, (4) general diffuse, (5) semiindirect, and (6) indirect. The percentage of light sharing by the upper and lower hemisphere is shown in Figure 2-7. Note that the direct-indirect classification, as shown in the figure, is accepted by IES only and not by CIE.

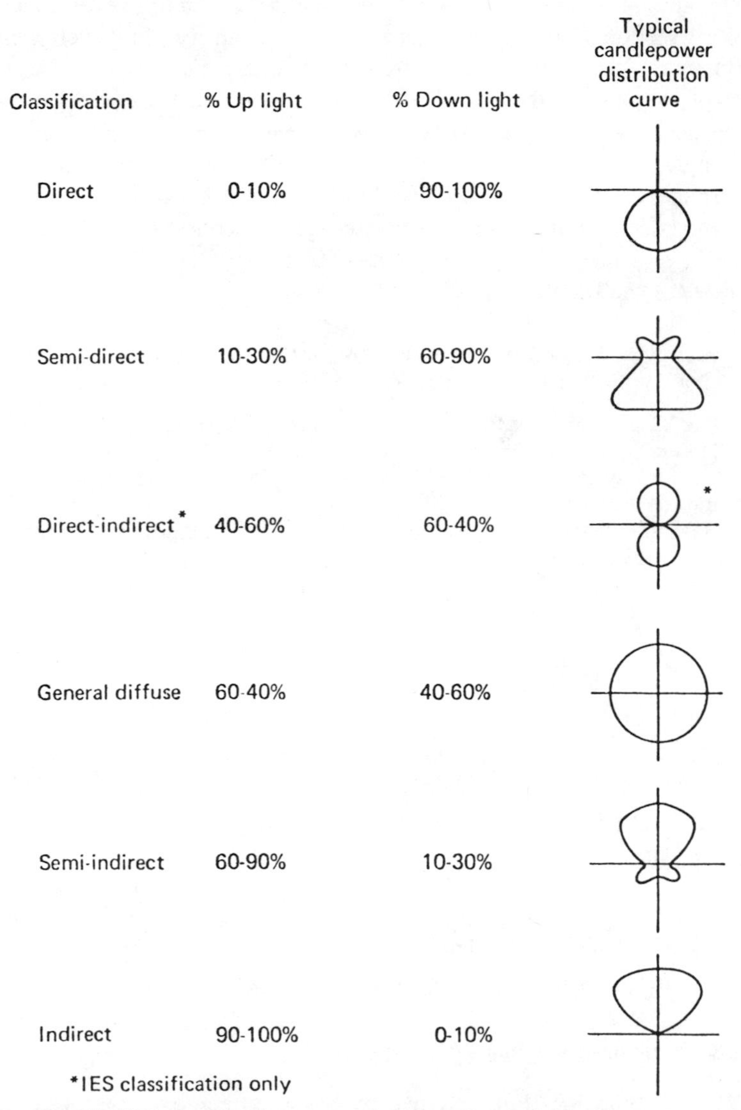

Classification
% Up light
% Down light
Typical candlepower distribution curve
Direct
0-10%
90-100%
Semi-direct
10-30%
60-90%
Direct-indirect*
40-60%
60-40%
*
General diffuse
60-40%
40-60%
Semi-indirect
60-90%
10-30%
Indirect
90-100%
0-10%
*IES classification only

Fig. 2-7. The CIE–IES classification system describes a luminaire by the percentage of light in the upper and lower hemispheres of the candlepower distribution curve.

To determine the CIE–IES classification, the percentage of total output in the upper and lower hemisphere has to be determined. This is given by:

$$\%\ \text{total output} = \frac{\text{Lumens produced in upper and lower hemisphere}}{\text{Total lumens produced from 0 to } 180^\circ} \times 100.$$

For our example, thus:

$$\%\ \text{total output (upper hemisphere)} = \frac{\text{Lumens at 90 to } 180^\circ \text{ zone} \times 100}{\text{Lumens at 0 to } 180^\circ \text{ zone}}$$

$$= \frac{0}{5793} = 0\%$$

and

$$\%\ \text{total output (lower hemisphere)} = \frac{\text{Lumens at 0 to } 90^\circ \text{ zone} \times 100}{\text{Lumens at 0 to } 180^\circ \text{ zone}}$$

$$= \frac{8313 \times 100}{8313}$$

$$= 100\%.$$

Thus the luminaire is CIE–IES classified as a direct type of luminaire.

Shielding Angle

Shielding a luminaire is the hiding of the source from direct viewing. A luminaire can be shielded at any angle. The shielding angle is actually the angle measured from a horizontal line drawn at the bottom of the luminaire and the line of first sight of the source. For an evenly illuminated source, such as frosted incandescent, phosphor-coated HID or fluorescent lamps, the shielding angle is measured between the horizontal line and the bottom of the bulb. This is shown in Figure 2-8. For sources with filaments, such as the incandescents, the angle is measured between the horizontal line and the filament. For a clear HID source, this angle is measured between the horizontal line and the bottom of the arc tube.

Spacing-to-Mounting-Height Ratio (S/MH) and Spacing Criterion (SC)

One of the most important and frequently used features in the photometric report is the spacing-to-mounting-height ratio, or simply the S/MH. It is an important tool in lighting design for all types of applications where uniformity

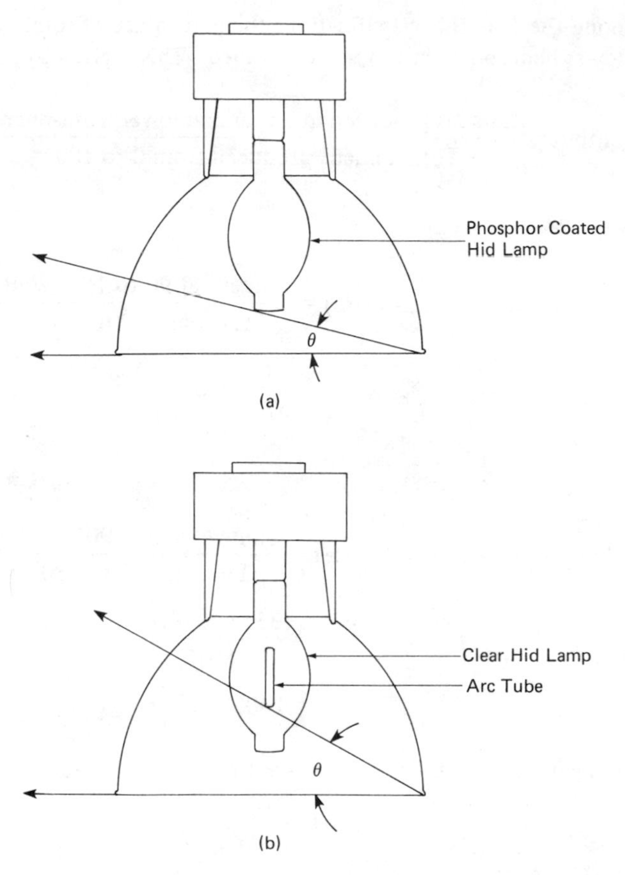

Fig. 2-8. Shielding angle of luminaires. (a) For luminaires with frosted lamps, such as the coated HID, incandescent, or fluorescent, the shielding angle is measured from a horizontal line at the bottom of the luminaire to the first sight of lamp. (b) For a luminaire with a filament or arc tube source, the shielding angle is measured from a horizontal line at the bottom of the luminaire to the first sight of the filament or arc tube of the lamp.

in illuminance is required. Uniformity of a lighting level in an indoor application rests on several factors. A maximum uniformity in lighting can be achieved if there is a homogeneous luminous ceiling throughout the room. However, most interiors contain only a finite number of luminaires placed uniformly throughout the room, causing some degree of nonuniformity. In general, illuminance level right below a luminaire is often higher than that in between two luminaires. As a measure of tolerance, in the past it was accepted that a room full of light would be considered uniform provided its point-to-point variation does not exceed plus or minus one-sixth of the average value.

Uniformity of a lighting level for an indoor application depends mainly on the luminaire spacings and the distance from the walls. The maximum allowable distance will vary with mounting height above work plane. Each luminaire, depending on its type of beam spread, has a maximum S/MH ratio, which, when multiplied by the mounting height from work level, determines the spacing limitation between luminaires. Illuminance uniformity determined by this method is a common practice and has been accepted as an official way of lighting design for years. However, the technique has been found to have several limitations and drawbacks.

One of its main drawbacks is its failure to take into account the possible unidirectional lighting and the shadow effect that might be produced if the luminaires were spaced at their farthest allowable distance. To demonstrate this, let us take two examples. Figure 2-9a shows the candlepower distribution curve of a luminaire whose intensity varies smoothly with angle up to 90°. With luminaires such as these, one would expect to have light contribution from all angles to each point of a horizontal surface and not to have abrupt changes in horizontal illuminance because of the absence of one luminaire. A direct application of the S/MH of these luminaires for uniformity probably will be justified.

Let us now consider Figure 2-9b, which shows the candlepower distribution curve for another type of luminaire, one having batwing-type intensity distribution. Note that the intensity distribution varies uniformly up to about 45° and then drops off abruptly. Unlike the previous example, this luminaire does not produce or contribute any light between 45 and 90 degrees. One cannot expect this luminaire to contribute appreciable direct horizontal illumination at large horizontal distances from the luminaire. As the luminaire spacing is increased, the uniformity will drop significantly and rapidly at some particular spacing. In addition, spacing these luminaires at their maximum points would result in unidirectional lighting, causing strong shadows that might be annoying and undesirable for certain types of applications.

Noting drawbacks such as these, the IES introduced the concept of spacing criterion (SC) to replace the previously accepted S/MH. This is shown in the Reference volume of the IES *Lighting Handbook*, 1981. The main purpose of its introduction was to erase the conventional assumption that the quality of lighting distribution would remain the same irrespective of luminaire spacing, as long as it did not exceed the S/MH.

The spacing criterion is also a numerical factor just like the S/MH ratio, and it has computational similarities with S/MH. However, as the name implies, this factor represents a number around which a criterion or an argument can be built to determine illuminance uniformity for a particular application. In many applications, SC can be directly used as S/MH—but not always. The scope of SC can be summarized as follows:

1. SC is a standard method of classifying luminaire spread, given numerically, that is multiplied to the luminaire mounting distance from the work plane, towards the determination of even illuminance.

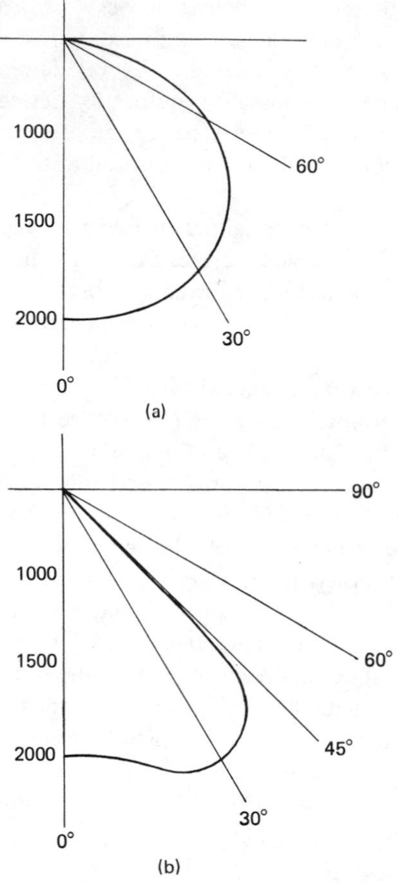

Fig. 2-9. (a) The candlepower distribution varies smoothly from nadir to 90°, assuring a certain amount of light at all angles. A direct application of S/MH for these luminaires will probably be justified. (b) Candlepower distribution varies smoothly up to 45° and then drops off abruptly, offering no light above 45°.

2. It only represents the spacings for uniform horizontal illuminance at a few selected points.
3. The effect is true only for nearby luminaires.
4. It only considers the direct component of illuminance and does not consider any reflections from the walls, ceiling, or floor.

In determination of SC, two typical types of luminaire orientations are considered. Figure 2-10a shows the first orientation where two luminaires are separated by a distance S_1. In this criterion, it is assumed that the maximum spacing S_1 between the two luminaires will be considered to provide uniform lighting if the

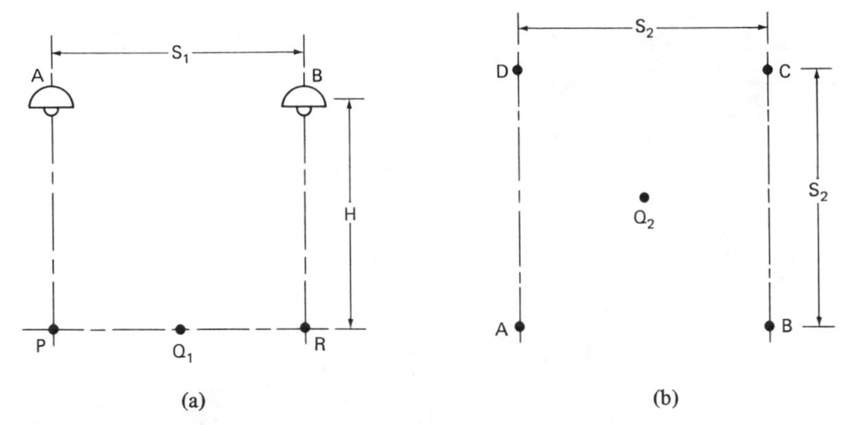

Fig. 2-10. (a) Luminaires A and B are located directly above points P and R. Q_1 is the midpoint between P and R. (b) Four luminaires located at each of the four corners of square A, B, C, and D. Q_2 is the midpoint.

points P and R situated directly under the luminaire have individually the same amount of light as that in Q_1, the midpoint between P and R. To achieve this, obviously each of the two luminaires has to provide half as much light at the midpoint Q_1 than directly underneath at points P and Q. For a given mounting height, H, the maximum spacing S_1 will thus create a situation where all points P, Q, and R will have uniform horizontal illuminance.

In the second orientation, four luminaires are arranged in a square array and the midpoint Q_2 is located right at the center of the square (see Figure 2-10b). In this situation, for the given mounting height, H, the maximum spacing S_2 will be considered in providing uniform lighting when the illuminance level at all points directly under luminaires and the midpoint Q_2 will be the same. Again, this is only possible if each luminaire provides one-fourth their downward light at midpoint Q_2. Both the criteria are analyzed, and the lower value of the two is accepted as the spacing criterion.

As can be imagined, SC determined by the four-luminaire-orientation concept can be successfully used for two-luminaire orientation, but not vice versa. Although IES recommendation for determining S/MH was to select the lower value of the two orientations, the S/MH published in older photometric data was based on the two-luminaire concept only. Obviously, this caused some error.

The computation of SC is very similar to S/MH. For luminaires whose candlepower distribution curve is normally symmetric about the nadir:

1. Plot the relative intensity (candlepower) of the luminaire in the chart shown in Figure 2-11. Smooth out any pronounced irregularities in the curve.

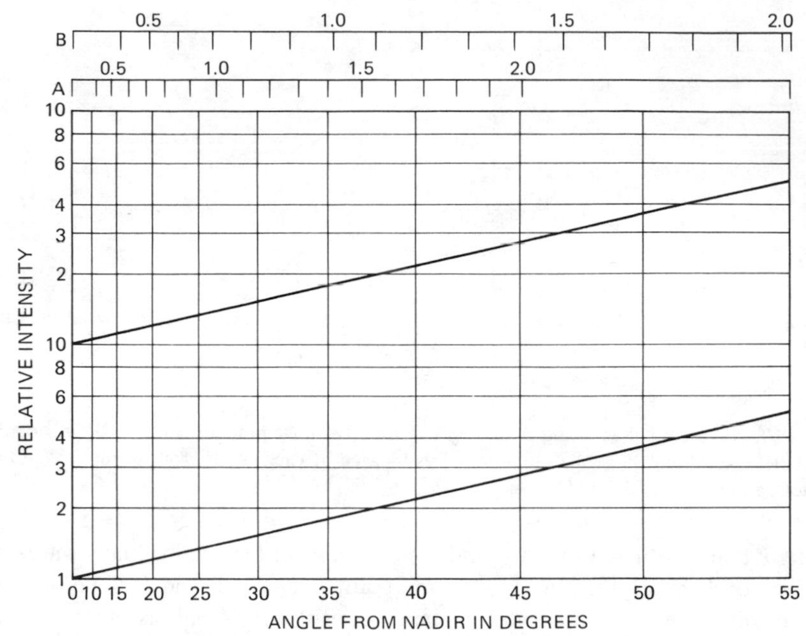

Fig. 2-11. Chart for determining the Luminaire Spacing Criterion. (Reprinted with permission from the IES Lighting Handbook, 1981, Reference Volume).

2. Locate the point of one-half the intensity at 0° on the ordinate (extreme left vertical line) and draw a line from this point parallel to the diagonal lines.
3. From the point where this line intersects the candlepower curve, draw a vertical line and read the value in scale A.
4. Repeat step 2 with one-quarter intensity at 0° on the ordinate.
5. Repeat the process and read the value in scale B.
6. The lower of these two values is the SC. The number can be rounded off to nearest tenth.

Note that in our example of the fluorescent luminaire, for a vertical plane across lamps, the SC found from scale A is 1.32 and that from B is 1.41. So the SC is 1.3, the smaller of the two rounded to nearest tenth.

Note that luminaire SC only suggests a maximum spacing at which the horizontal illuminance will possibly be reasonably uniform. It is just a beginning of the consideration where uniform lighting is required. When other criteria, such as shadowing, vertical illuminance, overlap between luminaries, etc., are considered, an S/MH ratio lower than the SC value may result in a more satisfactory result. While a true picture of uniformity can only be assessed by making point-to-point analysis, it is in general suggested that a maximum value of 1.5 be assigned as the luminaire SC, since the use of larger value usually does not produce acceptable lighting installation. Luminaires with higher SC would cause discomfort glare at higher angles, lowering the probability of visual comfort.

Luminaires that have vertically mounted lamps and have circular reflectors usually have a symmetrical candlepower distribution at all vertical planes. The SC for these luminaires is constant and can be used for all vertical planes. Others may have more than one SC, depending on their distribution pattern and symmetry. Fluorescent luminaires, such as in this example, producing much of its light from the sides of the lamps obviously have higher SC at a vertical plane across lamps than along them. With these luminaires a uniform illuminance can only be achieved if they are placed in a rectangular array or in continuous rows. If the luminaires are placed in continuous rows, then SC value at a vertical plane across the lamps is only required.

In the determination of SC, one of the major considerations is that all luminaires donate a substantial amount of their light directly underneath. This has a disadvantage in that in a group if any one of the luminaires burns out there will be a dark area right underneath the problem luminaire. This may cause some annoyance and difficulty in a certain type of work. For an application where uniformity of light is critical, luminaire orientation can be so arranged that illuminance level at any spot under a luminaire will not be fully dependent on the luminaire above and some amount of light will be always available from other sources nearby in case of the luminaire's failure. The amount of light to be obtained from other sources is dependent upon the nature of the work and the furniture or equipment layout in the room. A good compromise is 50%, where one half the total light underneath each luminaire may be obtained from nearby luminaires.

Spacing Criteria (SC) and Energy Saving. Energy saving is a comparative term. How much energy a lighting system saves can only be assessed by comparison with another existing or proposed system for the same amount of light. The amount of energy consumed, depends upon the total number of luminaires and the individual power consumption of each. For the same amount of light, obviously, the fewer the number of luminaires, the less energy consumption there will be. In determination of the total number of luminaires for a certain lighting level, it is necessary to know the CU of the luminaire under the applicable surrounding. As will be seen in Chapter 3, the higher the value of CU, the fewer the total number of luminaires for the required lighting level. Although the formula used to determine total number of luminaires is independent of the SC of the luminaires, it is dependent on CU. High SC usually means wide beam spread of the luminaire. Any time a luminaire scatters more and more light towards higher vertical angles, the direct light towards the nadir reduces, resulting in a drop in illuminance level on the task. The lumens at higher angles bounce off the reflecting room surfaces before reaching the task and hence there is a drop in the CU value. CU values, as will be seen later in this chapter, represent that percentage of bare lamp lumens that fall on the task. A higher CU value would mean more light for the same amount of energy use.

High SC values usually indicate good uniformity with fewer luminaires. Very high SC values, however, will cause direct glare at high angles. A higher SC value will also mean a reduction in CU values. A good energy-saving design would thus depend on a compromise between SC, CU, and the brightness of the luminaire. In a noncritical area, such as a storage room or a warehouse, where a high level of illuminance or reduced glare is not of prime importance, luminaires with high SC may be adequate. But in applications where reduced glare and higher CU are of prime importance, such as in offices and schools, a balanced compromise must be made. Luminaires with very low SC, e.g., 1.0 or smaller, require closer spacing for uniformity. This results in a high level of minimum illuminance, and may raise some design problems for areas of low ceiling height, where the illuminance-level requirement is low.

As will be seen, CU values are also dependent upon the luminaire quality in terms of its finish, the surface reflectance of the reflector, and the lamps' relative locations from each other. As a result, for the same type of luminaire, different manufacturers may have different CU values for the same room cavity ratio and room surface reflectances. The best selection can be made only by comparing the performance of a family of similar types of luminaires by different manufacturers and selecting the one with the highest CU value and highest SC value without causing glare. If two luminaires have the same acceptable SC value, select the one with highest CU.

Coefficient of Utilization: Zonal Cavity Method

As we have seen earlier, the efficiency of a luminaire tells us how many lumens are emitted by the luminaire compared to the total lumens produced by the bare lamps. However, a luminaire with high efficiency may not be very energy effective, since little is known about the direction in which the lumens actually fall. CU represents that portion of the total bare-lamp lumens that falls on the work plane. From this respect, it represents a better measure of energy effectiveness than the efficiency of the luminaire. In general, lumens on the work plane that represent the CU, are accumulated from two types of light rays. The first is the direct rays, and the second, the rays that hit the room surfaces, bounce back and forth, and finally reach the work plane. Thus, determination of CU depends not only on the candlepower distribution of the luminaire, but also on the room proportions and surface reflectances, height of work plane, and the height of the luminaire above the work plane. Item J of Figure 2-1, represents the CU table of our example. The values are expressed corresponding to the room cavity ratios and effective ceiling, wall, and floor reflectances. CU is an integral part of lighting calculations; a thorough understanding of its determination is extremely important.

Room Cavity and Cavity Ratios. The zonal cavity method of determining the CU is based on the concept that the area to be lighted has a series of cavities or spaces that have effective reflectances with respect to each other and the work plane. The space to be lighted is divided into three cavities; this is shown in Figure 2-12. The space between the bottom surface of the luminaire and the ceiling is known as the *ceiling cavity*. The distance is known as h_{cc}, or the height of ceiling cavity. The space between work plane and the bottom of the luminaire is known as the *room cavity*, and this distance is known as h_{rc}, or the height of room cavity. The space between the work plane and the floor is known as the *floor cavity*, and the distance is known as h_{fc}, or the height of floor cavity.

The cavity ratios represent the geometric proportions of the ceiling, room, and floor cavities and can be found by the following formula:

$$\text{Cavity ratio} = \frac{5\,h\,(\text{room length} + \text{room width})}{\text{room length} \times \text{room width}}$$

where

$h = h_{cc}$ for ceiling cavity ratio (CCR)
$= h_{rc}$ for room cavity ratio (RCR)
$= h_{fc}$ for floor cavity ratio (FCR)

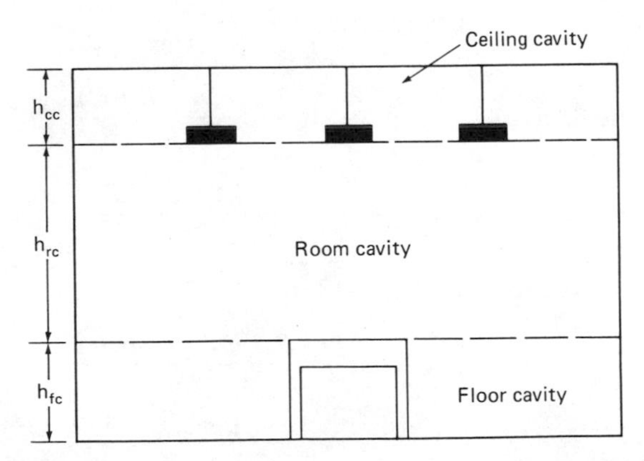

Fig. 2-12. Space to be lighted is divided into three spaces or cavities: ceiling cavity, room cavity, and floor cavity.

The formula is applicable for all rectangular or square-shaped rooms with flat ceilings. The formula can be modified as follows:

1. If the luminaires are installed flush with ceiling, or surface mounted, h_{cc} is 0; hence, CCR = 0.
2. If the work plane is the floor, $h_{fc} = 0$; then FCR = 0.
3. If the room is square, room length = room width, so the formula can be modified:

$$\text{Cavity ratio} = \frac{5\,h\,(\text{room length} + \text{room length})}{\text{room length} \times \text{room length}}$$

$$= \frac{5\,h\,(2)\,(\text{room length})}{(\text{room length})^2}$$

$$= \frac{10\,h}{\text{room length}}$$

4. If the area is of irregular shape, the following modifications are possible:
   a. If the room is of an L shape, such as shown in Figure 2-13a, the formula is given by:

$$\text{Cavity ratio} = \frac{2.5\,h\,(\text{perimeter of room in feet})}{\text{Area of cavity base}}$$

from the figure,

$$= \frac{2.5\,h \times 2(W + L)}{(WL - XY)}$$

   b. For a triangular room (Figure 2-13b):

$$\text{Cavity ratio} = \frac{2.5\,h \times \text{perimeter}}{\text{Area}}$$

from the figure,

$$= \frac{2.5\,h\,(A + B + C)}{\frac{1}{2}BC}$$

$$= \frac{5\,h\,(A + B + C)}{BC}$$

   c. For a circular room (Figure 2-13c):

$$\text{Cavity ratio} = \frac{2.5\,h \times \text{perimeter}}{\text{Area}}$$

$$= \frac{2.5\,h\,(2\pi R)}{\pi R^2}$$

$$= \frac{5\,h}{R}$$

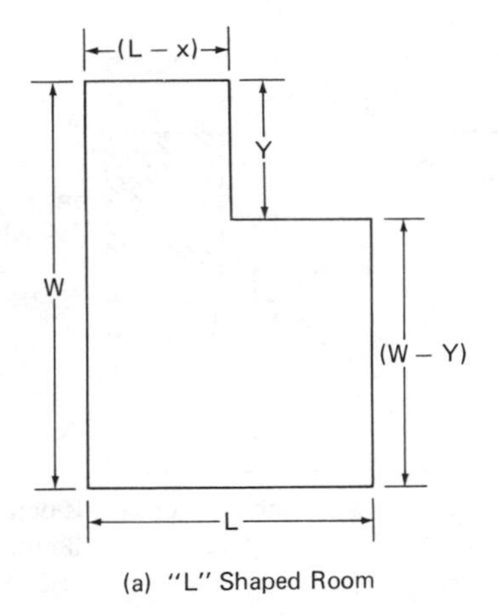

(a) "L" Shaped Room

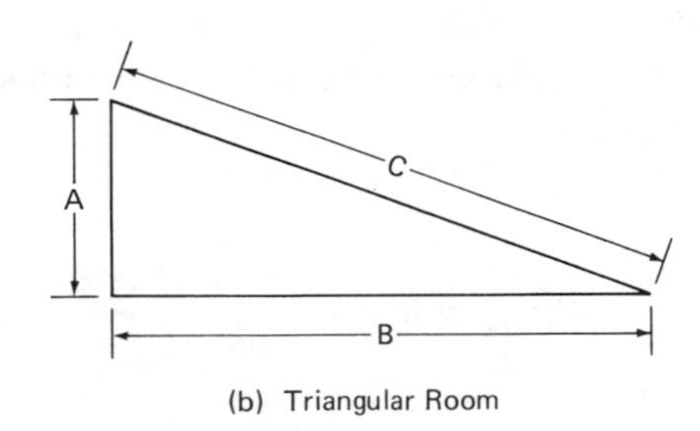

(b) Triangular Room

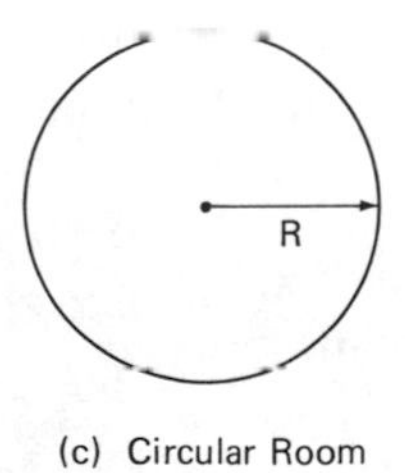

(c) Circular Room

Fig. 2-13. Cavity ratio determination for irregular-shaped rooms. (a) L-shaped room. (b) Triangular room. (c) Circular room.

In many applications, for example, in industrial facilities, more than one type of task is performed in a single room. These tasks may call for different illuminance levels. The area in which one type of task is performed should be considered as a separate room even though no physical barrier such as walls or partitions may exist between the rooms. The true cavity ratios can be obtained with the help of the formulas given above. This is shown in the following example.

The engineering office area shown in Figure 2-14 is a rectangular room with shaded area assigned for drafting purposes. The drafting area requires, say, 150

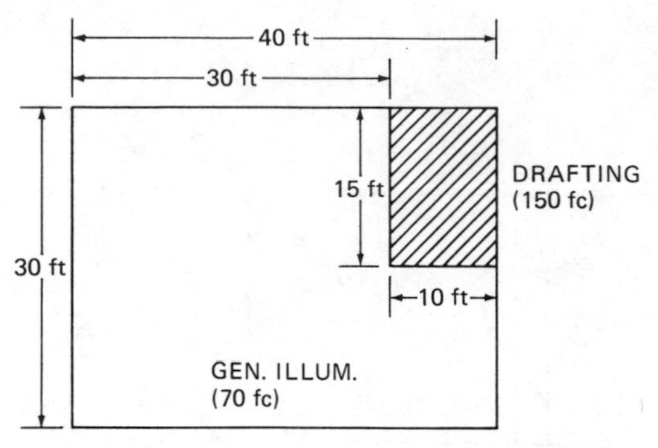

Fig. 2-14. In many applications, more than one type of task is performed in a single room, with separate illuminance level requirements. These areas should be treated as separate rooms for cavity ratio calculations.

fc, and the remaining area needs 70 fc. These two rooms, because of their illuminance level requirements and different room configurations, should be treated as two separate rooms.

For the drafting area the regular formula can be used:

$$\text{Cavity ratio} = \frac{5\,h\,(10 + 15)}{10 \times 15}$$

$$= \frac{5\,h}{6}$$

For the L-shaped remaining room, the formula for irregular shaped rooms can be used:

$$\text{Cavity ratio} = \frac{2.5\,h \times 2(W + L)}{WL - XY}$$

$$= \frac{2.5\,h \times 2(40 + 30)}{(40 \times 30) - (15 \times 10)}$$

$$= \frac{h}{3}$$

By substituting the values of h_{cc}, h_{rc}, h_{fc} for h, the CCR, RCR, and FCR can be found for each room. Note that the cavity ratios for the drafting area can also be found directly from the chart shown in Table 2-1.

Reflectances. CU values are much dependent on the reflecting surfaces of a room. In general, higher reflectance of the ceiling, walls, and floor would mean higher CU values. Thus, to evaluate the correct CU of a luminaire, one must know the true reflectance of the room surfaces. The reflectance of room surfaces largely varies with the type of work performed inside. An industrial facility working with fumes, smoke, or dirt would obviously have lower reflectances than an office room that is well maintained. The other difficulty in evaluating true reflectance arises when the surface has several different colors, for example, in the case of a multicolor wallpaper or a painted wall with partial wooden panels. Table 2-2 shows the typical reflectances of different colors that can be used directly when the color matches. For a multicolor surface, a weighted average of all the colors must be determined. The technique of determining such a value is as follows.

Suppose the walls in a specific room have 800 square feet of white paint (80% reflectance), 400 square feet of wooden panels (10% reflectance), 500 square feet of wallpaper (50% reflectance) and 200 square feet of bookshelves of approximately 30% reflectance. The average weighted reflectance of the whole wall should be determined as follows:

Area (sq. feet)		Reflectance		New Surface
800	X	0.80	=	640
400	X	0.10	=	40
500	X	0.50	=	250
200	X	0.30	=	60
1900				990

The total wall surface = 1900 square feet and the total new surface = 990 square feet, so the weighted average = 990/1900 = 0.52. Thus, the weighted average reflectance of the whole wall is 52%.

Reflectance of a surface can be determined easily by a reflectometer. A handy method of determining reflectance is with the help of a light meter and an 8 X 10 inch card or paper of known reflectance. Let us assume that the known reflectance of the card is 80%. The procedure of determining the reflectance of the surface is as follows:

1. Place the card against the surface whose reflectance is to be measured, with the known reflectance side outwards.
2. Hold the light meter at the center of the card, with the open slot facing the card 2 inches away. Read the meter. Suppose it is 50 fc.

Table 2-1. Cavity Ratio Table. For room and cavity dimensions other than below, the cavity ratio can be calculated by the formulas shown in this chapter. If smaller room and cavity dimensions are required divide width, length and cavity depth by 10. (Reprinted with permission from the *IES Lighting Handbook*, 1981, Reference Volume).

Room Dimensions		Cavity Depth																			
Width	Length	1.0	1.5	2.0	2.5	3.0	3.5	4.0	5.0	6.0	7.0	8	9	10	11	12	14	16	20	25	30
8	8	1.2	1.9	2.5	3.1	3.7	4.4	5.0	6.2	7.5	8.8	10.0	11.2	12.5	—	—	—	—	—	—	—
	10	1.1	1.7	2.2	2.8	3.4	3.9	4.5	5.6	6.7	7.9	9.0	10.1	11.3	12.4	—	—	—	—	—	—
	14	1.0	1.5	2.0	2.5	3.0	3.4	3.9	4.9	5.9	6.9	7.8	8.8	9.7	10.7	11.7	—	—	—	—	—
	20	0.9	1.3	1.7	2.2	2.6	3.1	3.5	4.4	5.2	6.1	7.0	7.9	8.8	9.6	10.5	12.2	—	—	—	—
	30	0.8	1.2	1.6	2.0	2.4	2.8	3.2	4.0	4.7	5.5	6.3	7.1	7.9	8.7	9.5	11.0	—	—	—	—
	40	0.7	1.1	1.5	1.9	2.3	2.6	3.0	3.7	4.5	5.3	5.9	6.5	7.4	8.1	8.8	10.3	11.8	—	—	—
10	10	1.0	1.5	2.0	2.5	3.0	3.5	4.0	5.0	6.0	7.0	8.0	9.0	10.0	11.0	12.0	—	—	—	—	—
	14	0.9	1.3	1.7	2.1	2.6	3.0	3.4	4.3	5.1	6.0	6.9	7.8	8.6	9.5	10.4	12.0	—	—	—	—
	20	0.7	1.1	1.5	1.9	2.3	2.6	3.0	3.7	4.5	5.3	6.0	6.8	7.5	8.3	9.0	10.5	12.0	—	—	—
	30	0.7	1.0	1.3	1.7	2.0	2.3	2.7	3.3	4.0	4.7	5.3	6.0	6.6	7.3	8.0	9.4	10.6	—	—	—
	40	0.6	0.9	1.2	1.6	1.9	2.2	2.5	3.1	3.7	4.4	5.0	5.6	6.2	6.9	7.5	8.7	10.0	12.5	—	—
	60	0.6	0.9	1.2	1.5	1.7	2.0	2.3	2.9	3.5	4.1	4.7	5.3	5.9	6.5	7.1	8.2	9.4	11.7	—	—
12	12	0.8	1.2	1.7	2.1	2.5	2.9	3.3	4.2	5.0	5.8	6.7	7.5	8.4	9.2	10.0	11.7	—	—	—	—
	16	0.7	1.1	1.5	1.8	2.2	2.5	2.9	3.6	4.4	5.1	5.8	6.5	7.2	8.0	8.7	10.2	11.6	—	—	—
	24	0.6	0.9	1.2	1.6	1.9	2.2	2.5	3.1	3.7	4.4	5.0	5.6	6.2	6.9	7.5	8.7	10.0	12.5	—	—
	36	0.6	0.8	1.1	1.4	1.7	1.9	2.2	2.8	3.3	3.9	4.4	5.0	5.5	6.0	6.6	7.8	8.8	11.0	—	—
	50	0.5	0.8	1.0	1.3	1.5	1.8	2.1	2.6	3.1	3.6	4.1	4.6	5.1	5.6	6.2	7.2	8.2	10.2	—	—
	70	0.5	0.7	1.0	1.2	1.5	1.7	2.0	2.4	2.9	3.4	3.9	4.4	4.9	5.4	5.8	6.8	7.8	9.7	12.2	—
14	14	0.7	1.1	1.4	1.8	2.1	2.5	2.9	3.6	4.3	5.0	5.7	6.4	7.1	7.8	8.5	10.0	11.4	—	—	—
	20	0.6	0.9	1.2	1.5	1.8	2.1	2.4	3.0	3.6	4.2	4.9	5.5	6.1	6.7	7.3	8.6	9.8	12.3	—	—
	30	0.5	0.8	1.0	1.3	1.6	1.8	2.1	2.6	3.1	3.7	4.2	4.7	5.2	5.8	6.3	7.3	8.4	10.5	—	—
	42	0.5	0.7	1.0	1.2	1.4	1.7	1.9	2.4	2.9	3.3	3.8	4.3	4.7	5.2	5.7	6.7	7.6	9.5	11.9	—
	60	0.4	0.7	0.9	1.1	1.3	1.5	1.8	2.2	2.6	3.1	3.5	3.9	4.4	4.8	5.2	6.1	7.0	8.8	10.9	—
	90	0.4	0.6	0.8	1.0	1.2	1.4	1.6	2.0	2.5	2.9	3.3	3.7	4.1	4.5	5.0	5.8	6.6	8.3	10.3	12.4
17	17	0.6	0.9	1.2	1.5	1.8	2.1	2.3	2.9	3.5	4.1	4.7	5.3	5.9	6.5	7.0	8.2	9.4	11.7	—	—
	25	0.5	0.7	1.0	1.2	1.5	1.7	2.0	2.5	3.0	3.5	4.0	4.5	5.0	5.5	6.0	7.0	8.0	10.0	12.5	—
	35	0.4	0.7	0.9	1.1	1.3	1.5	1.7	2.2	2.6	3.1	3.5	3.9	4.4	4.8	5.2	6.1	7.0	8.7	10.9	—
	50	0.4	0.6	0.8	1.0	1.2	1.4	1.6	2.0	2.4	2.8	3.1	3.5	3.9	4.3	4.5	5.4	6.2	7.7	9.7	11.6
	80	0.4	0.5	0.7	0.9	1.1	1.2	1.4	1.8	2.1	2.5	2.9	3.3	3.6	4.0	4.3	5.1	5.8	7.2	9.0	10.9
	120	0.3	0.5	0.7	0.8	1.0	1.2	1.3	1.7	2.0	2.3	2.7	3.0	3.4	3.7	4.0	4.7	5.4	6.7	8.4	10.1
20	20	0.5	0.7	1.0	1.2	1.5	1.7	2.0	2.5	3.0	3.5	4.0	4.5	5.0	5.5	6.0	7.0	8.0	10.0	12.5	—
	30	0.4	0.6	0.8	1.0	1.2	1.5	1.7	2.1	2.5	2.9	3.3	3.7	4.1	4.5	4.9	5.8	6.6	8.2	10.3	12.4
	45	0.4	0.5	0.7	0.9	1.1	1.3	1.4	1.8	2.2	2.5	2.9	3.3	3.6	4.0	4.3	5.1	5.8	7.2	9.1	10.9
	60	0.3	0.5	0.7	0.8	1.0	1.2	1.3	1.7	2.0	2.3	2.7	3.0	3.4	3.7	4.0	4.7	5.4	6.7	8.4	10.1
	90	0.3	0.5	0.6	0.8	0.9	1.1	1.2	1.5	1.8	2.1	2.4	2.7	3.0	3.3	3.6	4.2	4.8	6.0	7.5	9.0
	150	0.3	0.4	0.6	0.7	0.8	1.0	1.1	1.4	1.7	2.0	2.3	2.6	2.9	3.2	3.4	4.0	4.6	5.7	7.2	8.6

24	24	0.4	0.6	0.8	1.0	1.2	.5	1.7	2.1	2.5	2.9	3.3	3.7	4.1	4.5	5.0	5.8	6.7	8.2	10.3	12.4
	32	0.4	0.5	0.7	0.9	1.1	.3	1.5	1.8	2.2	2.6	2.9	3.3	3.6	4.0	4.3	5.1	5.8	7.2	9.0	11.0
	50	0.3	0.5	0.6	0.8	0.9	.1	1.2	1.5	1.8	2.2	2.5	2.8	3.1	3.4	3.7	4.4	5.0	6.2	7.8	9.4
	70	0.3	0.4	0.6	0.7	0.8	.0	1.1	1.4	1.7	2.0	2.2	2.5	2.8	3.0	3.3	3.8	4.4	5.5	6.9	8.2
	100	0.3	0.4	0.5	0.6	0.8	.9	1.0	1.3	1.6	1.8	2.1	2.4	2.6	2.9	3.1	3.7	4.2	5.2	6.5	7.9
	160	0.2	0.4	0.5	0.6	0.7	.8	1.0	1.2	1.4	1.7	1.9	2.1	2.4	2.6	2.8	3.3	3.8	4.7	5.9	7.1
30	30	0.3	0.5	0.7	0.8	1.0	.2	1.3	1.7	2.0	2.3	2.7	3.0	3.3	3.7	4.0	4.7	5.4	6.7	8.4	10.0
	45	0.3	0.4	0.6	0.7	0.8	.0	1.1	1.4	1.7	1.9	2.2	2.5	2.7	3.0	3.3	3.8	4.4	5.5	6.9	8.2
	60	0.3	0.4	0.5	0.6	0.7	.9	1.0	1.2	1.5	1.7	2.0	2.2	2.5	2.7	3.0	3.5	4.0	5.0	6.2	7.4
	90	0.2	0.3	0.4	0.6	0.7	.8	0.9	1.1	1.3	1.6	1.8	2.0	2.2	2.5	2.7	3.1	3.6	4.5	5.6	6.7
	150	0.2	0.3	0.4	0.5	0.6	.7	0.8	1.0	1.2	1.4	1.6	1.8	2.0	2.2	2.4	2.8	3.2	4.0	5.0	5.9
	200	0.2	0.3	0.4	0.5	0.6	.7	0.8	1.0	1.1	1.3	1.5	1.7	1.9	2.0	2.2	2.6	3.0	3.7	4.7	5.6
36	36	0.3	0.4	0.6	0.7	0.8	.0	1.1	1.4	1.7	1.9	2.2	2.5	2.8	3.0	3.3	3.9	4.4	5.5	6.9	8.3
	50	0.2	0.4	0.5	0.6	0.7	.8	1.0	1.2	1.4	1.7	1.9	2.1	2.5	2.6	2.9	3.3	3.8	4.8	5.9	7.2
	75	0.2	0.3	0.4	0.5	0.6	.7	0.8	1.0	1.2	1.4	1.6	1.8	2.0	2.3	2.5	2.9	3.3	4.1	5.1	6.1
	100	0.2	0.3	0.4	0.5	0.6	.7	0.8	0.9	1.1	1.3	1.5	1.7	1.9	2.1	2.3	2.6	3.0	3.8	4.7	5.7
	150	0.2	0.3	0.3	0.4	0.5	.6	0.7	0.9	1.0	1.2	1.4	1.6	1.7	1.9	2.1	2.4	2.8	3.5	4.3	5.2
	200	0.2	0.2	0.3	0.4	0.5	.6	0.7	0.8	1.0	1.1	1.3	1.5	1.6	1.8	2.0	2.3	2.6	3.3	4.1	4.9
42	42	0.2	0.4	0.5	0.6	0.7	.8	1.0	1.2	1.4	1.6	1.9	2.1	2.4	2.6	2.8	3.3	3.8	4.7	5.9	7.1
	60	0.2	0.3	0.4	0.5	0.6	.7	0.8	1.0	1.2	1.4	1.6	1.8	2.0	2.2	2.4	2.8	3.2	4.0	5.0	6.0
	90	0.2	0.3	0.3	0.4	0.5	.6	0.7	0.9	1.0	1.2	1.4	1.6	1.7	1.9	2.1	2.4	2.8	3.5	4.4	5.2
	140	0.2	0.2	0.3	0.4	0.5	.5	0.6	0.8	0.9	1.1	1.2	1.4	1.5	1.7	1.9	2.2	2.5	3.1	3.9	4.6
	200	0.1	0.2	0.3	0.4	0.4	.5	0.6	0.7	0.9	1.0	1.1	1.3	1.4	1.6	1.7	2.0	2.3	2.9	3.6	4.3
	300	0.1	0.2	0.3	0.3	0.4	.5	0.5	0.7	0.8	0.9	1.1	1.3	1.4	1.5	1.7	1.9	2.2	2.8	3.5	4.2
50	50	0.2	0.3	0.4	0.5	0.6	.7	0.8	1.0	1.2	1.4	1.6	1.8	2.0	2.2	2.4	2.8	3.2	4.0	5.0	6.0
	70	0.2	0.3	0.3	0.4	0.5	.6	0.7	0.9	1.0	1.2	1.4	1.5	1.7	1.9	2.0	2.4	2.7	3.4	4.3	5.1
	100	0.1	0.2	0.3	0.4	0.4	.5	0.6	0.7	0.9	1.0	1.2	1.3	1.5	1.6	1.8	2.1	2.4	3.0	3.7	4.5
	150	0.1	0.2	0.3	0.3	0.4	.5	0.5	0.7	0.8	0.9	1.1	1.2	1.3	1.5	1.6	1.9	2.1	2.7	3.3	4.0
	300	0.1	0.2	0.2	0.3	0.3	.4	0.5	0.6	0.7	0.8	0.9	1.0	1.1	1.3	1.4	1.6	1.9	2.3	2.9	3.5
60	60	0.2	0.2	0.3	0.4	0.5	.6	0.7	0.8	1.0	1.2	1.3	1.5	1.7	1.8	2.0	2.3	2.7	3.3	4.2	5.0
	100	0.1	0.2	0.3	0.3	0.4	.5	0.5	0.7	0.8	0.9	1.1	1.2	1.3	1.5	1.6	1.9	2.1	2.7	3.3	4.0
	150	0.1	0.2	0.2	0.3	0.3	.4	0.5	0.6	0.7	0.8	0.9	1.0	1.2	1.3	1.4	1.6	1.9	2.3	2.9	3.5
	300	0.1	0.1	0.2	0.2	0.3	.3	0.4	0.5	0.6	0.7	0.8	0.9	1.0	1.1	1.2	1.4	1.6	2.0	2.5	3.0
75	75	0.1	0.2	0.3	0.3	0.4	.5	0.5	0.7	0.8	0.9	1.1	1.2	1.3	1.5	1.6	1.9	2.1	2.7	3.3	4.0
	120	0.1	0.2	0.2	0.3	0.3	.4	0.4	0.5	0.6	0.8	0.9	1.0	1.1	1.2	1.3	1.5	1.7	2.2	2.7	3.3
	200	0.1	0.1	0.2	0.2	0.3	.3	0.4	0.5	0.5	0.6	0.7	0.8	0.9	1.0	1.1	1.3	1.5	1.8	2.3	2.7
	300	0.1	0.1	0.2	0.2	0.2	.3	0.3	0.4	0.5	0.6	0.7	0.7	0.8	0.9	1.0	1.2	1.3	1.7	2.1	2.5
100	100	0.1	0.1	0.2	0.2	0.3	.3	0.4	0.5	0.6	0.7	0.8	0.9	1.0	1.1	1.2	1.4	1.6	2.0	2.5	3.0
	200	0.1	0.1	0.1	0.2	0.2	.3	0.3	0.4	0.4	0.5	0.6	0.7	0.7	0.8	0.9	1.0	1.2	1.5	1.9	2.2
	300	0.1	0.1	0.1	0.2	0.2	.2	0.3	0.3	0.4	0.5	0.5	0.6	0.7	0.7	0.8	0.9	1.1	1.3	1.7	2.0
150	150	0.1	0.1	0.1	0.2	0.2	.2	0.3	0.3	0.4	0.5	0.5	0.6	0.7	0.7	0.8	0.9	1.1	1.3	1.7	2.0
	300	—	0.1	0.1	0.1	0.1	.2	0.2	0.2	0.3	0.3	0.4	0.5	0.5	0.6	0.6	0.7	0.8	1.0	1.2	1.5
200	200	—	0.1	0.1	0.1	0.1	.2	0.2	0.2	0.3	0.3	0.4	0.5	0.5	0.6	0.6	0.7	0.8	1.0	1.2	1.5
	300	—	0.1	0.1	0.1	0.1	.1	0.2	0.2	0.2	0.3	0.3	0.4	0.4	0.5	0.5	0.6	0.7	0.8	1.0	1.2
300	300	—	—	0.1	0.1	0.1	.1	0.1	0.2	0.2	0.2	0.3	0.3	0.3	0.4	0.4	0.5	0.5	0.6	0.7	0.8
500	500	—	—	—	—	0.1	.1	0.1	0.1	0.1	0.1	0.2	0.2	0.2	0.2	0.2	0.3	0.3	0.4	0.5	0.6

Table 2-2. Average Surface Reflectances of Various Commonly Used Colors.

Color	Surface Reflectance (avg)
White and very light tints	75–90%
Medium blue-gray, yellow-gray	50%
Dark gray, medium blue	30%
Dark blue, brown, Dark green, wood finish	10%

3. Remove the card and read the meter again, holding the meter in the exact same position. Suppose it reads 40 fc.
4. The ratio of the second reading over the first multiplied by the known reflectance is the reflectance of the surface. In this case it will be $(40 \times 80)/50 = 64\%$.

Obviously, the reflectance measured this way is not always absolutely correct since the meter condition, the difficulty of holding the meter in the exact same location, and even the accuracy of the known reflectance of the card may have a certain error factor. But for practical purposes, the method is fine.

The next move in determining the CU is to find the effective reflectances of the ceiling and floor. Basically, the effective reflectance takes into account the fact that because of the cavity depth, the total overall reflectance of the ceiling cavity or the floor cavity is different from the direct surface reflectance. Obviously, the deeper the cavity, the less light will be reflected. In addition, some of this reflected light at the ceiling or floor is actually the light bounced off the walls. Thus, the determination of the effective ceiling or floor reflectances would be dependent upon their cavity depths, their surface reflectances, and also the cavity ratios. Table 2-3 is used to obtain the effective reflectances for ceiling or floor cavities. To clarify the use of this table, let us take an example.

Let us suppose a room has 80, 50, and 30% ceiling, wall, and floor reflectances, respectively. Let us also suppose the CCR and FCR are 1.6 and 0.5, respectively. Thus, to determine the effective CCR, we select the column with 80% ceiling reflectance and 50% wall reflectance at the top of the table and read 60% effective ceiling-cavity reflectance at CCR = 1.6. Similarly, for the effective floor-cavity reflectance, we select the column with 30% floor reflectance and 50% wall reflectance at the top of the table and read 28% effective floor-cavity reflectance against FCR = 0.5.

Table 2-3. Per Cent Effective Ceiling or Floor Reflectances for Various Reflectance Combinations. (Reprinted from the *IES Lighting Handbook*, 1981, Reference Volume).

Effective Cavity Reflectance

BASE REFLECTANCE PER CENT	90									
WALL REFLECTANCE PER CENT	90	80	70	60	50	40	30	20	10	0
ROOM CAVITY RATIO										
0.2	89	88	88	87	86	85	85	84	84	82
0.4	88	87	86	85	84	83	81	80	79	76
0.6	87	86	84	82	80	79	77	76	74	73
0.8	87	85	82	80	77	75	73	71	69	67
1.0	86	83	80	77	75	72	69	66	64	62
1.2	85	82	78	75	72	69	66	63	60	57
1.4	85	80	77	73	69	65	62	59	57	52
1.6	84	79	75	71	67	63	59	56	53	50
1.8	83	78	73	69	64	60	56	53	50	48
2.0	83	77	72	67	62	56	53	50	47	43
2.2	82	76	70	65	59	54	50	47	44	40
2.4	82	75	69	64	58	53	48	45	41	37
2.6	81	74	67	62	56	51	46	42	38	35
2.8	81	73	66	60	54	49	44	40	36	34
3.0	80	72	64	58	52	47	42	38	34	30
3.2	79	71	63	56	50	45	40	36	32	28
3.4	79	70	62	54	48	43	38	34	30	27
3.6	78	69	61	53	47	42	36	32	28	25
3.8	78	69	60	51	45	40	35	31	27	23
4.0	77	69	58	51	44	39	33	29	25	22
4.2	77	62	57	50	43	37	32	28	24	21
4.4	76	61	56	49	42	36	31	27	23	20
4.6	76	60	55	47	40	35	30	26	22	19
4.8	75	59	54	46	39	34	28	25	21	18
5.0	75	59	53	45	38	33	28	24	20	16
6.0	73	61	49	41	34	29	24	20	16	11
7.0	70	58	45	38	30	27	21	18	14	08
8.0	68	55	42	35	27	23	18	15	12	06
9.0	66	52	38	31	25	21	16	14	11	05
10.0	65	51	36	29	22	19	15	11	09	04

BASE REFLECTANCE PER CENT	80									
WALL REFLECTANCE PER CENT	90	80	70	60	50	40	30	20	10	0
ROOM CAVITY RATIO										
0.2	79	78	78	77	[illegible]	76	76	75	74	72
0.4	79	77	76	75	[illegible]	73	72	71	70	68
0.6	78	76	75	73	[illegible]	70	68	66	65	63
0.8	78	75	73	71	[illegible]	67	65	63	61	57
1.0	77	74	72	69	[illegible]	65	62	60	57	55
1.2	76	73	70	67	[illegible]	61	58	55	53	51
1.4	76	72	68	65	[illegible]	59	55	53	50	48
1.6	75	71	67	63	[illegible]	57	53	50	47	44
1.8	75	70	66	62	[illegible]	54	50	47	44	41
2.0	74	69	64	60	[illegible]	52	48	45	41	38
2.2	74	68	63	58	[illegible]	49	45	42	38	35
2.4	73	67	61	56	[illegible]	47	43	40	36	33
2.6	73	66	60	55	[illegible]	45	41	38	34	31
2.8	73	65	59	53	[illegible]	43	39	36	32	29
3.0	72	65	58	52	[illegible]	42	37	34	30	27
3.2	72	65	57	51	[illegible]	40	35	33	28	25
3.4	71	64	56	49	[illegible]	39	34	32	27	24
3.6	71	63	54	48	[illegible]	38	32	30	25	23
3.8	70	62	53	47	[illegible]	36	31	28	24	22
4.0	70	61	53	46	[illegible]	35	30	26	22	20
4.2	69	60	52	45	[illegible]	34	29	25	21	18
4.4	69	60	51	44	[illegible]	33	28	24	20	17
4.6	69	59	50	43	[illegible]	32	27	23	19	15
4.8	68	58	49	42	[illegible]	31	26	22	18	14
5.0	68	58	48	41	[illegible]	30	25	21	18	14
6.0	66	55	44	38	[illegible]	27	22	19	15	10
7.0	64	53	41	35	[illegible]	24	19	16	12	07
8.0	62	50	38	32	[illegible]	21	17	14	11	05
9.0	61	49	36	30	[illegible]	19	15	13	10	04
10.0	59	46	33	27	[illegible]	18	14	11	08	03

BASE REFLECTANCE PER CENT	70									
WALL REFLECTANCE PER CENT	90	80	70	60	50	40	30	20	10	0
ROOM CAVITY RATIO										
0.2	70	69	68	68	67	67	66	66	65	64
0.4	69	68	67	66	65	64	63	62	61	58
0.6	69	67	65	64	63	61	59	58	57	54
0.8	68	66	64	62	60	58	56	55	53	50
1.0	68	65	62	60	58	55	53	52	50	47
1.2	67	64	61	59	57	54	50	48	46	44
1.4	67	63	60	58	55	51	47	45	44	41
1.6	67	62	59	56	53	47	45	43	41	38
1.8	66	61	58	54	51	46	42	40	38	35
2.0	66	60	56	52	49	45	40	38	36	33
2.2	66	60	55	51	48	43	38	36	34	32
2.4	65	60	54	50	46	41	37	35	32	30
2.6	65	59	54	49	45	40	35	33	30	28
2.8	65	59	53	48	43	38	33	30	28	26
3.0	64	58	52	47	42	37	32	29	27	24
3.2	64	58	51	46	40	36	31	28	25	23
3.4	64	57	50	45	39	35	29	27	24	22
3.6	63	56	49	44	38	33	28	25	22	20
3.8	63	56	49	43	37	32	27	24	21	19
4.0	63	55	48	42	36	31	26	23	20	17
4.2	62	55	47	41	35	30	25	22	19	16
4.4	62	54	46	40	34	29	24	21	18	15
4.6	62	53	45	39	33	28	24	21	17	14
4.8	62	53	45	38	32	27	23	20	16	13
5.0	61	52	44	36	31	26	22	19	16	12
6.0	60	51	41	35	28	24	19	16	13	09
7.0	58	48	38	32	26	22	17	14	11	06
8.0	57	46	35	29	23	19	15	13	10	05
9.0	56	45	33	27	21	18	14	12	09	04
10.0	55	43	31	25	19	16	12	10	08	03

BASE REFLECTANCE PER CENT	60									
WALL REFLECTANCE PER CENT	90	80	70	60	50	40	30	20	10	0
ROOM CAVITY RATIO										
0.2	60	59	59	59	58	57	56	56	55	53
0.4	60	59	59	58	57	55	54	53	52	50
0.6	60	58	57	56	55	53	51	51	50	46
0.8	59	57	56	55	54	51	48	47	46	43
1.0	59	57	55	53	51	48	45	44	43	41
1.2	59	56	54	51	49	46	44	42	40	38
1.4	59	56	53	49	47	44	41	39	38	36
1.6	59	55	52	48	45	42	39	37	35	33
1.8	58	55	51	47	44	40	37	35	33	31
2.0	58	54	50	46	43	39	35	33	31	29
2.2	58	53	49	45	42	37	34	31	29	28
2.4	58	53	48	44	41	36	32	30	27	26
2.6	58	53	48	43	39	35	31	28	26	24
2.8	58	53	47	43	38	34	29	27	24	22
3.0	57	52	46	42	37	32	28	25	23	20
3.2	57	51	45	41	36	31	27	23	22	18
3.4	57	51	45	40	35	30	26	23	20	17
3.6	57	50	44	39	34	29	25	22	19	16
3.8	57	50	43	38	33	29	24	21	19	15
4.0	57	49	42	37	32	28	23	20	18	14
4.2	56	49	42	37	32	27	22	19	17	14
4.4	56	49	42	36	31	27	22	19	16	13
4.6	56	49	41	35	30	26	21	18	16	13
4.8	56	48	41	34	29	25	21	18	15	12
5.0	56	48	40	34	28	24	20	17	14	11
6.0	55	45	37	31	25	21	17	14	11	07
7.0	54	43	35	30	24	20	15	12	09	05
8.0	53	42	33	28	22	18	14	11	08	04
9.0	52	40	31	26	20	16	12	10	07	03
10.0	51	39	29	24	18	15	11	09	07	02

BASE REFLECTANCE PER CENT	50									
WALL REFLECTANCE PER CENT	90	80	70	60	50	40	30	20	10	0
ROOM CAVITY RATIO										
0.2	50	50	49	49	48	48	47	46	46	44
0.4	50	49	48	48	47	46	45	45	44	42
0.6	50	48	47	46	45	44	43	42	41	38
0.8	50	48	47	45	44	42	40	39	38	36
1.0	50	48	46	44	43	41	38	37	36	34
1.2	50	47	45	43	41	39	36	35	34	29
1.4	50	47	45	42	40	38	35	34	32	27
1.6	50	47	44	41	39	36	33	32	30	26
1.8	50	46	43	40	38	35	31	30	28	25
2.0	50	46	43	40	37	34	30	28	26	24
2.2	50	46	42	38	36	33	29	27	24	22
2.4	50	46	42	37	35	31	27	25	23	21
2.6	50	46	41	37	34	30	26	23	21	20
2.8	50	46	41	36	33	29	25	22	20	19
3.0	50	45	40	36	32	28	24	21	19	17
3.2	50	44	39	35	31	27	23	20	18	16
3.4	50	44	39	35	30	26	22	19	17	15
3.6	50	44	39	34	29	25	21	18	16	14
3.8	50	44	38	34	29	25	21	17	15	13
4.0	50	44	38	33	28	24	20	17	15	12
4.2	50	43	37	32	28	24	20	17	14	12
4.4	50	43	37	32	27	23	19	16	13	11
4.6	50	43	36	31	26	22	18	15	13	10
4.8	50	43	36	31	26	22	18	15	12	09
5.0	50	42	35	30	25	21	17	14	12	09
6.0	50	42	34	29	23	19	15	13	10	06
7.0	49	41	32	27	21	18	14	11	08	05
8.0	49	40	30	25	19	16	12	10	07	03
9.0	48	39	29	24	18	15	11	09	07	03
10.0	47	37	27	22	17	14	10	08	06	02

Table 2-3. *(Continued)*

BASE REFLECTANCE PER CENT	40										30									
WALL REFLECTANCE PER CENT	90	80	70	60	50	40	30	20	10	0	90	80	70	60	50	40	30	20	10	0
ROOM CAVITY RATIO																				
0.2	40	40	39	39	39	38	38	37	36	36	31	31	30	30	29	29	29	28	28	27
0.4	41	40	39	39	38	37	36	35	34	34	31	31	30	30	29	28	28	27	26	25
0.6	41	40	39	38	37	36	34	33	32	31	32	31	30	29	28	27	26	26	25	23
0.8	41	40	38	37	36	35	33	32	31	29	32	31	30	29	28	26	25	25	23	22
1.0	42	40	38	37	35	33	32	31	29	27	33	32	30	29	27	25	24	23	22	20
1.2	42	40	38	36	34	32	30	29	27	25	33	32	30	28	27	25	23	22	21	19
1.4	42	39	37	35	33	31	29	27	25	23	34	32	30	28	26	24	22	21	19	18
1.6	42	39	37	35	32	30	27	25	23	22	34	33	29	27	25	23	22	20	18	17
1.8	42	39	36	34	31	29	26	24	22	21	35	33	29	27	25	23	21	19	17	16
2.0	42	39	36	34	31	28	25	23	21	19	35	33	29	26	24	22	20	18	16	14
2.2	42	39	36	33	30	27	24	22	19	18	36	32	29	26	24	22	19	17	15	13
2.4	43	39	35	33	29	27	24	21	18	17	36	32	29	26	24	22	19	16	14	12
2.6	43	39	35	32	29	26	23	20	17	15	36	32	29	25	23	21	18	16	14	12
2.8	43	39	35	32	28	25	22	19	16	14	37	33	29	25	23	21	17	15	13	11
3.0	43	39	35	31	27	24	21	18	16	13	37	33	29	25	22	20	17	15	12	10
3.2	43	39	35	31	27	23	20	17	15	13	37	33	29	25	22	19	16	14	12	10
3.4	43	39	34	30	26	23	20	17	14	12	37	33	29	25	22	19	16	14	11	09
3.6	44	39	34	30	26	22	19	16	14	11	38	33	29	24	21	18	15	13	10	09
3.8	44	38	33	29	25	22	18	16	13	10	38	33	28	24	21	18	15	13	10	08
4.0	44	38	33	29	25	21	18	15	12	10	38	33	28	24	21	18	14	12	09	07
4.2	44	38	33	29	24	21	17	15	12	10	38	33	28	24	20	17	14	12	09	07
4.4	44	38	33	28	24	20	17	14	11	09	39	33	28	24	20	17	14	11	09	06
4.6	44	38	32	28	23	19	16	14	11	08	39	33	28	24	20	17	13	10	08	06
4.8	44	38	32	27	22	19	16	13	10	08	39	33	28	24	20	17	13	10	08	05
5.0	45	38	31	27	22	19	15	13	10	07	39	33	28	24	19	16	13	10	08	05
6.0	44	37	30	25	20	17	13	11	08	05	39	33	27	23	18	15	11	09	06	04
7.0	44	36	29	24	19	16	12	10	07	04	40	33	26	22	17	14	10	08	05	03
8.0	44	35	28	23	18	15	11	09	06	03	40	33	26	21	16	13	09	07	04	02
9.0	44	35	26	21	16	13	10	08	05	02	40	33	25	20	15	12	09	07	04	02
10.0	43	34	25	20	15	12	08	07	05	02	40	32	24	19	14	11	08	06	03	01

BASE REFLECTANCE PER CENT	20										10										0									
WALL REFLECTANCE PER CENT	90	80	70	60	50	40	30	20	10	0	90	80	70	60	50	40	30	20	10	0	90	80	70	60	50	40	30	20	10	0
ROOM CAVITY RATIO																														
0.2	21	20	20	20	20	20	19	19	19	17	11	11	11	10	10	10	10	09	09	09	02	02	02	01	01	01	01	00	00	0
0.4	22	21	20	20	20	19	19	18	18	16	12	11	11	11	11	10	10	09	09	08	04	03	03	02	02	02	01	01	00	0
0.6	23	21	21	20	19	19	18	18	17	15	13	13	12	11	11	10	10	09	08	08	05	05	04	03	03	02	02	01	01	0
0.8	24	22	21	20	19	19	18	17	16	14	15	14	13	12	11	10	10	09	08	07	07	06	05	04	04	03	02	02	01	0
1.0	25	23	22	20	19	18	17	16	15	13	16	14	13	12	12	11	10	09	08	07	08	07	06	05	04	03	02	02	01	0
1.2	25	23	22	20	19	17	17	16	14	12	17	15	14	13	12	11	10	09	07	06	10	08	07	06	05	04	03	02	01	0
1.4	26	24	22	20	18	17	16	15	13	12	18	16	14	13	12	11	10	09	07	06	11	09	08	07	06	04	03	02	01	0
1.6	26	24	22	20	18	17	16	15	13	11	19	17	15	14	12	11	09	08	07	06	12	10	09	07	06	05	03	02	01	0
1.8	27	25	23	20	18	17	15	14	12	10	19	17	15	14	13	11	09	08	06	05	13	11	09	08	07	05	04	03	01	0
2.0	28	25	23	20	18	16	15	13	11	09	20	18	16	14	13	11	09	08	06	05	14	12	10	09	07	05	04	03	01	0
2.2	28	25	23	20	18	16	14	12	10	09	21	19	16	14	13	11	09	07	06	05	15	13	11	09	07	06	04	03	01	0
2.4	29	26	23	20	18	16	14	12	10	08	22	19	17	15	13	11	09	07	06	05	16	13	11	09	08	06	04	03	01	0
2.6	29	26	23	20	18	16	14	11	09	08	23	20	17	15	13	11	09	07	06	04	17	14	12	10	08	06	05	03	02	0
2.8	30	27	23	20	18	15	13	11	09	07	23	20	18	16	13	11	09	07	05	03	17	15	13	10	08	07	05	03	02	0
3.0	30	27	23	20	17	15	13	11	09	07	24	21	18	16	13	11	09	07	05	03	18	16	13	11	09	07	05	03	02	0
3.2	31	27	23	20	17	15	12	11	09	06	25	21	18	16	13	11	09	07	05	03	19	16	14	11	09	07	05	03	02	0
3.4	31	27	23	20	17	15	12	10	08	06	26	22	18	16	13	11	09	07	05	03	20	17	14	12	09	07	05	03	02	0
3.6	32	27	23	20	17	15	12	10	08	05	26	22	19	16	13	11	09	06	04	03	20	17	15	12	10	08	05	04	02	0
3.8	32	28	23	20	17	15	12	10	07	05	27	23	19	17	14	11	09	06	04	02	21	18	15	12	10	08	05	04	02	0
4.0	33	28	23	20	17	14	11	09	07	05	27	23	20	17	14	11	09	06	04	02	22	18	15	13	10	08	05	04	02	0
4.2	33	28	23	20	17	14	11	09	07	04	28	24	20	17	14	11	09	06	04	02	22	19	16	13	10	08	06	04	02	0
4.4	34	28	24	20	17	14	11	09	07	04	28	24	20	17	14	11	08	06	04	02	23	19	16	13	10	08	06	04	02	0
4.6	34	29	24	20	17	14	11	09	07	04	29	25	20	17	14	11	08	06	04	02	23	20	17	13	11	08	06	04	02	0
4.8	35	29	24	20	17	13	10	08	06	04	29	25	20	17	14	11	08	06	04	02	24	20	17	14	11	08	06	04	02	0
5.0	35	29	24	20	16	13	10	08	06	04	30	25	20	17	14	11	08	06	04	02	25	21	17	14	11	08	06	04	02	0
6.0	36	30	24	20	16	13	10	08	05	02	31	26	21	18	14	11	08	06	03	01	27	23	18	15	12	09	06	04	02	0
7.0	36	30	24	20	15	12	09	07	04	02	32	27	21	17	13	11	08	06	03	01	28	24	19	15	12	09	06	04	02	0
8.0	37	30	23	19	15	12	08	06	03	01	33	27	21	17	13	10	07	05	03	01	30	25	20	15	12	09	06	04	02	0
9.0	37	29	23	19	14	11	08	06	03	01	34	28	21	17	13	10	07	05	02	01	31	25	20	15	12	09	06	04	02	0
10.0	37	29	22	18	13	10	07	05	03	01	34	28	21	17	12	10	07	05	02	01	31	25	20	15	12	09	06	04	02	0

Coefficient of Utilization Table. Once the effective floor- and ceiling-cavity reflectances have been determined, the determination of CU of a luminaire is easy. Manufacturers provide charts or tables for CU values for each type of luminaire, a typical one of which is shown in the section J of Figure 2-1. The CU value can be easily determined now by knowing the values of the effective ceiling-cavity reflectance, the percentage of wall reflectance, and the room-cavity ratio. The effective floor-cavity reflectance of such tables is normally 20%. If the effective floor-cavity reflectance determined is other than 20%, a correction factor is used to modify the CU. Table 2-4 shows such multiplying factors to convert a 20% effective floor-cavity-reflectance value to 0%, 10%, or 30%. The intermediate values are interpolated.

Let us now take an example that will demonstrate all steps in the computation of the CU values for two situations: (I) when the luminaires are suspended and (II) when they are recessed in ceiling.

Given: Room dimensions: Length, $L = 50$ ft.; width, $W = 30$ ft.; height, $H = 8.5$ ft.
Work plane or task height = 2.5 ft. above floor
Ceiling/wall/floor reflectances = 80/50/20%
Luminaires: 2 X 4 ft., 4 lamps, acrylic prismatic lens
Find the appropriate CU values for (I) Luminaires are suspended 1.5 ft. below ceiling; (II) Luminaires are recessed in ceiling. Use CU table in photometric report of figure 2-1.

I. Suspended Luminaires:

Step 1. Ceiling-cavity height, $h_{cc} = 1.5$ ft.
Floor-cavity height, $h_{fc} = 2.5$ ft.
Room-cavity height, $h_{rc} = 8.5 - (1.5 + 2.5) = 4.5$ ft.

Step 2: Determination of CCR, RCR, and FCR: Using cavity ratio formula, $5\,h\,(L + W)/L \times W$,

$$\text{CCR} = \frac{5 \times 1.5\,(50 + 30)}{50 \times 30} = 0.40$$

$$\text{RCR} = \frac{5 \times 4.5\,(50 + 30)}{50 \times 30} = 1.2$$

$$\text{FCR} = \frac{5 \times 2.5\,(50 + 30)}{50 \times 30} = 0.67$$

These values of the cavity ratios could be directly obtained from Table 2-1.

Table 2-4. Multiplying Factors for other than 20% Effective Floor Cavity Reflectances. (Reprinted from permission from the *IES Lighting Handbook*, 1981, Reference Volume).

% Effective Ceiling Cavity Reflectance, ρ_{CC}	80				70				50			30			10		
% Wall Reflectance, ρ_W	70	50	30	10	70	50	30	10	50	30	10	50	30	10	50	30	10
For 30 Per Cent Effective Floor Cavity Reflectance (20 Per Cent = 1.00)																	
Room Cavity Ratio																	
1	1.092	1.082	1.075	1.068	1.077	1.070	1.064	1.059	1.049	1.044	1.040	1.028	1.026	1.023	1.012	1.010	1.008
2	1.079	1.066	1.055	1.047	1.068	1.057	1.048	1.039	1.041	1.033	1.027	1.026	1.021	1.017	1.013	1.010	1.006
3	1.070	1.054	1.042	1.033	1.061	1.048	1.037	1.028	1.034	1.027	1.020	1.024	1.017	1.012	1.014	1.009	1.005
4	1.062	1.045	1.033	1.024	1.055	1.040	1.029	1.021	1.030	1.022	1.015	1.022	1.015	1.010	1.014	1.009	1.004
5	1.056	1.038	1.026	1.018	1.050	1.034	1.024	1.015	1.027	1.018	1.012	1.020	1.013	1.008	1.014	1.009	1.004
6	1.052	1.033	1.021	1.014	1.047	1.030	1.020	1.012	1.024	1.015	1.009	1.019	1.012	1.006	1.014	1.008	1.003
7	1.047	1.029	1.018	1.011	1.043	1.026	1.017	1.009	1.022	1.013	1.007	1.018	1.010	1.005	1.014	1.008	1.003
8	1.044	1.026	1.015	1.009	1.040	1.024	1.015	1.007	1.020	1.012	1.006	1.017	1.009	1.004	1.013	1.007	1.003
9	1.040	1.024	1.014	1.007	1.037	1.022	1.014	1.006	1.019	1.011	1.005	1.016	1.009	1.004	1.013	1.007	1.002
10	1.037	1.022	1.012	1.006	1.034	1.020	1.012	1.005	1.017	1.010	1.004	1.015	1.009	1.003	1.013	1.007	1.002

For 10 Per Cent Effective Floor Cavity Reflectance (20 Per Cent = 1.00)

Room Cavity Ratio																	
1	.923	.929	.935	.940	.933	.939	.943	.948	.956	.960	.963	.973	.976	.979	.989	.991	.993
2	.931	.942	.950	.958	.940	.949	.957	.963	.962	.968	.974	.976	.980	.985	.988	.991	.995
3	.939	.951	.961	.969	.945	.957	.966	.973	.967	.975	.981	.978	.983	.988	.988	.902	.996
4	.944	.958	.969	.978	.950	.963	.973	.980	.972	.980	.986	.980	.986	.991	.987	.992	.996
5	.949	.964	.976	.983	.954	.968	.978	.985	.975	.983	.989	.981	.988	.993	.987	.992	.997
6	.953	.969	.980	.986	.958	.972	.982	.989	.977	.985	.992	.982	.989	.995	.987	.993	.997
7	.957	.973	.983	.991	.961	.975	.985	.991	.979	.987	.994	.983	.990	.996	.987	.993	.998
8	.960	.976	.986	.993	.963	.977	.987	.993	.981	.988	.995	.984	.991	.997	.987	.994	.998
9	.963	.978	.987	.994	.965	.979	.989	.994	.983	.990	.996	.985	.992	.998	.988	.994	.999
10	.965	.980	.989	.995	.967	.981	.990	.995	.984	.991	.997	.986	.993	.998	.988	.994	.999

For 0 Per Cent Effective Floor Cavity Reflectance (20 Per Cent = 1.00)

Room Cavity Ratio																	
1	.859	.870	.879	.886	.873	.884	.893	.901	.916	.923	.929	.948	.954	.960	.979	.983	.987
2	.871	.887	.903	.919	.886	.902	.916	.928	.926	.938	.949	.954	.963	.971	.978	.983	.991
3	.882	.904	.915	.942	.898	.918	.934	.947	.936	.950	.964	.958	.969	.979	.976	.984	.993
4	.893	.919	.941	.958	.908	.930	.948	.961	.945	.961	.974	.961	.974	.984	.975	.985	.994
5	.903	.931	.953	.969	.914	.939	.958	.970	.951	.967	.980	.964	.977	.988	.975	.985	.995
6	.911	.940	.961	.976	.920	.945	.965	.977	.955	.972	.985	.966	.979	.991	.975	.986	.996
7	.917	.947	.967	.981	.924	.950	.970	.982	.959	.975	.988	.968	.981	.993	.975	.987	.997
8	.922	.953	.971	.985	.929	.955	.975	.986	.963	.978	.991	.970	.983	.995	.976	.988	.998
9	.928	.958	.975	.988	.933	.959	.980	.989	.966	.980	.993	.971	.985	.996	.976	.988	.998
10	.933	.962	.979	.991	.937	.963	.983	.992	.969	.982	.995	.973	.987	.997	.977	.989	.999

Step 3: Determination of Effective Reflectances of Ceiling and Floor Cavities: Referring to Table 2-3, corresponding to ceiling and wall reflectances of 80 and 50% and CCR of 0.4, the effective ceiling cavity reflectance is found to be 74%. Similarly, with wall and floor reflectances of 50 and 20%, and FCR of 0.67, the effective floor cavity reflectance is found to be 19%, rounded off to 20%.

Step 4: Determination of Required CU Value: From the CU table, corresponding to effective ceiling-, wall- and floor-cavity reflectances of 80/50/20 and 70/50/20, the CU values for RCR = 1 and 2 are as follows:

RCR	80/50/20	70/50/20
1	0.69	0.68
2	0.62	0.61

The required CU value at 74% effective ceiling-cavity reflectance at RCR = 1.2 has to be interpolated. The technique is as follows. First, interpolate for the effective ceiling cavity reflectance. For RCR = 1, the difference in these reflectance values is 74 - 70 = 4; corresponding CU difference is [(0.69 - 0.68)/10 X 4] = 0.004. So, the new CU value is 0.68 + 0.004 = 0.684. Similarly, for RCR = 2, the new CU value is {[(0.62 - 0.61)/10] X 4} + 0.61 = 0.614. Rewriting these values:

RCR	74/50/20
1	0.684
2	0.614

Now interpolating for RCR = 1.2, the difference in RCR value from RCR = 1 is 1.2 - 1.0 = 0.2. Now 0.684 - 0.614 = 0.07, and 0.07 X 0.2 = 0.014; this must be subtracted from 0.684. So, the interpolated CU value for 74/50/20% reflectances and RCR of 1.2 = 0.684 - 0.014 = 0.67.

II. Recessed Luminaire:

Step 1: $h_{cc} = 0$ ft.
$h_{fc} = 2.5$ ft.
$h_{rc} = 8.5 - 2.5 = 6$ ft.

Step 2: $$\text{CCR} = \frac{5 \times 0 \times (50 + 30)}{50 \times 30} = 0$$

$$RCR = \frac{5 \times 6 \times (50 + 30)}{50 \times 30} = 1.6$$

$$FCR = \frac{5 \times 2.5 \times (50 + 30)}{50 \times 30} = 0.67$$

Step 3: Referring to Table 2-3, with ceiling and wall reflectances of 80 and 50% and CCR of 0, the effective ceiling-cavity reflectance = 80%; with floor and wall reflectances of 20 and 50% and FCR of 0.67, the effective floor-cavity reflectance = 19%, rounded off to 20%.

Step 4: Corresponding to effective surface reflectances of 80/50/20, from CU table,

RCR	80/50/20
1	0.69
2	0.62

For RCR = 1.6, the interpolated CU value is 0.69 - (0.69 - 0.62) × 0.6 = 0.648.

REFERENCES

Faucett, R. E., and Judge, J. R., "An Improved Method for S/MH Ratings of Luminaires with Direct Symmetrical Distributors," *Illuminating Engineering*, January 1971, p. 37.

IES. "Lighting Calculations." *IES Lighting Handbook, 1981 Reference Volume*, pp. 9-1 to 9-9.

IES, Design Practice Committee. "Classification of Interior Luminaires by Distribution: Luminaire Spacing Criteria." *Lighting Design & Application*, August 1977, pp. 20-21.

IES, Design Practice Committee, "Zonal-Cavity Method of Calculating and Using Coefficient of Utilization." *Illuminating Engineering*, May 1964, pp. 309-328.

Le Vere, Richard C., Levin, Robert E., and Primrose, William C., "Spacing Criteria for Interior Luminaires–The Practices and Pitfalls," *Journal of the Illuminating Engineering Society*, **3**, October 1973, pp. 41-49.

Lewin, Robert E., "Revision of S/MH Concept," *Lighting Design & Application*, August 1977, pp. 22-25.

Chapter Three
Basic Design Techniques

All artificial illumination design techniques are related to two major considerations: quality and quantity. While the quantity of light is an unavoidable requirement for any lighting design, its quality measurements bear equal importance in many.

DETERMINATION OF QUANTITY

There are two basic design techniques for the determination of quantity of illuminance: the point-by-point method and the lumen method.

Point-by-Point Method

The point-by-point method makes use of the inverse-square law, which states that the illuminance at a point on a surface perpendicular to the light ray is equal to the luminous intensity of the source at that point, divided by the square of the distance between the source and the point of calculation. Mathematically:

$$E = \frac{I}{D^2}$$

where

E = Illuminance in footcandles
I = Luminous intensity in candelas
D = Distance in feet between the source and the point of calculation.

If the surface is not perpendicular to the light ray, the appropriate trigonometrical functions must be applied to account for the deviation. This is shown in Figure 3-1.

Figure 3-1a shows the direct application of the formula, where the point of

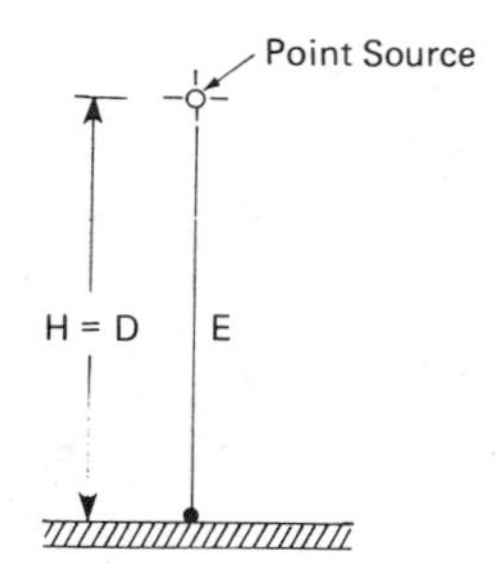
Point Source
H = D
E
(a) Luminaire is directly above the point of calculation – with the ray of light perpendicular to the surface. illuminance is given by:
$E = \frac{I}{D^2}$

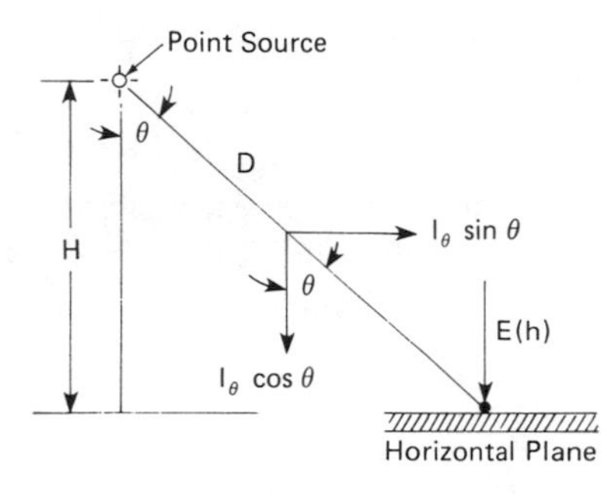
Point Source
θ
D
H
$I_\theta \sin\theta$
θ
E(h)
$I_\theta \cos\theta$
Horizontal Plane
(b) Light ray arrives at the point of calculation on a horizontal surface, at an angle θ. Illuminance is, then,
$E(h) = \frac{I_\theta \cos\theta}{D^2}$

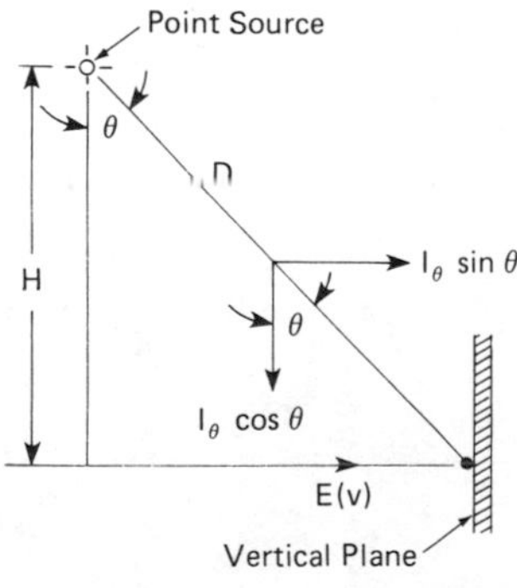
Point Source
θ
D
H
$I_\theta \sin\theta$
θ
$I_\theta \cos\theta$
E(v)
Vertical Plane
(c) Light ray arrives at the point of calculation on a vertical surface, at an angle θ. Illuminance is, then,
$E(v) = \frac{I_\theta \sin\theta}{D^2}$

Fig. 3-1. (a) Luminaire is directly above the point of calculation, with the ray of light perpendicular to the surface. Illuminance is given by $E = I/D^2$. (b) Light may arrive at the point of calculation on a horizontal surface, at an angle θ. Illuminance is then $E(h) = I_\theta \cos\theta / D^2$. (c) Light ray arrives at the point of calculation on a vertical surface, at an angle θ. Illuminance is then $E(v) = I_\theta \sin\theta / D^2$.

calculation lies directly beneath the luminaire. The luminous intensity used here is the quantity emerging at the nadir (0°). Figure 3-1b represents the situation where the ray of light arriving at the point of calculation on a horizontal surface is not perpendicular. The luminous intensity arriving at the angle

θ thus must be multiplied by cos θ, which will represent its component perpendicular to the horizontal surface. This modifies the basic formula as follows:

$$E(h) = \frac{I_\theta \cos \theta}{D^2}$$

Similarly, for the point of calculation lying on a vertical plane, the luminous intensity arriving at an angle θ must be multiplied by sin θ, which will represent its component perpendicular to the vertical surface. This modifies the basic formula to

$$E(v) = \frac{I_\theta \sin \theta}{D^2}$$

Let us illustrate this with an example. If the distance in each case is $D = 20$ feet, and the candlepower distribution curve of the luminaire is as shown in Figure 3-2, then for a situation as in Figure 3-1a the illuminance is

$$E = \frac{I}{D^2} = \frac{20{,}000}{20^2} = 50 \text{ fc.}$$

Similarly, illuminance for the point on a horizontal plane, as in Figure 3-1b, with angle $\theta = 30°$, is

$$E(h) = \frac{I_\theta \cos \theta}{D^2} = \frac{16{,}000 \times \cos 30°}{20^2} = 34.6 \text{ fc.,}$$

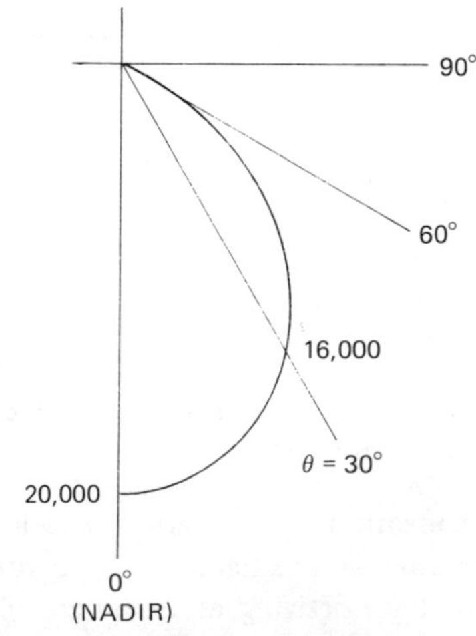

Fig. 3-2. Candlepower distribution of the luminaire in example.

and that for the point on a vertical plane, as in Figure 3-1c, with angle $\theta = 30°$, is

$$E(v) = \frac{I_\theta \sin\theta}{D^2} = \frac{16{,}000 \times \sin 30°}{20^2} = 20 \text{ fc}.$$

Where several luminaires contribute light to the point of calculation, each luminaire must be separately calculated and then added together for the net illuminance level. If luminaires are mounted in a horizontal plane, the distance D between each luminaire and the point will vary and can be eliminated by knowing the vertical distance (H) between the luminaire plane and the point of calculation. This modifies the formula to

$$E(h) = \frac{I_\theta \cos^3\theta}{H^2} \quad \text{and} \quad E(v) = \frac{I_\theta \sin\theta \cos^2\theta}{H^2},$$

since

$$D = \frac{H}{\cos\theta}$$

Although use of inverse-square law is basically simple, a number of assumptions limits is application:

1. The formula applies and holds true only for point sources. No lighting source is an ideal example of a point source. Clear lamps with filaments or smaller arc tubes are closer examples of a point source than the phosphor-coated bulbs, or lamps with larger arc tubes. In luminaires, a circular reflector with smaller opening is a closer example of a point source than any other type. As a rule of thumb, if the maximum dimensions of the source are no more than one-fifth the distance to the point of calculation, the source will be considered as a point source, since the error in such a case will be less than 1%, which is negligible.
2. Large luminaires or luminous ceilings can be considered a group of several point sources if they are divided into small sectors and calculations are made for each sector and added together. This method will be acceptable only if the luminance of the entire luminaire surface is uniform from all angles of view and the sectors are small enough to be qualified as point sources.
3. For the asymmetrical-distribution type of luminaire, the candlepower distribution curve of the required vertical angle must be used for calculations. If the point of calculation lies in an angle not shown in photometrics, interpolation of candlepower values must be used. This may cause unreliable or inaccurate results in some points.
4. Applications with rows of fluorescent lamps (linear sources), will be unsuit-

able, since the length of source in this case may greatly exceed the distance between the source and the point of calculation.

5. Application of point-by-point method for interior lighting with a great number of luminaires will necessitate the use of a computer rather than hand calculations, since donations of each luminaire at each location must be analyzed for all points in the room. Moreover, since the technique does not account for the room surface interreflectances, it will produce inaccurate results.

The point-by-point method of calculations, basically dealing with direct lighting only, is more suitable for outdoor lighting where a few luminaires are involved and where no surface reflectances are to be accounted for. This method is ideal for spot- or flood-lighting calculations. It may also be effectively used in interiors where task lighting is to be calculated on a surface lighted by a few luminaires and where the room has negligible or no surface reflectances.

Lumen Method

Development of the lumen method offers a much more simplified way of calculating an average uniform illuminance level on a plane in interiors. In many applications, a lack of prior knowledge on task locations, and also the possibility of multipurpose use of the room, it is necessary to provide an average uniform lighting throughout the room. Average uniform lighting, in essence, requires a number of luminaires spaced symmetrically on the ceiling to provide evenness in lighting distribution. To accomplish this with the point-by-point method would be tedious, time consuming, and expensive, since calculations would have to be made of a large number of points in the room. In addition, to account for the surface reflectances, a number of separate sets of calculations would have to be made by evaluating their luminance first and then treating them as a cluster of small sources. The lumen method eliminates all these problems and offers a much more simplified and accurate method.

The lumen method is developed from the basic definition of a footcandle, which states that "a footcandle is the illuminance on a surface of one square foot in area having a uniformly distributed flux of one lumen." Mathematically,

$$E = \frac{\ell}{A}$$

where

E = Illuminance in footcandles
ℓ = Lumen
A = Area in square feet.

If the area is in square meters, the illuminance is in lux.

If all luminous flux generated by the lamps falls on the work plane, the total lumens divided by the area will represent the average uniform footcandle. However, in reality this is never possible. A number of factors, such as the luminaire candlepower distribution, efficiency, room size and shape, surface reflectances, and luminaire mounting height, will affect the total number of lumens reaching the work plane. The formula thus must be multiplied by a coefficient of utilization (CU), that takes these into consideration. Thus,

$$E = \frac{\ell \times CU}{A}.$$

This, however, is the initial illuminance level. Luminaires and lamps will accumulate dirt and the lamps will depreciate in lumen output with time. To obtain a maintained illuminance level, the formula now must be multiplied by a light-loss factor (LLF), to account for this loss. Thus,

$$E = \frac{\ell \times CU \times LLF}{A}.$$

ℓ representing the total number of lumens produced by all luminaires, the formula now can be modified as follows:

$$E = \frac{(L \times N) \times CU \times LLF}{A}$$

where

L = Total initial lumens per luminaire
N = Number of luminaires

Expressing the formula in a more useful form:

$$N = \frac{A \times E}{L \times CU \times LLF}.$$

Examining the above formula, it is obvious that for an application where the area and the level of illuminance is already known, the only means of reducing the total number of luminaires is by using the highest values of L, CU, and LLF. The lower the total number of luminaires used, the less the power consumption will be. A designer, therefore, must look at all aspects in luminaire selection to determine which will offer the highest values of the three factors mentioned above.

Lumens (L). The only way of obtaining maximum lumens is by the proper selection of light source. Although the largest lumen-producing light source appears to be the immediate choice, this should not be the only criterion for light-source selection. Luminous efficacy (lumens per watt), color rendition, lumen maintenance, life, and cost are some of the other unavoidable considerations. For gaseous discharge sources, selection of the proper ballast is also an important consideration. A well-selected ballast will offer the greatest lumen output with less power consumption, the least noise output, and a high system-power factor. In selecting the right light source, the designer has to evaluate all these factors almost simultaneously. Chapter 1 takes a detailed look at these factors for the popular light sources of today.

Coefficient of Utilization (CU). Luminaire CU factors are the ratios of the lumens that reach the work plane to the lumens emitted by the lamps. In essence, it is not a fixed number but varies with room proportion, surface reflectances, mounting height above the work plane, and luminaire candlepower distribution. They combine the room characteristics with the luminaire characteristics to evaluate the efficiency of the interaction of the system. A higher CU value will offer more light for input power and hence will contribute to energy saving. The techniques and step-by-step procedures of determining the CU values of a luminaire have been discussed in Chapter 2.

Two of the most important factors involved in higher CU values of a luminaire are the efficiency of the luminaire, and the room proportion and its surface reflections. For indoor-type luminaires, a higher efficiency would usually mean a higher CU value, since no matter in which direction the light is emitted, it will finally arrive at the work plane after multiinterreflectances. The degree of improvement will depend upon the type of light distribution, quality of components, and engineering that has gone into the luminaire. This is discussed in Chapters 4 and 5.

Surface reflectances and room proportions will have significant effect on CU values. Which surface (ceiling, wall, floor) will have maximum impact on CU values varies with luminaires selected. Ceiling reflectance has the most significant effect on CU values of luminaires that produce light upwards. Indirect and direct-indirect luminaires fall into this category. Although no acceptable method of determining the CU values of indirect lighting systems has yet been developed, it is obvious that a highly reflective ceiling will send more light to work planes. The direct luminaires, on the other hand, will show insignificant change in CU values resulting from a major change in ceiling reflectance. Wall reflectance has significant effect for almost all luminaires, and particularly for luminaires with widespread distribution. Floor reflectances usually will have a low effect in CU values since most tasks are located above the work plane. Light reflecting from the floor will have to rebound from the ceiling to be of any use on the work plane. Most floors are intentionally of dark color because of dirt accumulation problems, and hence offer low reflectance. Highly reflecting floors are only useful in those limited applications where the task is located below the work plane, such as for air-craft repair work. A low-reflecting floor under such circumstances will offer lower levels of light on the task and possibly will require supplementary lighting.

To illustrate the impact of various surface reflectances, three graphs have been plotted to show the changes in CU values for the 2 X 4-foot grid-suspended troffer of the example in Chapter 2. These are shown in Figure 3-3. Figure

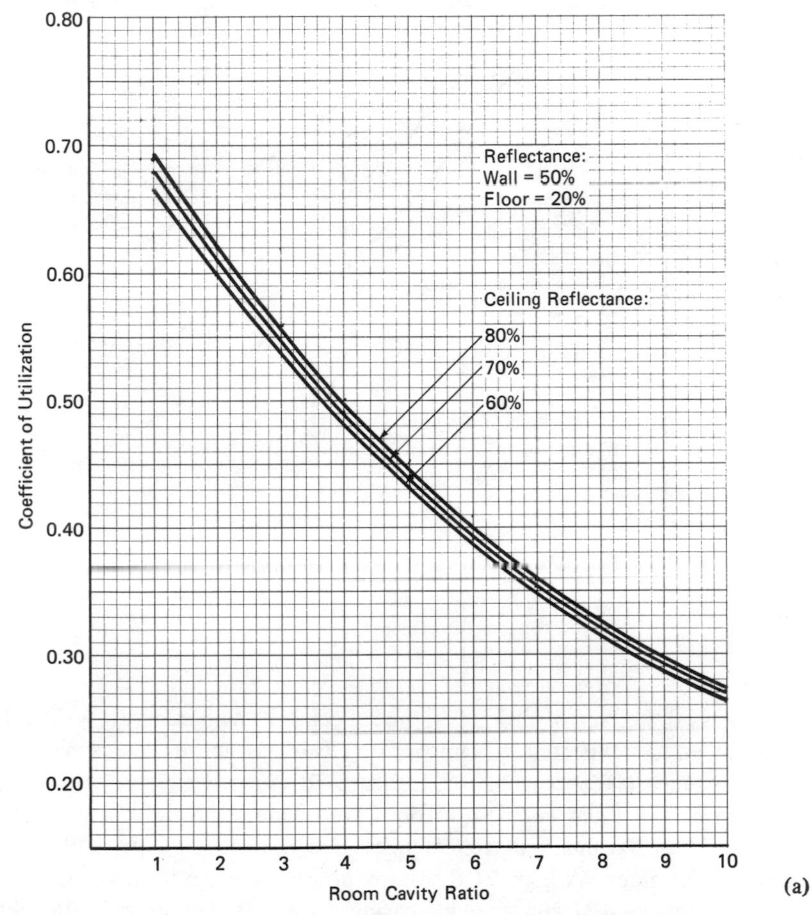

(a)

Fig. 3-3. (a) The effect of various ceiling reflectances with constant wall and floor reflectances on CU values. The change in CU values is insignificant, since the luminaire produces 100% downlight. (b) The effect of various wall reflectances with constant ceiling and floor reflectances. The change in CU values is significant, particularly for rooms with higher RCR values. A higher RCR value usually means a smaller room. (c) The effect of various floor reflectances with constant wall and ceiling reflectances. The change in CU values is slightly effected for larger rooms but quickly diminishes as the rooms get smaller.

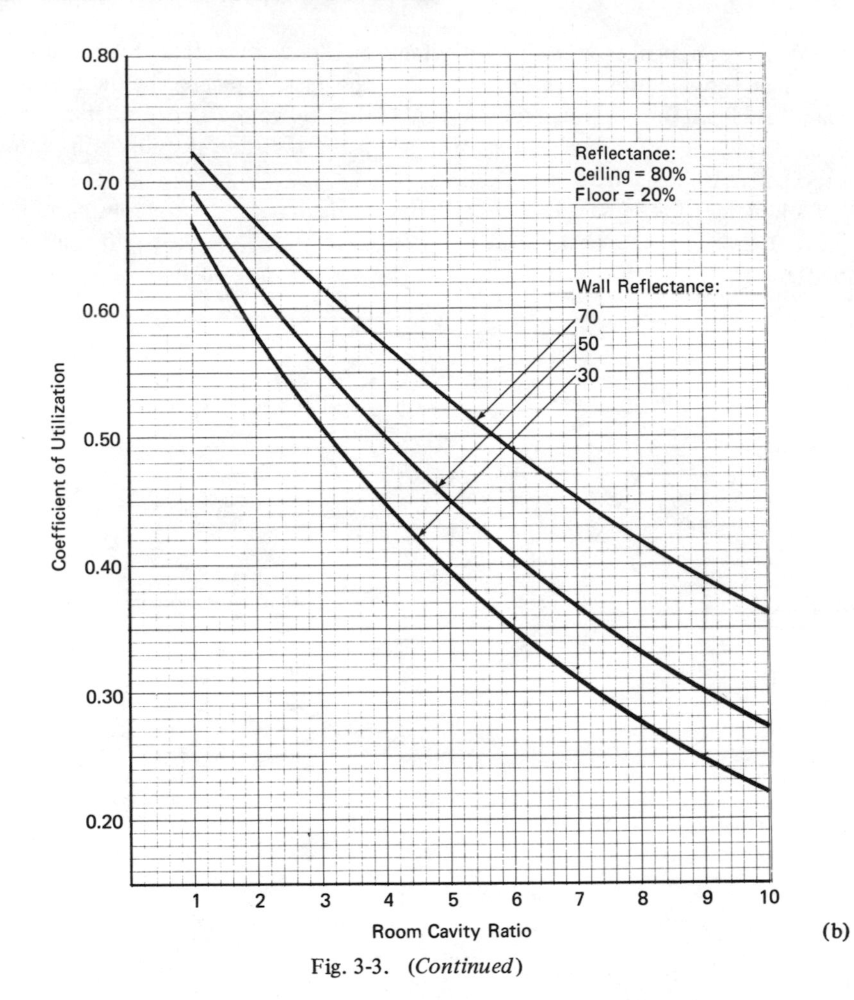

Fig. 3-3. *(Continued)*

3-3a shows the effect of various ceiling reflectances with constant wall and floor reflectances. As expected, the impact on CU is least in this case, since the luminaire produces 100% downlight. Figure 3-3b shows the effect of various wall reflectances and fixed ceiling and floor reflectances. Note that the change in CU values for various wall reflectances is significant, and particularly so for higher RCR values. A high RCR value will usually mean a small room. As the room gets smaller, CU values are increasingly dependent upon wall reflectances.

A light color, such as white or off-white, offers maximum reflectance. (Reflectance values of different popular colors are shown in Chapter 2). Selection of colors in a room, including those of the furniture, plays an important role on the overall working efficiency. A room with stark white surface colors and white furniture will definitely provide higher CU, but it may bring a monotonous

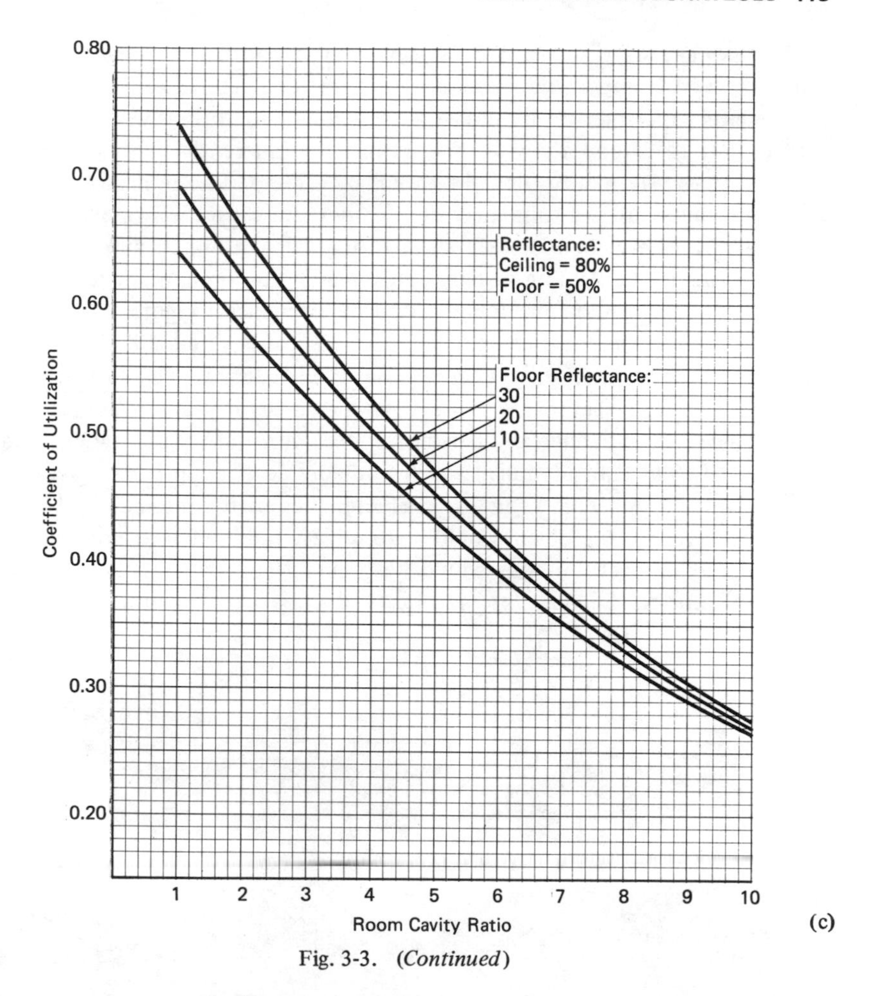

Fig. 3-3. *(Continued)*

and boring psychological effect on the workers. The total color scheme must be balanced to provide an efficient lighting system and yet a cheering atmosphere for maximum working efficiency.

Figure 3-3c shows the effect in CU values for a change in floor reflectances. The change in CU values is significant for large rooms, but quickly diminishes as the rooms get smaller. In developing the graph, proper correction factors were multiplied by the CU values to show the changes for 10%, 20%, and 30% floor reflectances.

Light Loss Factor (LLF). Without periodic scheduled and appropriate maintenance, no lighting system will be fully effective to retain the initial light output of the luminaire. An LLF would not be required in lighting calculations if the lighting system were to provide its light output constant throughout its life. This, however, is impossible. Inherent light loss characteristics of the luminaire components supplemented by unfavorable ambient conditions will force the net light output to deteriorate with time. While the inherently permanent reasons may systematically reduce the light output by 25% to 30%, an inadequate maintenance of luminaire and lamp, may drastically reduce it to 50% or even lower.

An LLF must be considered in the lighting calculations to make up for the expected loss in the lighting system. This guarantees a minimum required illuminance level until the time for a period maintenance. Figure 3-4 shows the curves of the concept. Inclusion of the LLF will cause a high initial illuminance level that drops in value with time and rises to initial value after each scheduled maintenance. Two factors are to be remembered here. First, even if a proper maintenance job is periodically performed, the regain in initial value as well as the maintained level of illuminance will gradually decrease, primarily because of the unavoidable, inherent light loss characteristics of the luminaire. The only way to get rid of this problem would be to replace old luminaires with new ones at a periodic interval, but this would be a costly and impractical solution for most applications. Second, the magnitude of initial value, and hence the total

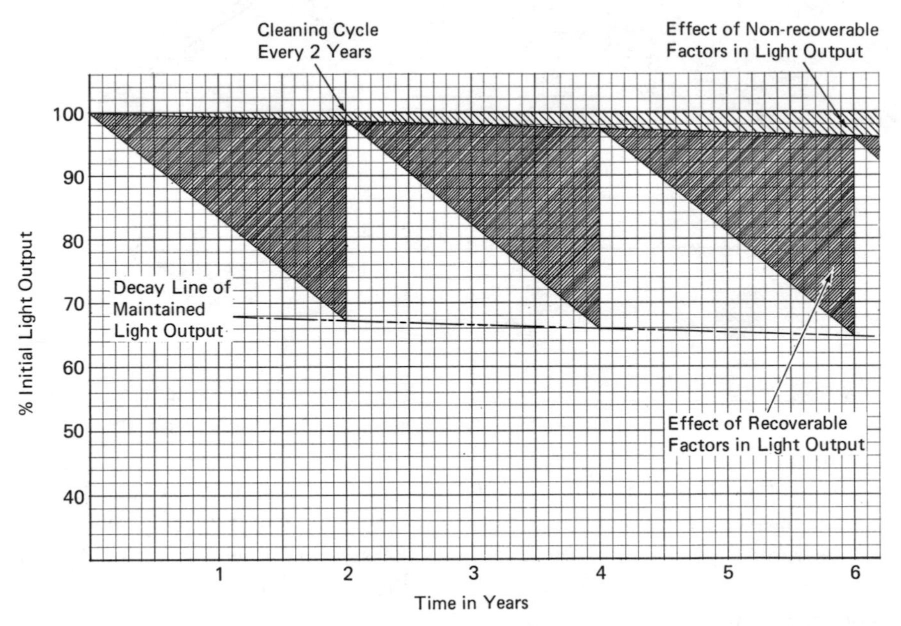

Fig. 3-4. Effect of light loss in illuminance level. An LLF must be taken into consideration, since initial illuminance level will continuously drop in value because of recoverable and nonrecoverable factors. In this diagram, it is assumed that luminaire cleaning and lamp replacement are done every 2 years. Note the gradual drop in initial light output due to nonrecoverable factors.

waste of energy shown by the shaded areas, will increase with longer maintenance intervals. Avoiding LLF is impossible, but a substantial portion of the wasted energy can be saved with an increase in maintenance intervals and with a good maintenance job. This means that the higher the LLF, the less the waste of energy.

We will now discuss the various individual factors that together determine the net LLF of a situation. The designer must look at the most practical means of achieving the optimized value of each of these factors. In addition, it is important that the designer inform the building owner of the parameters around which the LLF has been built and explain to the owner the importance of a good maintenance program. A failure to follow up on the maintenance schedule will defeat the purpose of introducing an LLF in calculations. Light output will continue to diminish, yet the power consumption will remain the same.

The various factors that constitute the net LLF fall into either of two types: the nonrecoverable and the recoverable.

$$LLF = \text{Nonrecoverable} \times \text{Recoverable factors}$$

$$LLF = (LAT \times VV \times BF \times LSD) \times (LDD \times RSDD \times LLD \times LBO)$$

where

LAT = Luminaire ambient temperature
VV = Voltage variation
BF = Ballast factor
LSD = Luminaire surface depreciation
LDD = Luminaire dirt depreciation
$RSDD$ = Room surface dirt depreciation
LDD = Lamp lumen depreciation
LBO = Lamp burn out

In explaining these factors, let us consider the same example of Chapter 2 and see how the LLF can be constructed.

Given: Room dimensions: $L = 50$ ft,; $W = 30$ ft.; $H = 8.5$ ft.
Office atmosphere = Clean
Luminaires: 4 lamps, 2 × 4 ft ceiling recessed fluorescent troffer with solid metal top and acrylic prismatic lens bottom. Maintenance of room surfaces and luminaires is normal. Maintenance interval is undecided.

Much of the information required in the construction of the LLF is unknown here, and purposely so, so as to simulate an average condition. Many such unknown quantities have to be guessed at, based on a reliable judgment. Although one example cannot represent all working conditions, we will discuss most conditions whenever possible.

Nonrecoverable Factors. Nonrecoverable factors represent those conditions of a lighting system that may reduce light output, when nothing in terms of periodic maintenance can be done to recover the loss.

Luminaire Ambient Temperature (*LAT*). About the only sources which are directly affected by ambient temperature are the fluorescents. A variation in temperature above or below the normally encountered operating temperature does not have much effect on incandescents or HID units. Indoor-type fluorescent lamps are usually designed to peek in light output at an ambient temperature of 77°F. Since the photometric tests of the indoor-type luminaires are carried out in an open space with still air at 77°F, the measured CU factors represent the system's efficiency under those specific thermal conditions only. In use, when the luminaires are operated at their respective mounting conditions, there may be a significant difference in light output, since mounting methodology has an impact on the ambient temperature. Recessed luminaires, for instance, will be affected since the plenum temperature is usually 10 to 20°F higher than the room ambient temperature. A surface mounted luminaire may be subject to a higher operating temperature due to the insulating effect of the ceiling material. This, and drastically different room ambient temperatures, such as in a cold storage facility with extremely low temperature and in a foundry or blacksmith facility with a very high temperature, will have a significant impact in luminaires' performance regardless of their mounting methodology. Each lamp-ballast-luminaire combination has its own distincitve characteristics of light output versus ambient temperature. This is particularly true of the heat-transfer-type luminaires.

LAT in essence represents the thermal factor of the situation which compensates for the light losses (or gains), and sometimes is regarded as a factor to renew the CU of the system. A method of incorporating this effect in calculations is shown in Chapter 4. In the depicted example, since no information is available as to the luminaire ambient temperature, LAT is assumed to be unity.

Voltage Variation (*VV*). A variation in incoming voltage will affect luminaire operating characteristics. For incandescents, small deviation from rated lamp voltage would cause about 3% change in lumens for each 1% voltage variation. For high-intensity discharge units, the effect will vary significantly for different lamps from one type of ballast to another (see Chapter 1). Variation in incoming voltage is usually an unpredictable occurrence. If the deviation is stable and of a known quantity, an appropriate VV factor may be used in calculations. Otherwise, it is assumed to be unity.

Luminaire Surface Depreciation (*LSD*). Adverse changes in metal, paint, and plastic components of a luminaire cause a reduction in light output. Insofar as reflective surfaces are concerned, most luminaires either have baked enamel or processed aluminum. Although enamels, in general, provide higher initial reflectances and are easier to clean, their inherent porous characteristics cause a constant depreciation. Processed aluminum, on the other hand, does not offer as much reflectance but has a slower rate of depreciation. Luminaires using glass, porcelain, or specially processed aluminum are most tolerant and have insignificant depreciation. Plastics, such as polystyrene and polycarbonate, tend to turn yellow with constant exposure to the ultraviolet rays of the lamps. Most luminaires contain several materials of different depreciating characteristics. The best source of information on this factor is the manufacturer. For the purpose of convenience, the LSD for this example is assumed to be unity.

Ballast Factor (*BF*). Each CBM approved ballast has a BF that is determined by ANSI technique. This factor represents the ratio of light output obtained by a commercial ballast to that obtained by a reference ballast. Mathematically,

$$BF = \frac{\text{Light output by a commercial ballast}}{\text{Light output by a reference ballast}}$$

If the BF of the ballast used in a luminaire is different from that of the ballast used in the photometric report, the light output will differ by the same proportion. Note that for a reliable BF, the ballast must be CBM certified. The best source of obtaining any information on BF is to consult the manufacturer. BF for all HID sources can be assumed to be unity. For fluorescent ballasts, however, they may vary substantially. A discussion on this is in Chapter 4.

All four nonrecoverable factors mentioned above will depreciate light output permanently and little can be done to recover them once they are installed. The best solution is to select the lamp-ballast-luminaire combination carefully in the first place.

Recoverable Factors

Luminaire Dirt Depreciation (*LDD*). The greatest loss of light output is mainly attributed to the dirt on lamps and luminaire reflecting surfaces. The amount and type of air differs from one area to another. The air in an office building located near a residential area may be clean, while that in another office building near an industrial belt may be considered dirty. Manufacturing areas produce relatively dirtier air and are not as dirty as areas in which there are foundries or welding shops. Figure 3-5 represents the depreciation in light output that can be expected if dirt and dust are allowed to accumulate on lamps and luminaires.

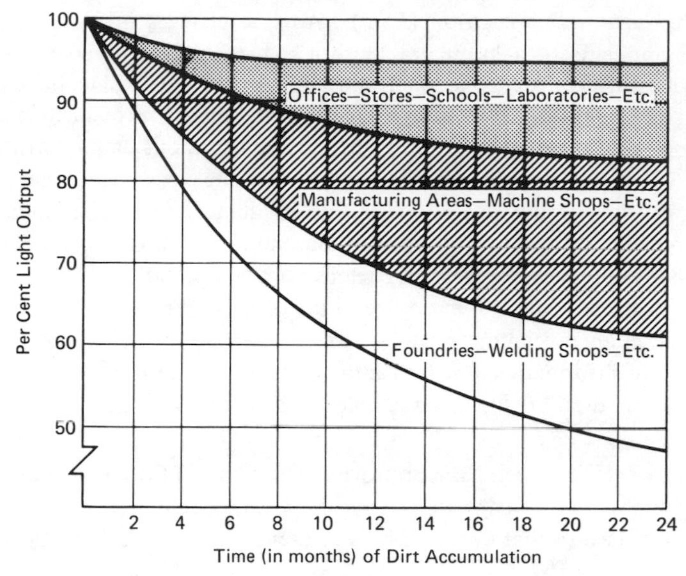

Fig. 3-5. Depreciation of light output, which can be expected if dirt and dust are allowed to accumulate on lamps and luminaires. (Courtesy of General Electric.)

The wide band of each of the three categories actually is the result of the wide variety of dirt conditions in the air and the type of luminaires in use. In general, all luminaires with nonventilated reflectors usually collect more dirt than those with ventilation; convection currents created by surrounding air carry out dust and dirt away from the luminaire.

In determining LDD, three factors are needed to be known: the type of luminaire, the atmospheric air condition, and the maintenance interval. If the first two are known, an LDD can be found from the chart corresponding to the maintenance interval. The steps are as follows.

Step 1: Knowing the type of luminaire used, select the proper maintenance category out of the six given in Figure 3-6. The luminaire used in the example fits in category V, since it produces 100% down-light and has an unaperatured opaque top and an unaperatured transparent bottom.

Step 2: Determine the type of atmosphere most appropriate from the following: (a) Very Clean, *VC*, (b) Clean, *C*, (c) Medium, *M*, (d) Dirty, *D* , (e) Very Dirty, *VD*. The office in our example has an atmosphere that is clean, or *C*.

Step 3: Knowing Steps 1 and 2 and the interval of maintenance, refer to the appropriate curve in Figure 3-7 and read the recommended LDD factor. In this example, the maintenance interval is unknown and has to be decided.

Maintenance Category	Top Enclosure	Bottom Enclosure
I	1. None.	1. None
II	1. None 2. Transparent, with 15% or more up-light through apertures. 3. Translucent, with 15% or more up-light through apertures. 4. Opaque, with 15% or more up-light through apertures.	1. None 2. Louvers or baffles
III	1. Transparent, with less than 15% upward light through apertures. 2. Translucent, with less than 15% upward light through apertures. 3. Opaque, with less than 15% up-light through apertures.	1. None 2. Louvers or baffles
IV	1. Transparent, unapertured. 2. Translucent, unapertured. 3. Opaque, unapertured.	1. None 2. Louvers
V	1. Transparent, unapertured. 2. Translucent, unapertured. 3. Opaque, unapertured.	1. Transparent, unapertured 2. Translucent, unapertured
VI	1. None. 2. Transparent, unapertured. 3. Translucent, unapertured. 4. Opaque, unapertured.	1. Transparent, unapertured 2. Translucent, unapertured 3. Opaque, unapertured

Fig. 3-6. Luminaire classification based on six maintenance categories. (Reprinted with permission from the IES Lighting Handbook, 1981, Reference Volume.)

Selection of a maintenance interval depends upon how clean the area really is. For air-conditioned offices, once a year or once in two years may be sufficient. Non-air-conditioned offices or schools may require cleaning at least once a year. In industrial areas, such as foundries or smoke-producing factories, the interval may be as frequent as once a month. For food-preparation areas, this should be increased to once a week because of the special cleanliness requirement.

For the purpose of comparing the effect on energy saving, let us assume that the office in this example is air-conditioned and clean and that a thorough cleaning will be done at every 2 or 3 years. Referring to curve C in category V, the LDD factors for these intervals are found to be 0.835 and 0.800, respectively.

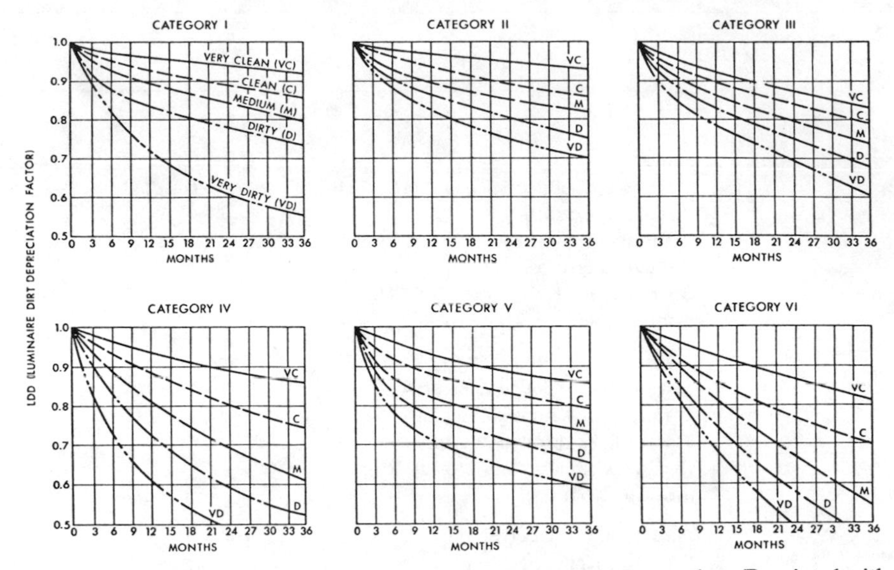

Fig. 3-7. Luminaire dirt depreciation factors for six luminaire categories. (Reprinted with permission from the IES Lighting Handbook, 1981, Reference Volume.)

LDD values recommended by luminaire manufacturers are usually too optimistic, since these values are based on the assumption that proper scheduled maintenance is done by all owners. In reality, this is seldom the case. A more suitable factor can be found from experience with similar luminaires and applictions, or by consulting with a luminaire maintenance company.

Room Surface Dirt Depreciation (*RSDD*). RSDD takes into account that dirt accumulates on room surfaces and reduces surface reflectances. A schedule must be maintained to clean the surface and possibly repaint. The steps of determining this factor are as follows.

Step 1: Knowing the cleaning interval and the type of atmospheric condition, determine the expected dirt depreciation in Figure 3-8. In this example, corresponding to cleaning intervals of 2 years and 3 years and an atmospheric condition *C*, the expected dirt depreciation factors are 18% and 21%, respectively.

Step 2: Knowing the expected dirt depreciation from Step 1, the type of luminaire distribution from Figure 2-7, and the RCR from Table 2-1, both in Chapter 2, determine the RSDD from Figure 3-9. In this example, the luminaire is a direct type and the RCR is 1.6, so the RSDD factors are 0.97

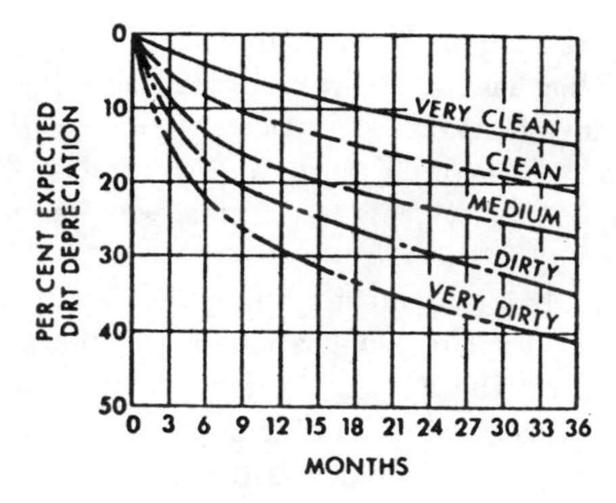

Fig. 3-8. Percent expected dirt depreciation because of cleaning interval and atmospheric conditions. (Reprinted with permission from the IES Lighting Handbook, 1981, Reference Volume.)

and 0.96, corresponding to maintenance intervals at 2 years and 3 years, respectively.

Lamp Lumen Depreciation (*LLD*) *and Lamp Burn-Out* (*LBO*). A depreciation in lamp lumens is an inherent characteristic of all types of light sources. Much of this has already been discussed in Chapter 1. If the lamps are allowed to burn the duration of their rated life, the lumen output gradually will decrease and terminate in a low value before burn-out. If it is assumed that all lamps burn out at the same time and that all burned-out lamps are instantly replaced with new ones, a factor representing the final LLD may be used in calculations to make up for the loss. This is seldom the case, however, since all lamps do not burn out at the same time. While a specific lamp's burn-out is impossible to predict, a group of lamps will fail in a predictable manner. This is shown in the

	Luminaire Distribution Type																			
	Direct				Semi-Direct				Direct-Indirect				Semi-Indirect				Indirect			
Per Cent Expected Dirt Depreciation	10	20	30	40	10	20	30	40	10	20	30	40	10	20	30	40	10	20	30	40
Room Cavity Ratio																				
1	.98	.96	.94	.92	.97	.92	.89	.84	.94	.87	.80	.76	.94	.87	.80	.73	.90	.80	.70	.60
2	.98	.96	.94	.92	.96	.92	.88	.83	.94	.87	.80	.75	.94	.87	.79	.72	.90	.80	.69	.59
3	.98	.95	.93	.90	.96	.91	.87	.82	.94	.86	.79	.74	.94	.86	.78	.71	.90	.79	.68	.58
4	.97	.95	.92	.90	.95	.90	.85	.80	.94	.86	.79	.73	.94	.86	.78	.70	.89	.78	.67	.56
5	.97	.94	.91	.89	.94	.90	.84	.79	.93	.86	.78	.72	.93	.86	.77	.69	.89	.78	.66	.55
6	.97	.94	.91	.88	.94	.89	.83	.78	.93	.85	.78	.71	.93	.85	.76	.68	.89	.77	.66	.54
7	.97	.94	.90	.87	.93	.88	.82	.77	.93	.84	.77	.70	.93	.84	.76	.68	.89	.76	.65	.53
8	.96	.93	.89	.86	.93	.87	.81	.75	.93	.84	.76	.69	.93	.84	.76	.68	.88	.76	.64	.52
9	.96	.92	.88	.85	.93	.87	.80	.74	.93	.84	.76	.68	.93	.84	.75	.67	.88	.75	.63	.51
10	.96	.92	.87	.83	.93	.86	.79	.72	.93	.84	.75	.67	.92	.83	.75	.67	.88	.75	.62	.50

Fig. 3-9. Room surface dirt depreciation factors. (Reprinted with permission from the IES Lighting Handbook, 1981, Reference Volume.)

lamp's mortality curves supplied by the manufacturers. Under these conditions, the owner of the building has two choices: (1) spot replacement of lamps as they burn out or (2) a group relamping.

Group relamping makes use of the mortality curve and is of two basic systems: (1) that with no interim replacements and (2) that with interim replacements.

System 1: Group relamping with no interim replacements is basically simple and particularly suitable for large applications where the burning out of a few lamps does not affect the lighting level and esthetics significantly. In this method, the individual burning out of lamps is ignored until a predetermined group of lamps goes out; this indicates that a group relamping is due. The quantity of lamp burn-out for group relamping is determined from the mortality chart. Such a system is more suitable to fluorescents and the group replacing is recommended not to exceed 70% of average life to make the best use of lumen depreciation. When 10 lamps out of 100 burn out, it usually represents the 70% of average life of fluorescent lamps, which indicates that it is time for a group relamping.

System 2: A more popular and esthetically pleasing way is to group relamp with interim replacements. Right after a group relamping at 80% of average life, 20% of the best remaining lamps are stored as replacements for the burn-outs before the next group relamping. These are used to replace the early burn-out lamps, and when they are exhausted, it indicates that it is time for a group relamping again.

For either of the two systems, the group relamping period for fluorescents and incandescents should be selected for an average of 50 through 80% of rate lamp life. For fluorescents, group relamping periods shorter than 50% usually will be uneconomical because of lamp cost, and those beyond 80% will be unfavorable, since not too many good lamps can be saved for interim replacements. For filament lamps, this is true after 85% of life.

The timing of group relamping is important. For fluorescents and incandescents, timing can be as follows:

1. Office buildings, schools, and some industrial plants have about 2500–3500 burning hours per year on single-shift operation. Group relamping every 3-5 years for fluorescents and 3-3½ months for incandescents will be economical.
2. Double-shift operation on factories and many stores have burning hours between 4000 and 5000 hours per year. Group relamping every 2–3 years for fluorescents and 2–2½ months for incandescents will be economical.
3. For other multishift operations where burning hours may be 6500–8000 hours per year, the relamping may be done once a year for fluorescents and about every 6 weeks for incandescents.

Assuming that the luminaires in the example are left on for 12 hours a day and that there are 250 working days per year, the total number of burning hours is $250 \times 12 = 3000$ hours per year.

The rated life of a typical indoor-type fluorescent lamp is 20,000 hours. Suppossing that group relamping is done at 70% of rated life, the total burning hours per lamp is $20{,}000 \times 0.7 = 14{,}000$ hours. This means a total of $14{,}000/3000 = 4.6$ years of continuous operation before lamps will be replaced. Under this condition, with group relamping done at 70% rated life, the LLD will be 0.80 from the curve in Figure 1-19b in Chapter 1.

The LLF for a 3-year cleaning and a 4.6-year relamping interval is:

$$(LAT \times VV \times BF \times LSD) \times (LDD \times RSDD \times LLD \times LBO) = (1 \times 1 \times 1 \times 1) \times (0.80 \times 0.96 \times 0.80) = 0.614$$

If relamping is done at 60% of rated life, the total burning hours will be $20{,}000 \times 0.6 = 12{,}000$ hours, representing a total of $12{,}000/3000 = 4$ years of continuous operation before lamps are replaced. From the curve in Figure 1-19b, the LLD for this interval at 60% rated life is 0.83.

The LLF for a 2-year cleaning and a 4-year relamping interval is then,

$$(1 \times 1 \times 1 \times 1) \times (0.835 \times 0.97 \times 0.83) = 0.67$$

The benefit of a 2-year cleaning interval over one of 3 years is obvious. This can be established in the following calculations.

Let us suppose a maintained amount of 70 fc. is required for the example. The total number of luminaires required in each case is as follows:

Given area = 50×30 sq. ft.
Lumens per lamp = 3200
Required illumination = 70 fc.
CU value = 0.648

For 2-year cleaning interval:

$$N = \frac{A \times E}{L \times CU \times LLF} = \frac{50 \times 30 \times 70}{(4 \times 3200) \times 0.648 \times 0.67} = 19 \text{ luminaires.}$$

For 3-year cleaning interval:

$$N = \frac{50 \times 30 \times 70}{(4 \times 3200) \times 0.648 \times 0.614} = 21 \text{ luminaires.}$$

Use of a high LLF will result in a lower number of luminaires and will save energy. It will be the duty of the lighting designer to inform the owner about all parameters used in constructing the LLF and about the benefit of regular maintenance. For convenience, instructions like those in Figure 3-10 may either be included in the lighting specifications or separately presented to the owner.

Luminaire Cleaning. Luminaires are made of different types of material. Each material has a special type of compound that cleans best. The best source of information on this subject is the luminaire manufacturer. The following applies to most luminaire finishes:

Plastics	Use destaticizer. Do not wipe dry after the application of rinse solution.
Glass	Use nonabrasive cleaners or detergent.
Synthetic enamel	Use detergent. Do not use alcohol or abrasive cleaners.
Porcelain	Use nonabrasive cleaners.
Aluminum	Use soap and cleaners, and then rinse clean with clear water. Do not use strong alkaline cleaners or acid solution.

Lighting Maintenance Schedule

Project: XYZ Office Building
123 Golden Avenue
Lakewood, Colorado

Elec. Eng. By: Butterweck–Sorcar Engineering Co.
Denver, Colorado

In Designing Lighting Systems of the Above Project the Following Parameters Have Been Used in Calculations. To Retain the Calculated Illuminance Level, it is Recommended that the Following Maintenance Schedule be Observed.

	Room No.									
	101	102	103	104						
Luminaire and Lamp Cleaning Interval (In Years)	2	2	1	2						
Room Surface Cleaning or Painting Interval (In Years)	2	3	2	X						
Group Relamping Interval (In Years)	3	4	2	X						

Fig. 3-10. Maintenance schedule form.

In large applications, such as factories, schools, shopping centers, or high-rise commercial buildings, the greater quantity and out-of-reach luminaires usually prevent the in-house maintenance man or janitor from doing a good upkeep job on the luminaires. Proper care of the products requires the right type of equipment for mobility, compounds, and plenty of room for lamp storage. With these, along with the necessity of periodic positive cleaning, usually it is more convenient and economical to hire professional cleaning companies to do the job. Their experience with luminaire cleaning, possession of proper equipment, and swiftness offer good quality and a dependable maintenance job within a short time.

DETERMINATION OF QUALITY

Quality determination methods relate to the measurement of the glare present in a lighting system. Glare can be loosely divided into two types: the direct and indirect. The direct glares are associated with tasks with heads-up position, and the indirect glares are associated with tasks in head-down position. Direct glare mainly results from two types of sources. The first is an excessively bright luminous source shining into a person's eyes so as to reduce his or her ability to see the task. This is also known as a "disability glare." Looking straight at the high beam of an oncoming car is an example of disability glare. The other type of direct glare is due to peripheral stray light that is not necessarily aimed specifically at the eye. This is known as "discomfort glare." Indirect glare is the result of the mirror effect of the tasks that reflect the light source from tasks to the eyes of a viewer, resulting in a loss in visibility. Direct glare is measured in terms of VCP and in essence measures the "discomfort factor of direct glare." The indirect glare is measured in terms of equivalent sphere illumination (ESI).

In an effort to provide good-quality design, it is necessary that VCP and ESI values of a lighting system be as high as possible. This presents a problem since their principles of operation oppose each other. The direct glare zone is between 45° and 90°. For a high value of VCP, it is necessary that the luminance associated with this zone be as low as possible and that the light emerge through the 0–45° zone. The indirect glare zone, on the other hand, is between 0° and 45°. For a high value of ESI, it is necessary that the luminance associated with this zone be as low as possible and that light emerge between 45° and 90°. A lighting design will thus evaluate the importance of each in the application and create a design that will be most suitable for the purpose (see Figure 3-11).

Visual Comfort Probability. VCP is one of several methods available in evaluating the direct glare. This is the only method currently accepted by the IES for evaluating direct glare in a room. While two people may have a difference of opinion as to the lighting glare of a space, the majority of a group of people

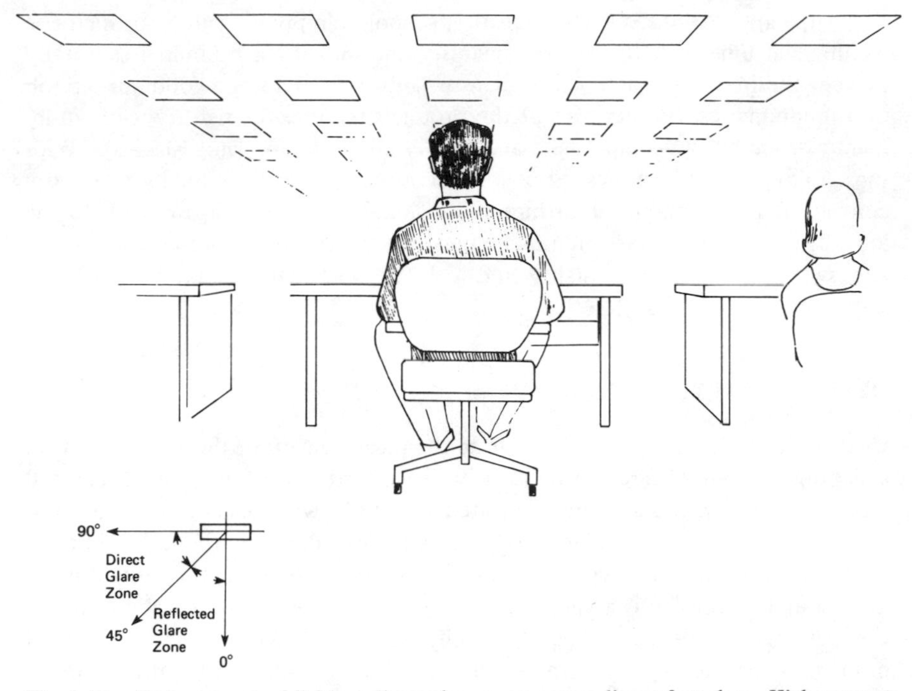

Fig. 3-11. High amount of light at direct-glare zone causes discomfort glare. High amount of light at indirect-glare zone causes a loss in visibility.

should have the same opinion. VCP represents that percentage of people who probably will find the glare produced by luminaires in a space just comfortable. The method of computing VCP is explained in the IES handbook. Virtually all formulas related to VCP computation take into account the location of luminous elements, luminous-element brightness, viewing angle, and overall room brightness. The formula is as follows:

$$\text{Luminaire visual comfort} \propto \frac{\text{Apparent luminance} \times \text{apparent source size}}{\text{Source position} \times \text{overall room luminance}}$$

All visual-comfort values of the luminaires in the room are added together and modified by an empirical relationship (the discomfort glare rating) and then converted into a useful VCP rating for the condition. The evaluation of VCP values are based on some predetermined standard conditions. Any deviation from these standard conditions may have considerable effect on the resulting VCP values. The accepted standard conditions are as follows:

1. Initial illuminance level from the overhead light is 100 fc.

2. Effective cavity reflectances of room surfaces: 80% ceiling, 50% walls, and 20% floor.
3. Mounting height above floor: 8.5, 10, 13, and 16 feet.
4. A range of room dimensions (floor area) so as to include square, long-narrow, and short-wide rooms.
5. A standard layout involving luminaires uniformly distributed throughout the space.
6. An observation point 4 feet in front of the center of the rear wall and 4 feet above the floor.
7. A horizontal line of sight directly forward.
8. A field of view limited to 53° above and directly forward from the observer.

Figure 3-12 shows a typical layout of luminaires in a room for calculating VCP values. The VCP values can be obtained either by hand calculations or by computer programming. Hand calculations are tedious and time consuming. If a request is made, manufacturers usually will supply a table of the VCP values of their luminaires along with the photometric report. Table 3-1 shows the VCP values of our sample luminaire equipped with four lamps and an acrylic prismatic lens. Note that the VCP values are given for different room dimensions and mounting sizes that match the standard conditions.

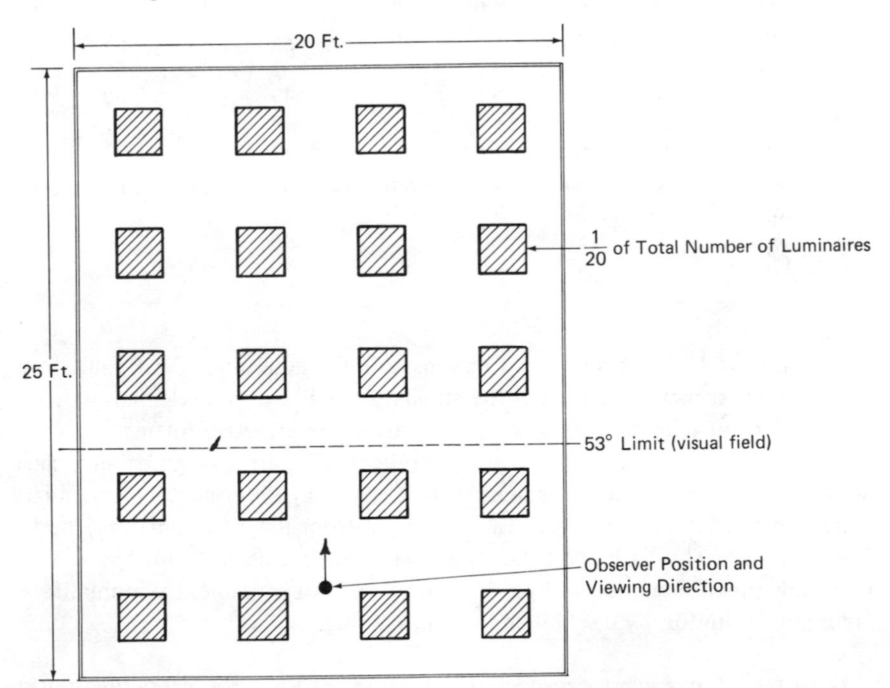

Fig. 3-12. Typical arrangement for calculating VCP tables.

Table 3-1. VCP Values of the Depicted Example.

Room Dimensions		VCP Lengthwise				VCP Crosswise			
		Mounting Height (ft)				Mounting Height (ft)			
W	L	8.5	10.0	13.0	16.0	8.5	10.0	13.0	16.0
20	20	62	67	74	80	59	64	70	76
20	30	55	59	65	70	52	56	61	66
20	40	51	54	59	63	48	51	55	59
20	60	48	50	54	57	46	48	50	54
30	20	63	68	73	78	61	65	70	75
30	30	56	59	64	68	54	57	60	65
30	40	51	54	57	61	49	52	54	57
30	60	48	50	52	55	46	48	49	52
30	80	45	48	49	52	44	46	47	49
40	20	65	69	74	77	63	67	72	76
40	30	57	60	64	68	56	59	62	65
40	40	53	55	58	61	51	53	55	58
40	60	49	50	52	55	48	49	50	52
40	80	46	48	49	51	45	47	47	49
40	100	46	47	47	49	44	46	45	47
60	30	59	62	65	68	58	60	63	66
60	40	54	56	58	61	53	55	56	59
60	60	50	51	52	55	49	50	50	52
60	80	47	49	49	51	46	47	47	49
60	100	46	47	47	48	45	46	45	46
100	40	58	59	60	63	57	58	59	62
100	60	53	54	54	56	53	53	53	55
100	80	50	51	50	52	49	50	49	51
100	100	49	49	48	50	48	49	47	48

Limitations of VCP. A number of reasons limit the determined VCP value from being totally accurate. Although the standard conditions are selected to match the majority of indoor lighting conditions, these are effective for making a preliminary evaluation only. The layout, luminaire mounting height, illuminance level, and the room dimensions used in the design may never match the standard conditions; hence, the "nearest" value received from the VCP table may be far from precise. The VCP evaluation method is applicable only for spaces with uniformly distributed luminaires; its values are not suitable for nonuniform layout or for uniformly distributed indirect lighting.

Effect of VCP on Energy Consideration. Although VCP is a technique of measuring the direct-glare aspects of quality design, its impact on energy conservation can be significant. Bear in mind that the technique works on the assumption of uniformly distributed luminaires; such a layout usually represents more energy consumption than nonuniform task lighting. The other factor is that luminaires with high VCP usually have low efficiency. Highly efficient luminaires offer more light per watt, with a substantial amount of light emerged through the direct-glare zones. This causes opposition to achieving high VCP values. The usual way to increase the VCP of such luminaires is with the use of special non-glaring louvers or modified lenses that are physical barriers to light emerging from the direct glare zone; this reduces efficiency substantially.

Dimming may be considered as a solution to this problem. A continuous dimming of these luminaires will gradually decrease the glare all around and increase VCP with insignificant change in the luminaire's efficiency. For fluorescents and mercury vapor, the percentage of light output is almost directly proportional to the input power. This is discussed in Chapter 7.

Comparison of the VCP of a dimmed luminaire producing the same amount of light as a luminaire with a special louver can be determined mathematically, but this will require extensive calculations. Figure 3-13 shows an empirical technique of making a quick comparison, which will be reasonably accurate for the purpose. To evaluate the resultant VCP of a dimmed luminaire compared to that of a high VCP luminaire, two factors must be known: (1) the efficiency or CU of both the luminaires and (2) the VCP of the luminaire to be dimmed. Let us suppose the dimming luminaire under consideration (luminare *A*) is a standard 2 X 4-foot troffer with 4 lamps and an acrylic prismatic lens, having a 65% VCP and 65% efficiency. The low-brightness luminaire (luminaire *B*) is a 2 X 4-foot troffer with 4 lamps and parabolic louvers, having an 85% VCP and 48% efficiency. Referring to Figure 3-13, the following are the steps involved to determine the resultant VCP of luminaire "*A*."

Step 1: Determine the efficiency ratio.

$$\frac{\text{Efficiency of luminaire } B}{\text{Efficiency of luminaire } A} = \left(\frac{48}{85}\right) = 0.73$$

Step 2: Determine resultant VCP. Referrring to Figure 3-13, start from 0.73 on the horizontal axis and go to the curve representing 65 VCP of luminaire *A*. Draw a horizontal line and read the resultant VCP = 70% on the vertical axis.

What this means is that for the same amount of light, the dimmed luminaire *A* will yield a VCP of approximately 70, compared to a VCP of 85 for the low-brightness luminaire *B*, using only 73% energy. The light output is almost directly proportional to power consumption for fluorescent lamps, and the same is true for mercury vapor down to about 50% of full light output.

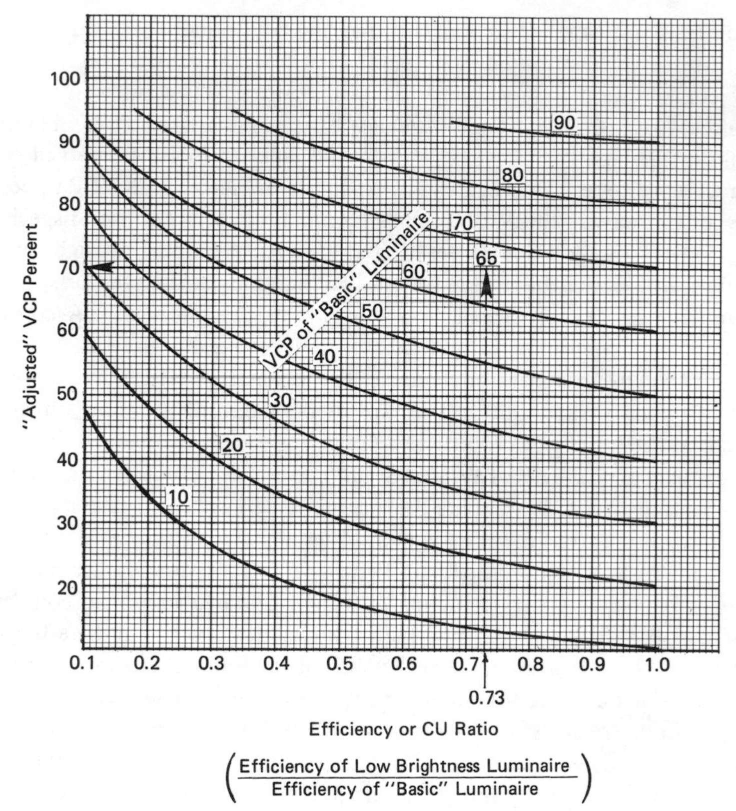

Fig. 3-13. Visual comfort can be increased by dimming luminaires, with significant savings in energy. Graph provides simple, empirical method for determining resultant VCP of dimmed luminaire. (Courtesy of Widelite.)

VCP ratings should be compared only if applicable areas and situations are identical and if luminaires under consideration are the only variables. It should be remembered that standard VCP data are calculated with a preset group of parameters, including the observer's looking straight ahead and sitting at the center of the rear of the room. These conditions are more applicable for office or school-like areas, rather than for industrial interiors. Visual comfort is one of the desirable factors in lighting systems and should be kept in proper perspective in relation to other objectives, such as energy conservation, cost, and productivity.

Indirect Glare and the Veiling Reflection. Indirect or reflected glare mainly results from light emitted in the 0–45° zone, reflecting off a shiny or glossy surface and pointed towards the eyes. The patches of light visible on the surface are the images of the light source. The glossier or more specular the surface of reflection, the more distinct the image of the source, causing a loss in visibility.

Depending on the angle and direction of the light rays, the reflected glare can be very distracting and annoying and the main reason for loss in visibility.

A ray of light arriving from an angle and direction such that it hinders visibility is termed a *poor ray*. In contrast, another ray arriving at an angle and direction that improves visibility is a *good ray*. Depending on various angles and directions of arrival on the task, the rays can be graded in different types of rays; see Figure 3-14. Figure 3-14a illustrates the grading of light rays with respect to direction of arrival, and Figure 3-14b shows grading with respect to angles of arrival. The angle and direction at which the viewer views the task is known as the *viewing angle* and *viewing direction*. Note that as the rays of light striking the task move away from the front of the viewer towards the back, visibility improves. The worst visibility occurs at an angle and direction of light rays that reflect off the task and head directly to the viewer's eyes. The combination of all such poor and bad rays hinders the visibility by creating a "veil" on the task; this is commonly known as the *veiling reflection*.

Veiling reflection reduces the contrast of the task, resulting in a loss of visibility and visual performance. This basically means that room lighting may actually hinder visibility rather than improve it, if light rays are striking the task from bad angles and directions. The complete lighting system, including types of luminaires, their locations, and room-surface reflectances, among others, will interact with the microstructure of the task, resulting in a reduction of contrast and hence adversely affecting visibility. See Figure 3-14c. This effect of loss in contrast or veiling reflection is measured in terms of ESI. The root of the ESI technique is based on the Blackwell Contrast Formula:

$$C = \frac{L_o - L_b}{L_b}$$

where

C = Contrast
L_o = Luminance of the object
L_b = Luminance of the background

Various factors affect contrast:

Room dimension
Room-surface reflectances
Luminaire layout
Candlepower distribution of luminaire and polarization
Viewing angle
Viewing direction
Physical characteristics of the task
Test location

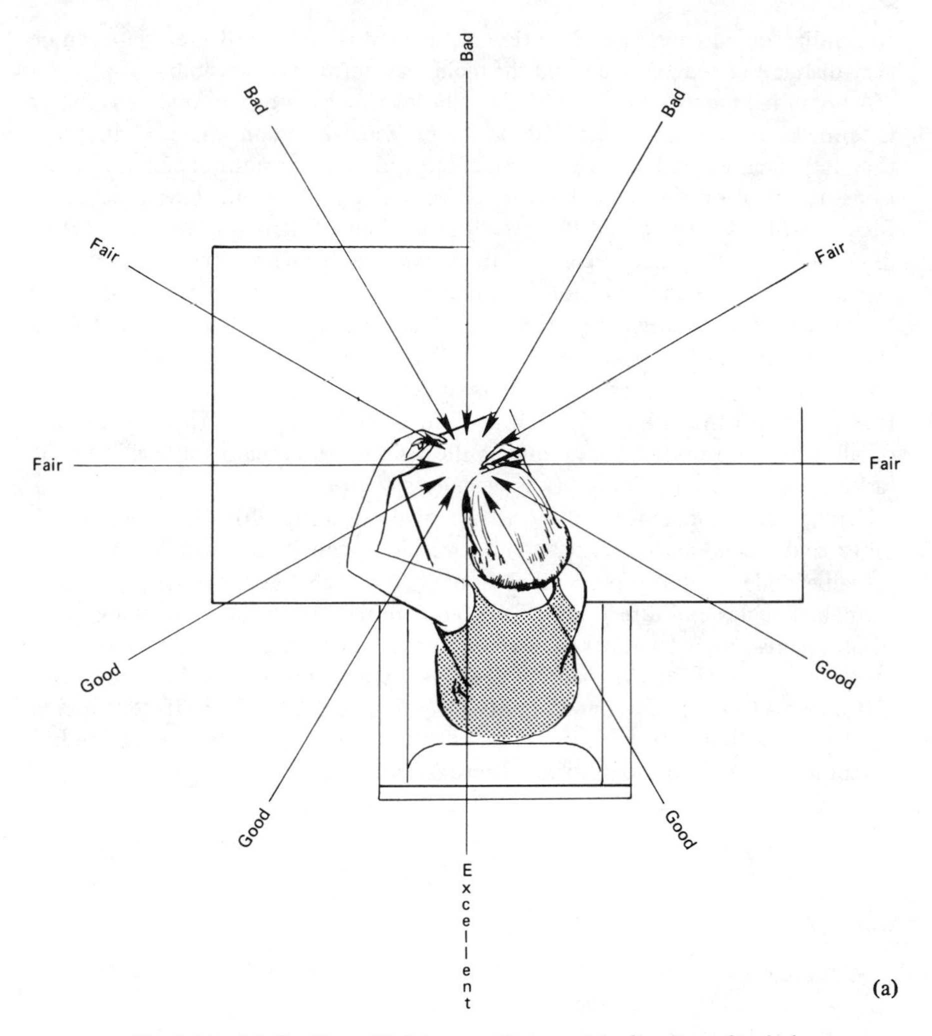

Fig. 3-14. (a) Grading of light rays with respect to *direction* of arrival.

In relating the concept of ESI to real environments, the following hypothetical example may be referred to. Figure 3-14b shows a real, everyday lighting environment, where light rays, either directly from luminaires or bouncing from room surfaces, strike the task at various angles and directions. Under these conditions, typical tasks, such as a book or a letter with black print on white pages, may not be satisfactorily visible, although a light meter near the task may read as much as 100 fc. The task is now placed inside a photometric sphere that provides the same visibility of the same task at the same viewing angle as was encountered in the room. A light meter reading the illuminance level inside

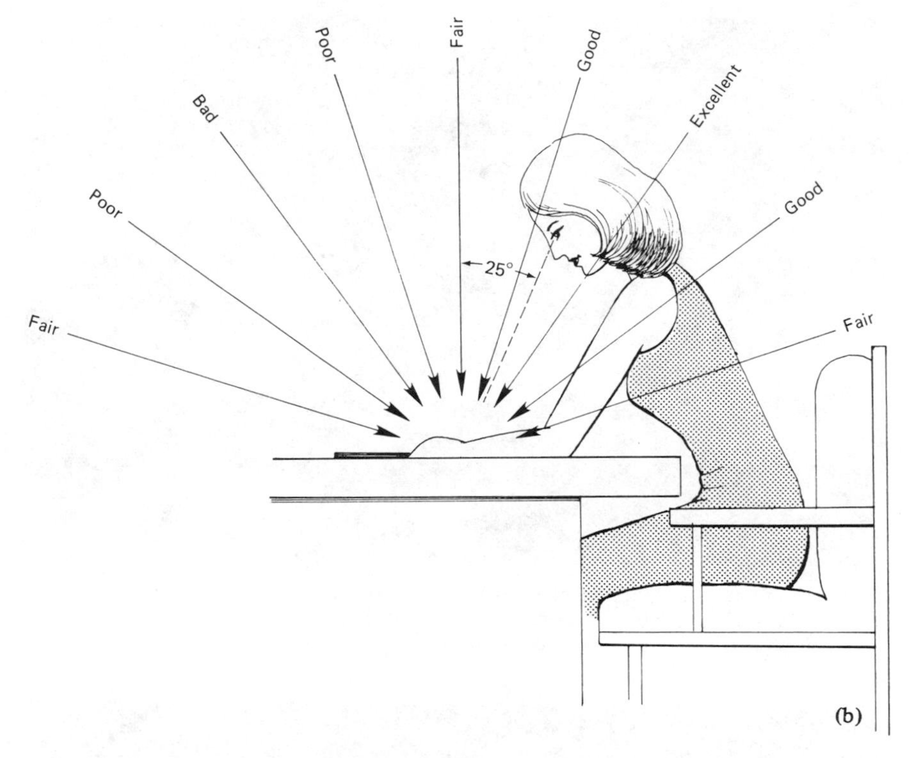

Fig. 3-14. (*Continued*). (b) Grading of light rays with respect to angle of arrival.

the sphere now may read a considerably smaller footcandle value—say, 50 fc. (see Figure 3-15). The ESI of the task in the room now can be claimed to have 50 ESI fc. since this is the equivalent of 100 fc. in the room.

A sphere is used as the reference mainly because it is easily reproducible, and experiments have shown that a proper placement of the task within an evenly illuminated spherical surface produces good task contrast. However, the important thing to remember is that a photometric sphere does not necessarily represent the ideal or the best lighting system; it is merely used as a reference. In addition, the ESI footcandle value determined for an application is valid only for the specific type of task, viewed at a specific angle and direction, with specific luminaires and orientation, and with given room-surface reflectances. A change in viewing direction or angle or a change of the type of task (which may commonly happen) will alter the ESI significantly. Based on these reasons, an average ESI value determined from various ESI values at different parts of the room will be meaningless.

Determination of an ESI value for an application is a complex process since there are so many contributing factors that are to be accounted for. Most

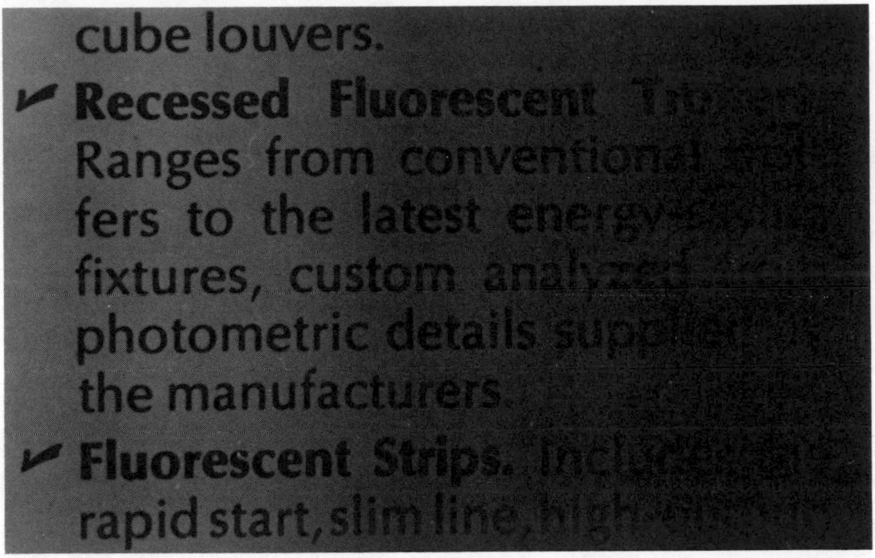

(i) Effect of contrast on printed material. The photograph above is purposely developed to create a background that darkens from left to right. Note that reading becomes increasingly difficult as the contrast between letters and background reduces.

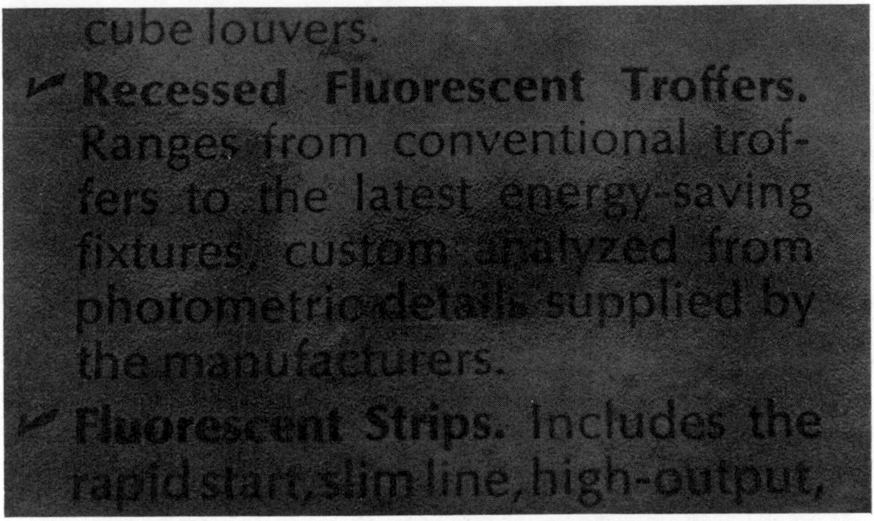

(ii) Black printing on glossy white paper. In this photograph, fluorescent troffer is located above and in front of the task in the offending zone. Note that the specular reflections occur in different positions on different letters as well as the background. This is a typical example of the veiling reflection which affects contrast.

(c)

Fig. 3-14. (*Continued*)

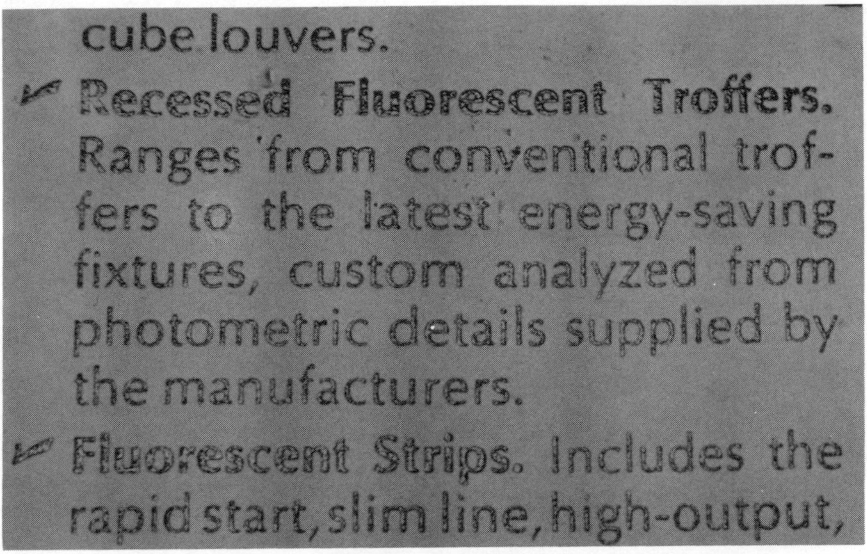

cube louvers.
Recessed Fluorescent Troffers. Ranges from conventional troffers to the latest energy-saving fixtures, custom analyzed from photometric details supplied by the manufacturers.
Fluorescent Strips. Includes the rapid start, slim line, high-output,

(iii) Black printing on a matte-white paper. In this photograph, fluorescent troffer is located above and in front of the task in the offending zone. The task is a dry-copy of the material used earlier. Note that the specular reflections mainly occur on the letters since these have a more shiney surface than the background, thereby reducing contrast and making it difficult to read.

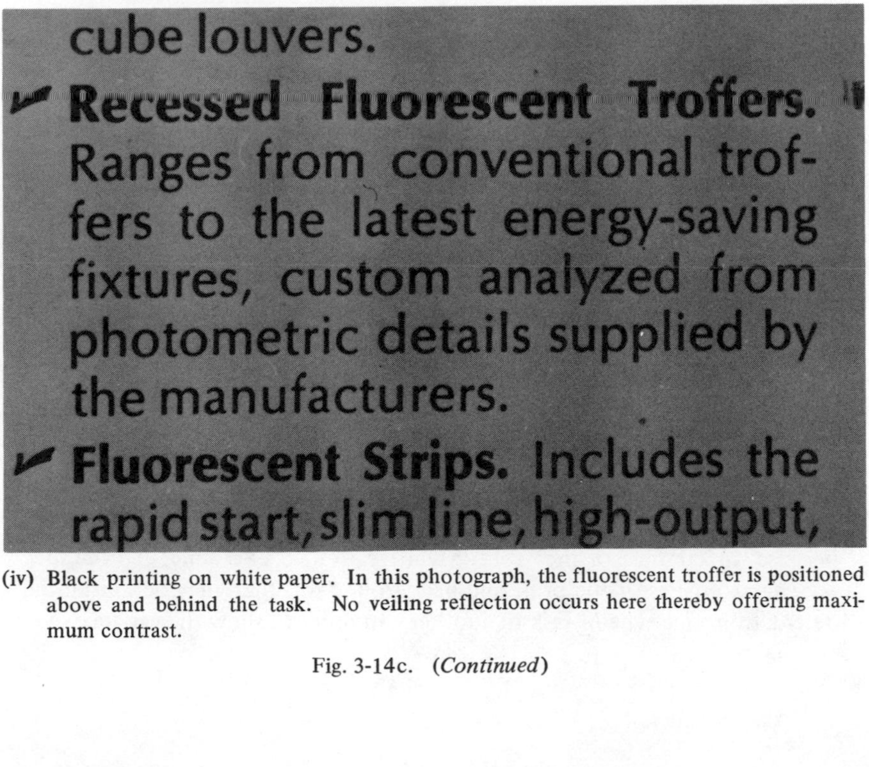

cube louvers.
Recessed Fluorescent Troffers. Ranges from conventional troffers to the latest energy-saving fixtures, custom analyzed from photometric details supplied by the manufacturers.
Fluorescent Strips. Includes the rapid start, slim line, high-output,

(iv) Black printing on white paper. In this photograph, the fluorescent troffer is positioned above and behind the task. No veiling reflection occurs here thereby offering maximum contrast.

Fig. 3-14c. (*Continued*)

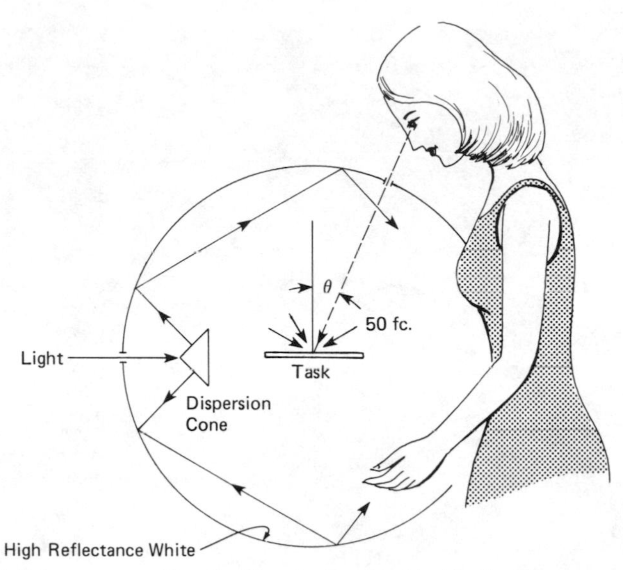

Fig. 3-15. A light meter reading the illumination level inside the sphere may read a considerably smaller footcandle level, say, 50 fc.

calculations are based on 25° viewing angle, since this appears to be the most common angle of viewing by an average person. The type of task may also vary significantly. The most common task used in ESI determination is a "pencil task": representing concentric rings drawn in pencil on white bond paper. There are a number of commercial computer programs available for ESI determination. Each program is usually capable of calculating with different types of common tasks. For instance, one popular programming offers the following types of tasks:

Pencil on white bond paper
Typewritten on white bond paper
Ballpoint pen on blue-lined tablet paper
Pencil on mylar drafting film
Offset printing on glossy paper
Dry copier on matte white paper
Felt-tip pen on matte white paper

It is important to remember that the ESI of a point may vary significantly from one type of task to another, even if all other variables are unchanged. This is because the contrast of the "print" relative to its background, and the specularity characteristics of the print and background together, may be substantially different from one type of task to another. In order to show the variance in ESI values that result from changes in types of tasks, the following table was developed using a popular programming, with all other variables unchanged:

Task	Avg. of All Points	Max. Value	Min. Value
Pencil	49	96	5
Typewritten	90	122	13
Ballpoint	82	120	11
Drafting	66	162	3
Offset	96	197	4
Dry copier	85	135	9
Felt tip	89	125	12

Note that an improvement in maximum or average ESI does not necessarily mean an improvement in minimum value. Compare the values for the pencil and drafting tasks.

Determination of task contrast is the main problem since its value is dependent upon so many factors, as discussed earlier. This is the main reason a computer must be used instead of hand calculations. If all the hard-to-find parameters are available, the remaining mathematics are quite easy in determining the ESI value. A number of programs are commercially available through time-sharing terminals. The following are the major steps involved in determining the ESI.

Step 1: Determination of L_b, L_o, ρ_o, E_t, and C_o. All of these can be measured directly if the ESI has to be determined in reality, in an existing situation. Since most applications are in a design stage, they must be calculated. Determination of L_b (luminance of the background of the task, e.g., the white page) and L_o (luminance of the object on the task, e.g., the back printing on the paper) is dependent upon E_t (illuminance on task at that location), which must be calculated. This is possibly the most complicated part of the whole procedure, since illuminance at a point E_t in the room is attributable to (1) direct light from luminaires and (2) interreflected light from room surfaces. While illuminance attributed from luminaires can be directly calculated, the room surfaces must be considered as a cluster of several small sources to be qualified for using the point-by-point method. For this, the luminance of surfaces must be determined first, and then they must be broken into several small areas to simulate the clustered light sources. Most of the information required in determining the ESI values is used at this stage, and a computer must be used for the complex calculations involved. ρ_0 (diffuse reflectance of the task background in sphere) and C_o (contrast of task in sphere) are predetermined figures and are constants for a specific type of task.

Step 2: C (contrast of task in real environment) is calculated by the Blackwell Contrast Formula, since L_o and L_b are known from Step 1.

Step 3: CRF (the contrast rendition factor) is determined by dividing C by C_o.

Step 4: RCS (the relative contrast sensitivity) in the real environment is determined from the table or curve, corresponding to the L_b found in Step 1. The curves can be found in the IES handbook.

Step 5: RCS_e (the relative contrast sensitivity) in the sphere is determined by multiplying RCS and CRF. This step in essence takes into account the veiling reflections.

Step 6: Using the same table or curve of Step 4, L_e (the equivalent background luminance of the task in sphere) is found corresponding to RCS_e.

Step 7: ESI value is determined by dividing L_e by ρ_o, the predetermined constant value for the diffuse reflectance of the task background in sphere.

A lighting effectiveness factor (LEF) is the ratio of ESI to E_t and represents a percentage of the illuminance found in a real environment, which is capable of producing the same glare-free visibility as can be found in the reference sphere.

As mentioned earlier, a number of programs are commercially available for determining ESI and VCP. Many luminaire manufacturers offer free service of such programs if they are related to their own products. Whatever route is taken, all programs will require the initial information in design criteria and parameters, which may be presented in a format of input data. To illustrate this, let us use the preceding example and determine ESI values at some arbitrary testing locations.

Model Example. The preceding sample problem required 19 luminaires for a maintained illuminance level of 70 fc. For the purpose of a symmetrical layout, let us suppose the total number of luminaires required is 18. Figure 3-16 is the lighting layout.

Let us suppose we are interested in finding the ESI values at points *A*, *B*, *C*, and *D*, as shown in the figure. The task is assumed to be "pencil." For the purpose of comparative study, let us suppose we also need the conventional illuminance and VCP reports on these points. Viewing directions are North, South, East, and West.

Input Data Form. In order to explain to the programmer exactly what is being requested, it is usually convenient to fill in an Input Data Form like the one shown in Figure 3-17. This will eliminate the necessity of sending a lighting

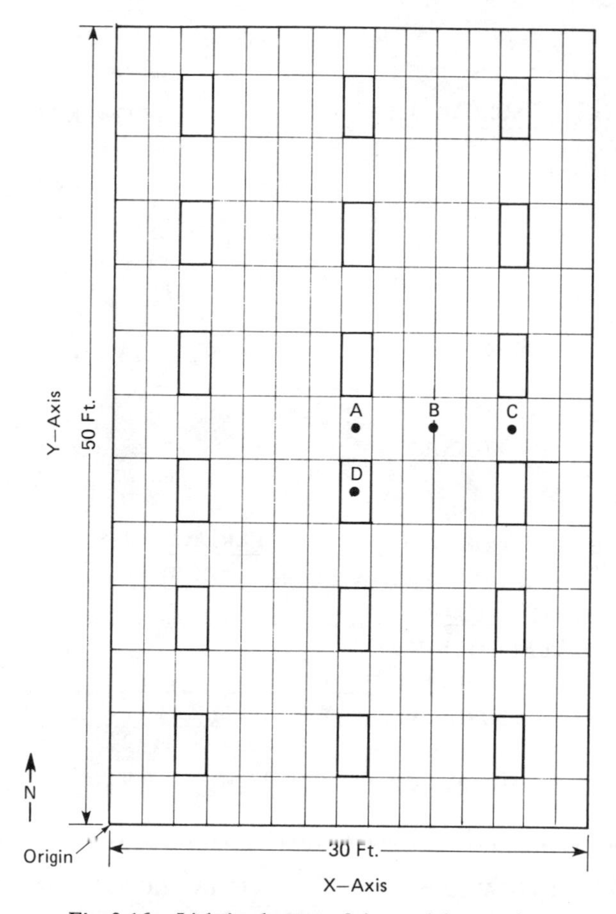

Fig. 3-16. Lighting layout of the model example.

floor plan and other information separately. Most luminaire manufacturers who offer ESI-determining services usually have a form of this kind.

Computer Output. Commercially available ESI programs will produce enough output to cover the requirements requested. One popular program, which performs a complete interreflectance calculation, taking into account all direct and reflected light within a room, will provide the following information:

ESI-Related Information

1. ESI, as viewed at North, South, East, and West
2. ESI minimum, maximum, and average

INPUT DATA FORM

PROJECT NAME SORCAR OFFICE BLDG. ENGINEER P.C.S.

ROOM DIMENSIONS

L 50 ft.
W 30 ft.
H 8.5 ft.

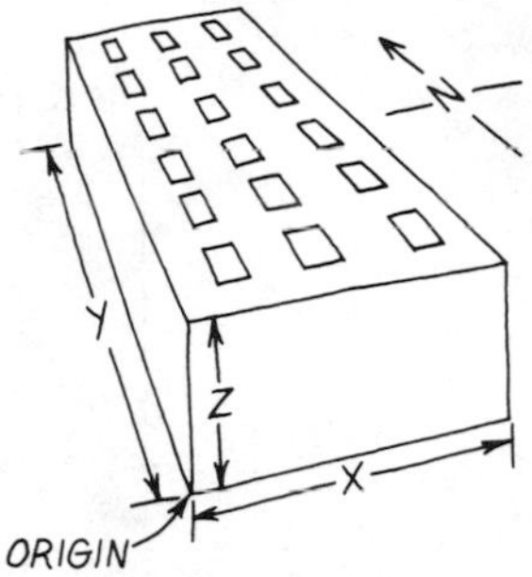

REFLECTANCES

CEILING 80%
WALL 50%
FLOOR 20%

LIGHTING SYSTEMS

DESCRIPTION 2 × 4 ft. TROFFER, ACR. PRSM. LENS.
MANUFACTURER/CATALOG NO. XYZ, #1234
LAMPS (4) F40 T12 CW LLF 0.67
FIXTURE LOCATIONS
X-COORDINATE 5, 15, 25
Y-COORDINATE 5, 13, 21, 29, 37, 45
Z-COORDINATE 8.5
LAMPS PARALLEL TO LENGTH OF ROOM (50 ft.)

CALCULATE ESI √ VCP √ ILLUMINATION √

TEST LOCATIONS TOTAL NO. OF TEST LOCNS. 4

X-COORDINATE	15	20	25	15
Y-COORDINATE	25	25	25	21
Z-COORDINATE	2.5	2.5	2.5	2.5

DIRECTION OF VIEW TASK 2.5 ft.
EAST √ WEST √
NORTH √ SOUTH √

Fig. 3-17. Typical input data form for ESI calculations.

3. Illuminance, with body shadow
4. Lighting effectiveness factors
5. Contrast rendition factors
6. Background luminaires
7. ESI bar chart

VCP-Related Information

8. VCP
9. VCP bar chart

Classical Illumination

10. Illuminance

Room- and Luminaire-Related Information

11. Room characteristics
12. Luminaire descriptions
13. Layouts
14. Room dimensions
15. Reflectances
16. Luminances

Table 3-3 is prepared from the computer output (Table 3-2) for a comparative study of the four points of interest. Reviewing the table of ESI values, it is obvious that, as expected, visibility in all directions is good if the viewer is sitting between rows and viewing between luminaires, as at point B. This provides a comfortable visibility since light arrives from both sides of the direction of view and there are no luminaires directly in the offending zone. Points A and C, on the other hand, produce low ESI at North and South directions since the luminaires are in the offending zones. Since there are no luminaires in the offending zones at East and West, the ESI values at these viewing directions are better, as would be expected. The worst location of the four points is D, since this is directly underneath a luminaire. ESI at all directions is poor, but more so for North and South, since these luminaires are closer to the offending zones than the others.

It is important to note that, although classical illuminance at point B is 63.7 fc, the ESI readings at this point are much higher in all four viewing directions. In contrast, although point D has the maximum classical illuminance, the average ESI value at this location is the least.

Good locations from the standpoint of high ESI and VCP can be determined by examining the computer output of such commercial programs. However, a reverse process, to determine the luminaire locations for predetermined ESI readings and work locations is almost impossible at this time because of the com-

Table 3-2. Computer Output.

EQUIVALENT SPHERE ILLUMINATION

SORCAR OFFICE BUILDING

LUMINAIRE: XYZ 1234
8′ × 10′ SPACING, 2.5 WORK PLANE
STANDARD LAMPS AND BALLASTS

TARGET DESCRIPTION: PENCIL TARGET-CONCENTRIC RINGS @ 25 DEGREE VIEWING ANGLE
SPHERE CONTRAST: 0.1675

	North	East	South	West	Total
Avg.	48.534	49.318	47.306	51.020	49.045
Min.	5.602	7.808	5.523	8.313	5.523
Max.	86.198	96.174	85.028	96.699	96.699
Mean Deviation	23.958	20.889	23.428	20.223	22.202

Abs. Y Coor.		Absolute X-Coordinate(s) 15.0	16.0	17.0	18.0	19.0	20.0	21.0	22.0	23.0	24.0	25.0
25.0	N	7.7	13.4	38.1	64.0	78.4	82.7	76.7	62.0	35.6	11.9	6.5
	E	75.8	89.4	96.2	93.8	81.9	68.6	53.9	41.1	41.7	53.7	68.7
	S	7.6	13.4	38.1	64.0	78.4	82.7	76.7	62.0	35.6	11.9	6.5
	W	76.3	58.6	45.3	43.5	56.8	70.6	83.4	94.9	96.7	89.0	76.0
24.0	N	17.7	26.6	51.6	72.2	82.4	85.1	80.8	70.1	48.4	23.8	14.1
	E	68.8	84.9	93.0	88.0	74.0	59.0	43.2	32.2	33.0	44.7	62.1
	S	6.5	10.3	30.5	57.5	74.7	79.8	73.4	55.8	28.4	9.2	5.6
	W	69.2	49.3	35.7	34.1	46.0	61.0	75.7	89.4	93.5	84.9	68.6
23.0	N	31.7	40.6	61.9	77.4	84.7	86.2	83.1	75.4	58.5	37.0	27.8
	E	55.0	75.4	82.5	72.1	55.4	38.9	25.3	17.7	18.9	29.9	49.4
	S	8.1	11.9	30.3	55.8	72.9	77.6	71.5	54.2	28.4	10.8	7.1
	W	55.4	32.8	20.5	18.9	27.0	40.5	57.0	73.6	83.1	75.1	54.6
22.0	N	35.8	43.3	60.3	75.0	82.8	83.9	81.1	73.1	57.1	39.8	32.0
	E	44.2	67.3	72.0	56.4	39.0	25.3	14.0	9.8	10.8	20.7	39.7
	S	13.0	18.8	36.1	59.8	74.6	78.3	73.4	58.3	34.1	17.1	11.6
	W	44.4	22.8	11.7	10.5	16.1	26.6	40.6	58.0	72.7	66.9	43.6
21.0	N	26.0	31.5	48.2	67.7	78.9	81.0	77.4	65.8	45.0	28.9	23.3
	E	40.6	64.2	69.3	49.9	32.0	20.1	11.2	8.1	9.2	18.0	36.4
	S	25.8	31.2	47.7	67.0	78.1	80.5	77.0	65.3	44.7	28.7	23.1
	W	40.8	19.8	9.9	8.7	12.0	21.3	33.6	51.3	69.9	63.9	40.0
20.0	N	13.1	19.0	36.5	60.5	75.3	78.9	73.7	58.6	34.3	17.3	11.7
	E	44.0	67.1	72.0	56.3	38.9	25.2	14.0	9.8	10.8	20.7	39.6

Table 3-2. *(Continued)*

Abs. Y Coor.		Absolute X-Coordinate(s)										
		15.0	16.0	17.0	18.0	19.0	20.0	21.0	22.0	23.0	24.0	25.0
	S	35.4	42.8	59.7	74.1	81.8	83.3	80.5	72.3	56.6	39.5	31.6
	W	44.3	22.7	11.7	10.5	16.0	26.5	40.5	57.7	72.7	66.9	43.6
19.0	N	8.2	12.1	30.7	56.6	73.2	77.8	71.7	54.8	28.6	10.9	7.2
	E	54.4	74.9	82.0	71.8	54.5	38.3	25.0	17.6	18.6	29.5	49.1
	S	31.2	40.3	61.6	77.0	83.6	85.0	82.3	75.0	57.9	36.6	27.5
	W	54.6	32.5	20.3	18.8	26.6	39.9	56.3	73.0	82.5	74.6	54.2
18.0	N	6.5	10.4	30.8	58.2	74.9	79.9	73.4	56.4	28.6	9.3	5.7
	E	67.8	84.2	92.1	87.5	72.9	57.9	42.4	31.9	32.5	44.0	61.6
	S	17.3	26.3	51.1	71.6	81.1	83.8	79.6	69.7	47.9	23.5	13.9
	W	68.1	48.7	35.1	33.7	45.2	59.9	74.5	88.7	92.7	84.2	67.9
17.0	N	7.7	13.5	38.4	64.5	78.4	82.7	76.8	62.5	35.7	11.9	6.5
	E	74.5	88.3	95.0	92.6	80.3	67.1	52.8	40.6	40.8	52.8	67.8
	S	7.4	13.2	37.5	63.2	76.8	81.2	75.4	61.5	35.1	11.6	6.3
	W	74.8	57.5	43.9	42.7	55.6	69.2	82.0	93.9	95.6	88.2	74.7
16.0	N	17.8	26.8	52.0	72.8	82.5	85.2	80.9	70.6	48.5	23.9	14.2
	E	67.6	83.9	91.9	87.2	72.6	57.8	42.2	31.7	32.4	43.9	61.4
	S	6.3	10.1	30.1	57.0	73.3	78.5	71.9	55.2	28.0	9.1	5.5
	W	67.9	48.4	35.0	33.6	45.0	59.8	74.2	88.4	92.5	84.0	67.8
15.0	N	31.8	41.0	62.3	78.0	84.8	86.2	83.1	75.7	58.6	37.2	28.1
	E	53.9	74.2	81.3	71.1	53.9	37.9	24.7	17.4	18.4	29.3	48.7
	S	7.9	11.7	29.9	55.1	71.4	76.1	70.1	53.4	28.0	10.6	7.0
	W	53.9	32.1	20.0	18.5	26.3	39.5	55.6	72.2	82.0	74.2	53.9
14.0	N	35.8	43.5	60.5	75.2	82.6	83.7	81.0	73.2	57.1	39.8	32.1
	E	42.6	65.4	70.1	54.5	37.3	24.0	13.4	9.5	10.5	20.0	38.7
	S	12.5	18.2	35.2	58.3	72.2	75.7	71.0	56.7	33.1	16.6	11.3
	W	42.8	22.0	11.3	10.1	14.3	25.3	38.7	55.8	70.7	65.2	42.5
13.0	N	26.0	31.6	48.4	67.9	78.7	80.6	77.1	66.1	45.0	28.9	23.4
	E	39.0	62.2	67.2	47.8	30.1	18.9	10.6	7.8	8.8	17.3	35.4
	S	24.7	30.2	46.3	65.2	75.4	77.5	74.2	63.5	43.3	27.8	22.5
	W	39.1	19.1	9.6	8.3	11.4	20.0	31.5	49.1	67.7	62.0	38.7
12.0	N	13.1	19.0	36.6	60.7	75.0	78.6	73.4	58.7	34.4	17.3	11.7
	E	42.3	64.9	69.6	53.9	36.8	23.8	13.2	9.4	10.4	19.8	38.4
	S	33.9	41.4	57.9	72.1	78.9	80.3	77.5	70.2	54.7	38.1	30.7
	W	42.4	21.8	11.2	10.0	14.2	25.1	38.3	55.1	70.0	64.6	42.1
11.0	N	8.1	12.0	30.7	56.5	73.2	78.0	71.9	54.7	28.6	10.9	7.2
	E	52.6	72.6	79.4	69.1	52.3	36.7	23.9	16.8	17.9	28.5	47.5
	S	29.8	38.7	59.3	74.4	80.9	82.4	79.6	72.3	55.7	35.2	26.4
	W	52.7	31.3	19.5	18.0	25.5	38.4	54.1	70.3	79.7	72.3	52.5

Table 3-2. *(Continued)*

Abs. Y Coor.		Absolute X-Coordinate(s)										
		15.0	16.0	17.0	18.0	19.0	20.0	21.0	22.0	23.0	24.0	25.0
10.0	N	6.4	10.3	30.7	58.1	74.9	79.9	73.5	56.4	28.5	9.2	5.6
	E	65.0	81.0	88.6	83.8	69.7	55.2	40.3	30.2	30.9	42.2	59.0
	S	16.1	24.8	48.8	68.7	78.0	80.6	76.5	66.3	45.1	22.2	13.0
	W	65.1	46.4	33.4	32.0	42.6	57.2	71.4	85.0	88.9	80.8	65.2
9.0	N	7.5	13.2	38.1	64.4	78.3	82.6	76.9	62.3	35.4	11.7	6.4
	E	70.7	83.9	90.7	88.3	76.2	63.5	50.0	38.1	38.4	50.1	64.6
	S	6.8	12.1	35.5	60.2	73.3	77.4	72.1	58.2	33.0	10.8	5.9
	W	71.0	54.3	41.5	40.3	52.4	65.5	78.1	89.3	90.9	84.2	71.3

plexity of optimization. Determination of good work locations are mostly dependent upon experience and judgment gathered from computer outputs like the one in the example. In an existing building, a good judgment of favorable locations can be determined with the help of a mirror. Use a mirror about the size of a typical task ($8\frac{1}{2} \times 11$ inches) and place it on the work plane. If no luminaires are visible in the viewing angle, it is likely to have a high ESI.

The average, classical illuminance level that is found by the lumen method usually is proportional to the amount of input power. The values of ESI, how-

Table 3-3. Short Table Prepared from the Computer Output for Comparative Study of the Four Points of Interest.

	Locations			
	A	B	C	D
ESI (fc)				
North	7.7	82.7	6.5	26.0
East	75.8	68.6	68.7	40.6
South	7.6	82.7	6.5	25.8
West	76.3	70.6	76.0	40.8
Classical Illumination (fc)	83.6	63.7	80.4	92.3
VCP (%)				
North	64.8	64.8	69.5	58.7
East	72.2	78.8	X	74.2
South	64.8	64.8	69.5	64.8
West	72.2	68.1	59.1	74.2

ever, cannot be accounted for in the same way. While for a good lighting design high ESI value is important, an energy-saving utilization will only be accomplished if a high ESI value is obtained from a low "raw" or "average" or "classical" level of illuminance.

Although ESI values are important for good lighting judgment, they should only be used where quality of visibility is important, such as in offices, schools, laboratories, or drafting areas. For other noncritical areas, such as storerooms, lobbies, or industrial areas, their application is practically useless. ESI is a good tool for analyzing a lighting system or making comparison between different luminaires for the same visibility; however, its values should not be used as the only criterion for the final decision. It is one of many criteria to be considered in lighting design.

REFERENCES

DiLaura, D. L. "On the Computation of Visual Comfort Probability." *Journal of the Illuminating Engineering Society*, 5, July 1976, p. 207.

DiLaura, D. L. "On the Computation of Equivalent Sphere Illumination." *Journal of the Illuminating Engineering Society*, 4, January 1975, p. 129.

DiLaura, D. L., and Stannard, S. M. "An Instrument for the Measurement of Equivalent Sphere Illumination." *Journal of the Illuminating Engineering Society*, 7, April 1978, pp. 183-187.

Florence, N. "Comparison of the Energy Effectiveness of Office Lighting Systems." A paper presented at the 1976 IES Annual Technical Convention, Cleveland, Ohio.

General Electric Company. *Lighting Maintenance*, TP-105-R. January 1969.

Guth, S. K. "Computing Visual Comfort Ratings for a Specific Interior Lighting Installation." *Illuminating Engineering*, October 1966.

Helm, Ronald N. "Energy and Lighting Design–Part Two." *Electrical Construction and Maintenance*, December 1979, pp. 61–67.

Helm, Ronald N. "Energy and Lighting Design–Part One." *Electrical Construction and Maintenance*, November 1979, pp. 62–70.

IES. "Lighting Calculations." *IES Lighting Handbook, 1981 Reference Volume*. pp. 9-1 to 9-12 and 9-60 to 9-74.

IES, Lighting Design Practice Committee. "General Procedure for Calculating Maintained Illumination." *Illuminating Engineering,* 65, October 1970, p. 602.

McNamara, A. C., and Andy Willingham. "The Concept of Visual Comfort Probability." *Plant Engineering*, June 12, 1975, pp. 141–144.

Mangold, S. A. "Lighting Economics Based on Proper Maintenance." *Lighting Design & Application*, August 1974, pp. 26–27.

Ngai, P. Y. "Veiling Reflections and Design of the Optimal Intensity Distribution of a Luminaire in terms of Visual Performance Potential." *Journal of the Illuminating Engineering Society*, 4, October 1974, pp. 53–59.

Ngai, P. Y., Zeller, R. D., and Griffith, J. W. "The ESI Meter–Theory and Practical Embodiment." *Journal of the Illuminating Engineering Society*, 5, October 1975, pp. 58–65.

Chapter Four
Commerical Lighting With Fluorescent Luminaires

For decades, the fluorescent has been the most popular light source for interior lighting systems. Relative lower cost and ease in lighting control for lower ceiling applications are the main reasons for this popularity.

The term *commercial lighting* used here is intended to include all types of nonresidential and nonindustrial interiors that use fluorescent lighting systems as applied in offices, schools, libraries, etc., where seeing ability is dependent not only upon quantity, but also upon quality. Adequacy of illuminance here is linked to both factors.

Commercial interiors with critical seeing areas can be loosely divided into two groups: (1) areas where task types and locations are known and (2) areas where task types and locations are unknown. Institutional buildings with interiors, such as classrooms, libraries, and conference rooms, are of the first kind. A prior knowledge of the type of tasks and locations in these areas enables the designers to create adequate lighting with minimum energy consumption. Office buildings, on the other hand, are of the second kind: Work-station locations, types of tasks, and occupancy may not be known until the buildings have been constructed. The lighting system for most of these applications, however, is to be designed before construction begins and, such being the case, it is virtually impossible to design a system that will be ideal for all purposes. Under these conditions, the only practical approach is to provide a flexible wiring system with low-power consuming luminaires that can be rearranged to a broad range of illuminance requirements.

Selection of a luminaire depends on several factors. They are as follows:

NUMBER OF LAMPS AND THEIR RELATIVE LOCATIONS

The total number of lamps and their relative locations from each other play an important role in the determination of candlepower distribution and efficiency. A bare lamp has the maximum efficiency (100%), since all of its light is able to come out in 360°. When the lamps are installed in an enclosure or on a mounting device, the overall efficiency of the system decreases. Efficiency of a single-lamp fluorescent strip is higher than that with two or more lamps, primarily because of the light trapped and the heat generated between lamps, as well as between the lamps and reflecting surface. As a general rule, the greater the number of lamps and the closer they are, the lower the efficiency. The phenomenon is shown in Figure 4-1, and it applies for all luminaires, irrespective of whether they are strips, surface-mounting enclosed, or recessed.

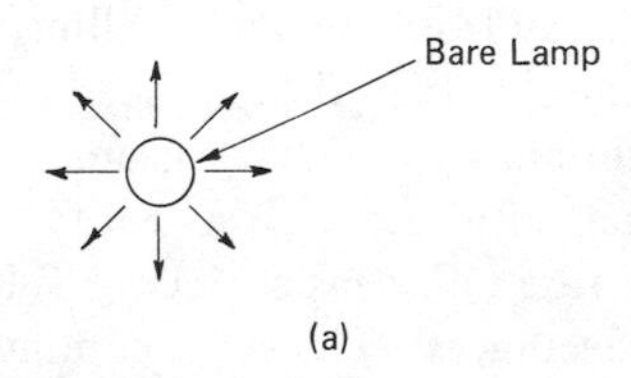

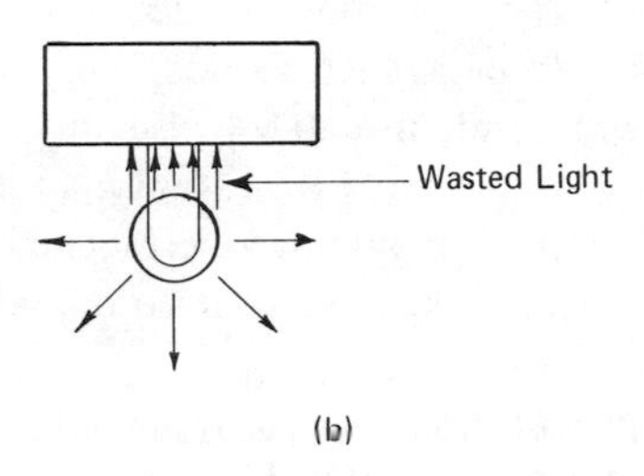

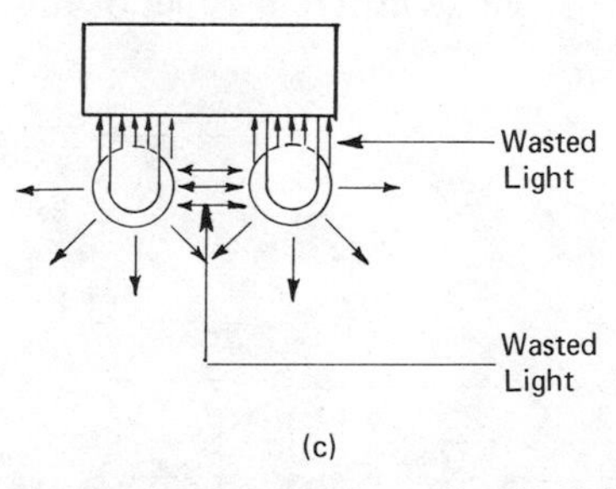

Fig. 4-1. (a) A bare lamp has 100% efficiency since all of its light is emitted at 360 degrees, uninterrupted. (b) A single lamp strip light has high efficiency, but not 100%. Light between the lamp and the channel is wasted. (c) Efficiency is further reduced because of additional light loss between lamps. The greater the number of lamps and the closer they are to each other, the lower the efficiency becomes.

REFLECTOR CONTOUR AND REFLECTANCES

Luminaire housing primarily has two functions: (1) protecting the lamps and ballast(s) from dirt build-up and mechanical damage and (2) direct the lights to useful areas. Figure 4-2 shows the cross-section of a typical indoor-type, fluorescent troffer. As can be seen, while a major portion of the light directly falls on the lens or louver, the remaining portion strikes first the reflecting surface and then reflects on the lens or louvers. In general, the reflecting surface of such luminaires are of the diffuse type, which in reality contains minute crystals or pigment particles. Each single ray from the lamps falling on an infinitesimal particle obeys the law of reflection, but as the surfaces of particles are in different planes, they reflect the light at many angles, causing a total diffusion.

Flat paints and other matte finishes such as white plaster or white terra-cotta are typical examples of diffuse reflecting surfaces. Table 4-1 lists diffuse reflecting materials and their reflectances. The most common reflecting material used in today's fluorescent luminaires is white paint, either dried in air or baked for durability. Some luminaires are made of prepainted steel, which in general has an air-dried thin coating of white paint, resulting in much lower reflectance than the baked white enamel. The reflecting surface is a main source of luminaire efficiency, so its value should be as high as possible. On the average, a 1% rise in painted reflectance will result in a 0.9% rise in CU values and efficiency. Large variation in quality of products from one manufacturer to another may result in as much as 15% variance in luminaire efficiency with the same reflector, design, and material. Before making a decision, compare different manufacturers' versions of a product.

Enamels are organic pigmented coatings applied for protection, decoration, and reflectance. They cure by oxidation by means of air or force-drying, or by polymerization by means of baking or catalytic action. This results in a very tough finish that mainly protects the reflecting surface from scratches and other physical damage. Table 4-2 shows different types of finishes and their properties

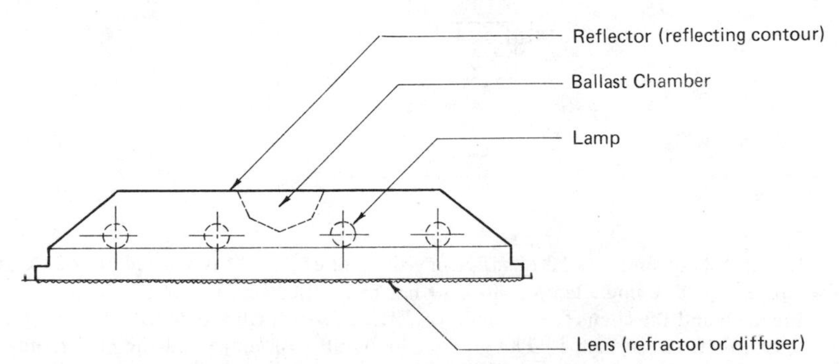

Fig. 4-2. Typical cross-section of an indoor-type, fluorescent troffer.

Table 4-1. Percent Reflectance or Transmittance of Materials. (Reprinted with permission from the IES Lighting Handbook, 5th Edition).

Material	Reflectance or Transmittance (%)	Characteristics
Specular		
Mirrored glass	80–90	Provide directional control of light and brightness at specific viewing angles. Effective as efficient reflectors and for special decorative lighting effects.
Metalized plastic	75–85	
Processed aluminum	75–85	
Polished aluminum	60–70	
Chromium	60–65	
Stainless steel	55–65	
Black structural glass	5	
Spread		
Processed aluminum (diffuse)	70–80	General diffuse reflection, with a high specular surface reflection of from 5 to 10 percent of the light.
Etched aluminum	70–85	
Satin chromium	50–55	
Brushed aluminum	55–58	
Porcelain enamel	60–90	
Aluminum paint	60–70	
Diffuse		
White plaster	90–92	Diffuse reflection results in uniform surface brightness at all viewing angles. Materials of this type are good reflecting backgrounds for coves and luminous forms.
White paint (mat)	75–90	
White terra-cotta	65–80	
White structural glass	75–80	
Limestone	35–65	

as they are applied for reflecting surfaces. The reflector contour design should be well balanced with lamp locations, so that a mutual heat build-up between lamps and between reflecting surface and lamps can be minimized.

MECHANICAL CONSTRUCTION

Mechanical construction is an important consideration, particularly for applications with vibrating environment or in areas where luminaires are handled very frequently to meet users' requirements. Selection of a luminaire solely on the basis of low initial cost usually ends up with bad-quality construction that may increase maintenance costs substantially over the years of operation. Most fluorescent luminaires used today are made with 22-gauge steel, although use of 20-gauge is not uncommon. A sturdy, well-constructed luminaire is important not only for the reasons above, but also for an assurance of safety for unusual conditions. The door frame usually is the weakest part of an indoor-type fluo-

Table 4-2. Properties of Finishes. (Reprinted with permission from the IES Lighting Handbook, 5th edition)

Type of Finish	Method of Application[a]	Principal Uses[b]	Possible Colors	Character of Reflected Light	Percent Reflectance[d]	Resistance[c]					
						Heat	Corrosion	Abrasion	Impact	Stability[c]	Flammability
Organic coatings											
Lacquers	D, B, S	A, P	Colorless of any color	Mixed to diffuse	10-90	F	F	P	F	F	Slow burn
Emulsions	D, B, S	A, P	All colors	Mixed to diffuse	10-90	G	G	G	G	G	Slow burn
Enamels	D, B, S	A, P, R	All colors	Mixed to diffuse	10-90	G	G	G	G	G	Slow burn
Baked clear coatings	D, B, S	A, P	Colorless, clear color	Diffuse to specular	0	G	G	G	G	G	Slow burn
Organisols	D, S	A, P	All colors	Mixed to diffuse	10-90	F	E	G	G	F	None
Ceramic coatings											
Vitreous enamels	D, S	A, P, R	All colors	Diffuse to specular	10-90	E	E	E	P	E	None
Ceramic enamels	D, S, B	A, R	All colors	Mixed to specular	10-90	E	E	E	P	E	None
Metallic coatings											
Chrome plate	Electrochemical	A, P		Specular to diffuse	60-88	E	E	E	E	E	None
Nickel plate	Electrochemical	A, P	Fixed; depending	Specular to diffuse	55	E	G	E	E	E	None
Cadmium plate	Electrochemical	P	on color of	Specular to diffuse	85	G	G	F	P	E	None
Brass plate	Electrochemical	A	plated metal	Specular to diffuse	55-80	P	P	F	P	F	None
Silver plate	Electrochemical	A, R		Specular	85-95	P	P	F	P	E	None
Laminates	Laminate	A, P, R	All colors of metallic effects	Mixed	10-90	Depends on nature of laminate					Slow burn
Conversion coatings											
Anodized aluminum	Electrochemical	A, P, R	Natural aluminum (or a wide variety of colors)	Diffuse to specular	60-90	E	E	E	E	E	None
Vacuum deposition	Vacuum chamber	A, R	Natural aluminum (or a wide variety of colors)	Specular	10-70	Depends on nature of protective coating					None

[a] D–dip, S–spray, B–brush.
[b] A–appearance, P–protection, R–reflectance.
[c] P–poor, F–fair, G–good, E–excellent.
[d] Depends upon color.

rescent luminaire, which is normally designed to hold a plastic lens. When a nonstandard lens (e.g., glass instead of plastic) or louver grid is specified, the specifier should make sure that the luminaire is capable of handling the additional weight. All exposed metal edges should be rounded off for easier handling and to avoid injury. Vibrating environments may cause the lamps to come loose from lampholders. This can be avoided by selecting spring-loaded lampholders. A full urethane foam gasket around the luminaire door will prevent light leakage and also reduce vibrations.

RADIO FREQUENCY INTERFERENCE

All arc-discharge lamps and their auxiliary components, such as starters, ballasts, or phase-control devices, produce electromagnetic pulses. The difference of potential across the electrodes of a fluorescent lamp results in electron flow, which generates a considerable amount of electromagnetic interference. This interference is both radiated directly from the lamp and conducted away through the supply leads. Depending on the age, conduction, and type of lamps and ballast, the frequency and magnitudes of the radiation vary considerably and are difficult to predict.

The interference can be substantially minimized and sometimes totally eliminated by properly grounding the luminaire, by bonding it either to a grounding wire or to the metal conduit of the circuit. The housing should be entirely metal, except for the opening for light output. Grounding between the door frame and the enclosure may be accomplished through a flexible conductive connection. A radio frequency suppression type of lens should be used for these areas. These lenses are equipped with a grid pattern of thin silver coating and metallic bussing around the edges. Two or three good contacts with the help of ground clips can bring sufficient grounding between lens bars and enclosure to avoid direct radio frequency interference. All precautions should be taken to make the luminaire totally free of any light leaks, since this is a potential for radio frequency interference.

APPEARANCE AND COST

Although luminaire appearance does not have any direct relation to efficiency and it is not necessary to sacrifice performance for esthetics, any consideration of long-term appearance must take luminaire surface depreciation factors into account. The durability of the finish and its maintainability should be complimentary to the overall esthetic effect intended in the space.

Cost is, no doubt, the most-weighed factor in the selection of a luminaire. While the majority of owners are more concerned about the first cost of luminaires, the net savings, if any, cannot be judged until the related upkeeping and maintenance costs have been evaluated. Popular luminaires are manufactured by dozens of companies, with significant difference in cost; which usually relates directly to their quality. Superiority of one over another cannot be judged until the quality of design, durability, and photometric distribution have been compared. A luminaire with a polystyrene lens, for instance, is less expensive than one with an acrylic lens. Although at first appearance and in performance they may seem to be identical, the polystyrene will turn yellow within a few years. Lower reflecting surfaces, inferior ballasts, improper lamp-holding arrangement, exposed or uncovered ballast wiring, thinness of metal and lens, and an overall weak structural condition are other major reasons for their inferior performance.

Speculation builders often are most concerned about first costs and select the lowest-costing luminaires in an attempt to reduce construction costs. Owners who build to occupy the building wholly or partially, usually recognize the benefit of life-cycle costing approach over that of the initial cost.

DIFFUSING AND SHIELDING MEDIA

There are many factors to consider when specifying a shielding or diffusing medium. The main purpose of a diffusing medium, commonly known as a lens, is to hide the lamps from direct vision, spread the lighting intensity uniformly over a larger area, and control the light output in a predetermined manner. Shielding medium is a physical barrier mainly to cut direct glare. Other important factors involved in selecting a lens or a louver are luminaire efficiency, candlepower distribution, S/MH ratio, and VCP. Specifying a lens or a louver without considering all of these factors will lead to an unsatisfactory result.

Lenses

From a photometric distribution standpoint, all lenses can be divided into two general types: diffusers and refractors.

The main purpose of diffusers is to hide the lamps and spread the brightness of the source over a larger area. They can be used where brightness control is required but there is no need for precise photometric control. Luminaires with diffusers have lower efficiency than open-bottom, or refractor, types.

Refractor design is based on the law of refraction, which states that a ray passing from a rare to denser medium is bent toward the normal to the interface, while a ray passing from a dense to rarer medium is bent away from the normal. This is shown in Figure 4-3. Refraction is commonly done with prisms. Prismatic light directors may be designed to provide a variety of light distribution using the principles of refraction. The majority of refractors used with fluorescent luminaires have a prismatic pattern. The purpose of these refractors is to intercept as many light rays as possible and redirect them away from the glare

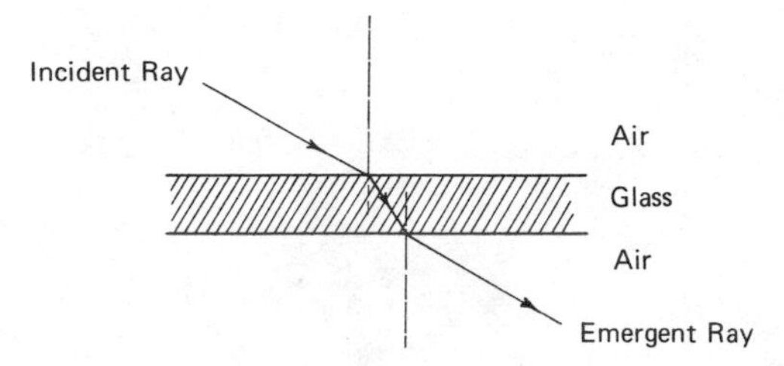

Fig. 4-3. Refraction of light rays at a plane surface causes bending of the incident rays and displacement of the emergent rays. A ray passing from a rare to a denser medium is bent toward the normal to the interface, while a ray passing from a dense to a rare medium is bent away from the normal.

zone to more useful directions. The bending of light at each prism is a function of the refractive index of the medium and the prism angle. The design of these prisms therefore determines the candlepower distribution curve, S/MH ratio, and VCP. This was discussed in detail in Chapters 2 and 3.

A proper functioning of the refractor is greatly dependent on the material and the quality of the overall luminaire design. A large number of smaller prisms produce greater accuracy of light control. However, if special precautions are not taken, they may "round-off" in the manufacturing process, and thus deteriorate performance. Retaining the sharp edges is important for optimum results. Refractors having prisms at the bottom surface and flat at the top are preferable from a maintenance standpoint.

Refractors are available in several types of lens design. Figure 4-4 shows a variation of such design, each specifically intended to transmit light in a unique pattern. While most commercial lenses have one consistent pattern throughout the panel, others combine many to produce a unique control of light.

Figure 4-5 shows the composite candlepower distribution of a 2 X 4-foot fluorescent troffer equipped with different types of popular lenses. Curve A represents a nonprismatic, flat-white acrylic diffuser, commonly known as an *opal lens*. These are mostly used where brightness control is required but there is no need for precise photometric control; the efficiency is low. Curve B represents a typical clear, acrylic prismatic lens, commonly used with standard fluorescent troffers. This type has female prisms at the bottom and a flat surface on the top; it is usually referred to as the *standard lens* of the industry. These are available in different thicknesses; the most popular is 0.125 inches. The basic concept behind this type of lens is to produce much of the light between 0 and 45° and as little as possible at the higher angles (direct-glare zone). This results in a better light control with higher efficiency than the diffuser in Curve A.

Curve C represents a unique lens that has a combination of female, flute, evaluating, and depressing prisms, each serving a special purpose. Figure 4-6 shows a cross-section of this lens as applied to a three-lamp luminaire. The

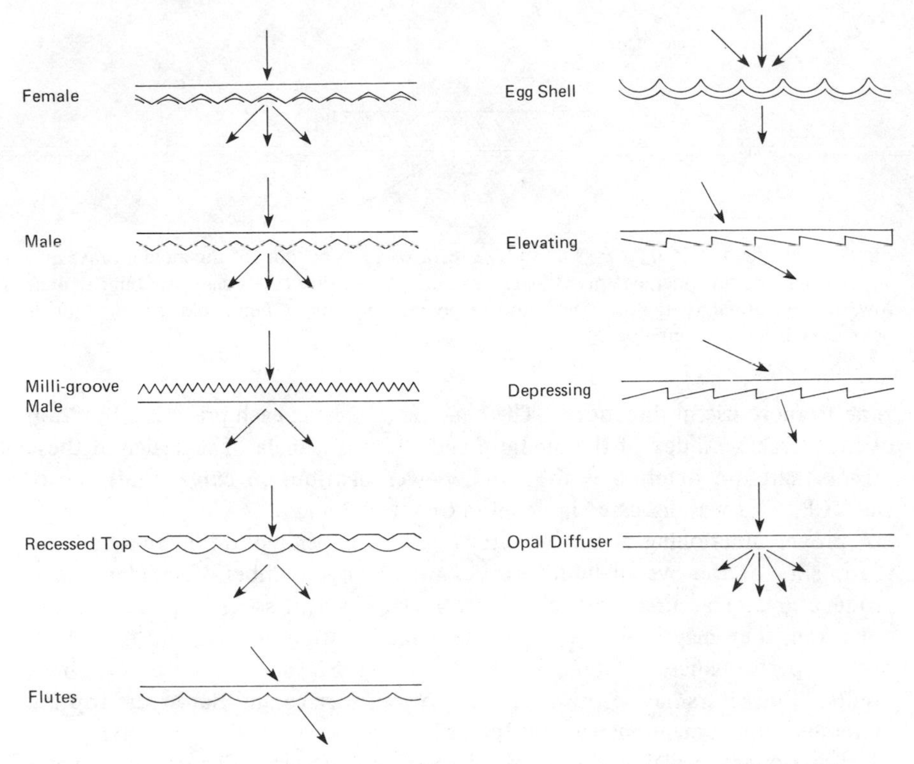

Fig. 4-4. Examples of various types of lens design. A commercial lens panel may have one consistent pattern or a combination of several.

female prism directly underneath the lamps spreads light just as in a standard lens, offering much downlight. Use of flutes between lamp areas directs light in a preferred direction relatively undisturbed. Elevating prisms raise light at low angles to increase the S/MH ratio, and depressing prisms at the end control the lights at higher angles, directing them to even lighting at far ends of spacing, minimizing direct glare. The net result is a uniform blanket of lighting from one end to the other, with high efficiency and S/MH ratio. A high S/MH ratio does increase the direct-zone lens brightness, resulting in a lower VCP. A substantial increase in VCP with equal or higher efficiency and S/MH ratio is achieved when the luminaire is equipped with two lamps instead of three. Omitting the center lamp tends to "even" lighting further by eliminating the slight bulge that occurs in the middle of the candlepower distribution of three lamps. With more light in the glare zone, the contrast rendition factor (CRF) and ESI increases with lower VCP, as discussed earlier.

The main disadvantage of these lenses is their inferior light distribution at a vertical plane parallel to the lamps. A good result can be obtained if the lumi-

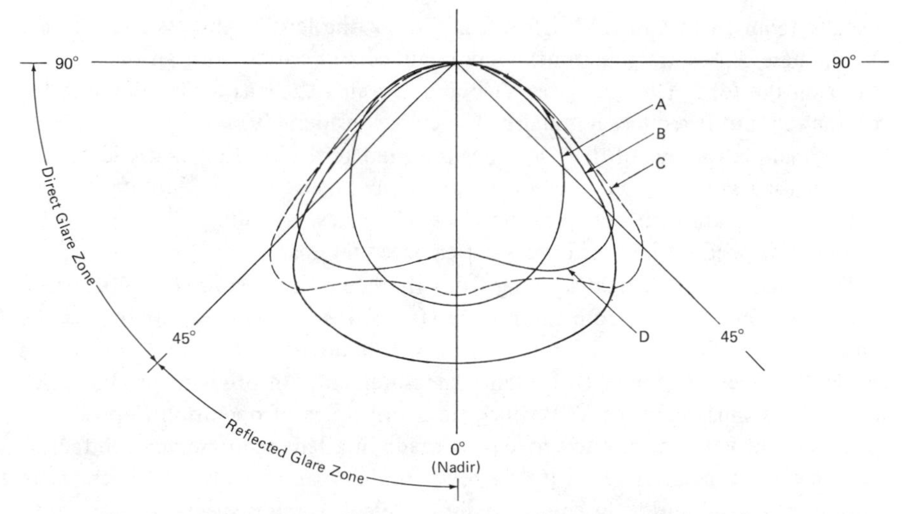

Fig. 4-5. Composite candlepower diagram of four types of popular lenses.

naires with such lenses are installed in a continuous row or in a checkered pattern, separated widely apart. The other disadvantage is that these are suitable for luminaires with two or three lamps only, precisely located to match the lens pattern. Multiprismatic pattern gives a distinct striated appearance, and at some angles, the lamps and hardware are visible. This may cause some unfavorable esthetics.

Other lenses are available that offer high-quality light with widely spread radial batwing distribution as shown by Curve D. These are either extruded or injection molded, each having a different but consistent and uniform prismatic pattern throughout the panel. The injection-molded lens obtains its distribution by a recess pattern on the top mating with a male prism on the bottom. The extruded lens has smaller male prismatic pattern on the top and a flat surface on the bottom. Both of these lenses produce radial batwing distribution with slightly high upper-angle brightness and a high S/MH ratio. The extruded lens suffers a low VCP, especially because they have no prisms on the bottom (prisms are

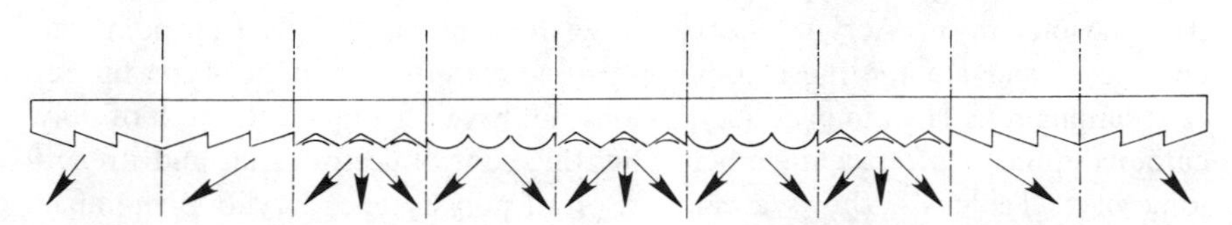

Fig. 4-6. Cross-section of a popular three-lamp lens that uses a combination of four different types of lenses.

usually required to control brightness). To cut the lens brightness and increase VCP, these lenses are sometimes recommended to be used with an overlay diffuser on the top. This improves VCP and prevents the lens from collecting dirt on the top, but it reduces luminaire efficiency considerably.

The main advantage of these lenses is that, unlike the multiprismatic lens, they can be used with any number of lamps at any position in the luminaire. The appearance is uniform and pleasing from all angles. Although their efficiency is low, they produce high ESI and can be spaced far apart.

The concept of task light arriving from the side of a luminaire (45-90° direct-glare zone) rather than from the middle (0-45°) is clearly contradictory to the basic concept of standard lenses. However, it increases ESI, and that is the presently accepted concept of quality measurement. In order to produce high ESI, a lens must sacrifice VCP, since their principles of operation oppose each other. What is more, any improvement made in a lens to preserve both factors in respectable position ends up slashing the luminaire's overall efficiency—an important consideration in energy savings. A lens which is specially designed to produce high ESI and VCP may have an efficiency as low as 40-45%. The designer must find a suitable application for such products to justify their high costs and the energy spent in producing the good-quality light. Blindly specifying such products in the anticipation of providing high-quality lighting in open spaces may be a sheer waste of money and energy until the types of tasks and their locations are exactly known and luminaires are precisely located for the best effect. Many times, the cure to offending zone lights, veiling reflections, and lack-of-contrast problems may be something no more complex than moving the task or desk by 90 degrees or moving the luminaires to a more suitable location. Table 4-3 shows the relative costs and test results using some popular plastic lenses in the same 2 × 4-foot troffer luminaire.

Other factors affecting luminaire efficiency are the thickness and the type of material used for the lens. A loss in efficiency and a good light control usually occur simultaneously with the use of thicker material that has been chosen for lens strength. Polystyrene, acrylic, high-impact acrylic, and polycarbonate are the different types of plastics frequently used for lenses. Polystyrene absorbs ultraviolet rays from lamps and turns yellow with age. Polycarbonate has relatively lower transmission efficiency but extremely high impact strength; this, unfortunately, degrades rapidly and becomes brittle with continuous exposure to ultraviolet rays. Acrylic plastics show the best results when all points are considered and are the most commonly used material for indoor-type fluorescent luminaires. High-impact acrylic does not have the impact strength of polycarbonate, but it offers a much better lighting-control performance and strength longevity. Table 4-4 shows a comparison of polystyrene, acrylic, high-impact acrylic, and polycarbonate plastics.

Table 4-3. Test Results Using Plastic Prismatic Lenses in the Same 2 X 4-Foot Troffer.

Type Lens (Acrylic)	Thickness	Luminaire Efficiency	Lens Brightness	Light Control	Relative Cost
A. Extruded					
Opaque diffuser	0.125	Low	High	Low	0.9
B. Extruded					
Clear prismatic	0.100	High	High	Medium	1.0
	0.125	High	High	Medium	1.9
Extruded					
Tinted prismatic	0.100	Low	Low	Medium	1.2
	0.125	Low	Low	Medium	2.3
Injection molded					
Clear prismatic	0.150	Medium	Medium	Medium	3.5
	0.175	Medium	Medium	Medium	4.0
C. Extruded					
Clear refractive grid, flutes, elevators, etc.	0.130	High	High	High	3.1
D. Extruded					
Clear prismatic with overlay	0.200	Low	High	High	4.4
E. Injection molded					
Clear refractive grid with overlay	0.280	Low	Low	Low	9.2
Injection molded					
Clear refractive grid without overlay	0.240	Medium	Low	Medium	7.5

Polarized Lenses

Polarized lenses adopt the basic principles of polarization of light. Light is composed of an infinite number of waves that run on planes perpendicular to the axis of the ray of light. When an unpolarized ray of light is passed through a polarized material, only certain waves are emitted, depending on its orientation. Figure 4-7 shows the phenomena. Experiments show that vertical vibration of polarized light drastically reduces veiling reflection and increases contrast of tasks. Working on the reflux principle, the blocked horizontally polarized rays

Table 4-4. Operating and Other Characteristics of Various Types of Plastic.

	Polystyrene	Acrylic	High-Impact Acrylic	Polycarbonate
Light transmission efficiency	90%	92%	90%	88%
Light stability age	4–7 yr.	15–20 yr.	10–15 yr.	3–4 yr.
Impact strength	0.5	1	10	30, fast degrade
Haze	Under 3%, degrades	Under 3%	Under 3%	Under 3%, degrades
Scratch resistance	Excellent	Excellent	Good	Good
Burning character (U.L.)	Slow	Slow	Slow	Self-extinguish
Smoke generation	High	Slight	Slight	High
Resistance to heat	170°–200°F	140°–190°F	130°–180°F	250°F
Type of smoke	Toxic	Nontoxic	Nontoxic	Toxic
Relative cost	0.67	1	2	4

bounce back inside luminaire and are re-reflected, with the horizontal portion blocked again. The process contains infinitely. The net result is approximately 10% less light output than that with standard acrylic lens, but there is a significant improvement in task contrast because of vertically polarized light.

Louvers

The use of louvers in fluorescent luminaires is mainly to improve the VCP. They are physical barriers to light output mainly in the glare zone that allow the light

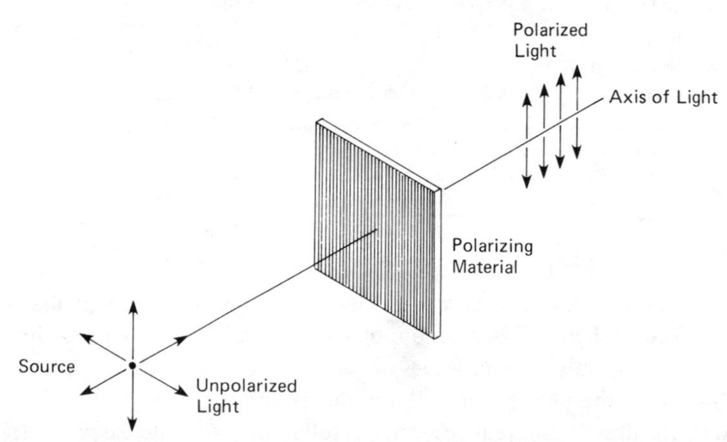

Fig. 4-7. When an unpolarized ray of light passes through a polarized material, it blocks all horizontal rays. The polarized vertical rays reduce veiling reflections and increase visibility.

Table 4-5. Comparative View of a Range of VCP and Efficiency of Various Types of Lens and Shielding Material.

Shielding/Lens	VCP (range)	Shielding/Lens	Efficiency (range)
Parabolic louver	85-95	Clear Lens	55-80
Toned lens	65-95	Polarizer	45-70
White metal, louver	65-85	Diffuser	40-60
Polarizer	60-70	Plastic louver	45-55
Clear lens	50-85	Parabolic louver	45-55
Plastic louver	50-70	White metal, louver	35-45
Diffuser	40-50	Dark metal louver	25-40

to travel relatively freely downwards. Louvers can be of different shapes, sizes, and colors. The more common patterns are square or rectangular.

Because much of the light is obstructed by louvers, luminaires with louvers usually have much lower efficiency than open-bottom types or those with lenses. The smaller the physical size of the cells and the darker the color, the lower the efficiency. Louvers can be made of metal or plastic. In recent years luminaires with parabolic-shaped aluminum louvers have become quite popular. These louvers have large square cells and blades that are about 3-5 inches high. The louvers maintaining a physical barrier between the lamps and the eyes, provide excellent visual comfort. Specular (or satin) finish of the parabolic blades reflect light rays at desired angles. The result is a widespread light distribution with minimized direct glare. However, because of the property of louvers, the efficiency of these luminaires is low. Table 4-5 provides a comparative view of VCP versus efficiency performance of luminaires equipped with louvers or lenses.

BALLAST SELECTION CRITERIA

The various types of ballasts available for fluorescent lamps have been discussed in detail in Chapter 1. Unless there is a special instruction, most luminaires are provided with standard ballasts and standard lamps as selected by the manufacturer. However, since the ballast has a considerable impact on the lamp's performance, the designer should investigate the many different options available and select a product that is most suitable for the application.

Sound

All ballasts produce a humming noise commonly known as the "60-cycle hum." The magnetic field in the ballast transformer core expands and collapses with input AC frequency agitating the metal laminations. This produces the audible humming noise.

Manufacturers produce different types of ballasts with various noise levels. A well-constructed ballast with a balanced amount of potting compound reduces noise. Luminaire construction, the location of ballast, and its secureness to the luminaire surface are the other reasons for the amplification of the noise produced by the ballast. The best technique to eliminate the noise is to remote-mount the ballasts. This method is expensive, but it may prove to be well worth the price, if controlling noise is critical, e.g., in music schools or libraries.

Temperature

The performance of a ballast in excessive temperatures is the most frequent reason for premature ballast failure. Underwriters' Laboratory- (U.L.-) listed ballasts must abide by the following temperature limitations: (1) 194°F (90°C) maximum case temperature, (2) 158°F (70°C) maximum capacitor temperature, and (3) 149°F (65°C) maximum temperature rise of the coil. In order to take precautions against excessive heat build-up, manufacturers take a variety of different design approaches that affect temperature rating. Locating the capacitor, the most heat-sensitive component, away from maximum heat-collecting areas, use of class H insulation and wiring with 356°F (180°C) rating, and positioning the core and coil against the housing are some of the frequently adopted means of avoiding excessive heat build-up. However, no matter how well it has been constructed, a ballast will not live up to its rated life if it is forced to operate under conditions for which it is not designed. For instance, operating a ballast against a ceiling may build up excessive heat if it was designed for suspended luminaires. One should always observe the minimum plenum depth, distance from ceiling, and minimum distance between units; these are usually specified on luminaires by manufacturers. Impact of voltage variation should also be observed, since each 1% increase in operating voltage will raise the ballast operating temperature by 2.1°F (1.2°C).

In order to obtain proper heat conduction from ballast to luminaire, the ballast should be secured with metallic fasteners on the metallic surface of the housing, preferably on the top area. Room ambient temperature, the number of lamps, the lamp cavity, luminaire mounting methodology, the type of ceiling material, etc. are some of the other factors directly involved in this matter. Surface-mounted luminaires against a low-density ceiling may experience a temperature as much as 18°F (10°C) higher than a similar unit that has been suspended. Each 1°C increase in ambient air temperature will raise the ballast temperature by about the same proportion. The impact of heat build-up because of luminaire-mounting methodology is shown elsewhere in this chapter.

Only ballasts carrying the U.L. and CBM labels should be used, since a U.L. label assures its safety criteria and CBM indicates its meeting the performance standards set by ANSI. Another important precautionary measure against excessive heat build-up inside a ballast, which may happen as a result of improper connections, abnormal voltage, or internal faults, is the inclusion of a self-resetting disconnecting means inside the ballast that trips with the excessive heat build-up. This is a requirement by NEC, and all ballasts meeting this requirement are classified by a letter *P* by U.L. (this only applies for ballasts for indoor use).

Impact of Ballast Factor and Thermal Factor in Lighting Design

The implementation of accurate ballast factor (BF) and thermal factor (TF) has a significant influence in energy-saving lighting design, since both numbers are contributing factors in determining the total number of luminaires. In the past, when energy was cheap, broad assumptions in such factors were sufficient to determine a reasonably correct result. With the introduction of newer low-power-consuming lamps and ballasts (see Chapter 1), both these factors have gained special interest in lighting calculations.

Ballast Factor. In evaluating the energy-saving performance of these newer products, manufacturers often compare them with the performance of traditional "standard" lamps and ballasts. The procedure for determining the BF is in accordance with the specifications set by ANSI standard C82.2, and is performed on a so-called bench test. The lamps are first seasoned by a reference ballast, which is a special inductive type that has certain prescribed characteristics that make it suitable for laboratory measurement of light output and input power. The test is done in open space with still air at controlled ambient temperature (77°F) and regulated power supply. After the lamps stabilize, a reading in light output is taken and then switched over to a cold commercial ballast, whose BF is to be determined. A reading of light output is taken within 30 seconds and then compared to determine the BF. Note that the procedure for BF determination applies only for ballasts having CBM certification. Presently, the accepted minimum ANSI/CBM BF for a commercial ballast operating standard F40T12 rapid-start lamps is 92.5% (in the past the accepted value was 95%). This basically means that the light output produced by a standard ballast may be as low as 92.5% of that produced by the reference ballast and yet be CBM-certified. ANSI procedure, not getting involved in setting standards for lamp lumen output, and basically confined only to electrical characteristics, holds that the determined BF value merely indicates a comparative number and does not imply that the reference condition light output is equal to rated lamp output or otherwise. However, it is important to remember that present ANSI/CBM specifications apply for all ballasts operating standard lamps only; they do not apply for lower-wattage lamps. The eagerness of users for saving energy and cutting operating costs, triggered by the promotional ventures of manufacturers, often results in the overlooking of the inherent adverse results of some combinations of such products. In many instances, the BF determined with these products may fall well below the required minimum of 92.5%. Ballast manufacturers, under these circumstances, do not publish the test results; they simply indicate that the products are not CBM certified.

For calculation purposes, BF can be assumed to be unity (100%), if the ballasts used in practice and photometrics have the same BF, and the changes due to BF are automatically encountered in the CU values. This, however, is rarely the case. CU values are relative terms which show the percentage of bare lamp lumens that arrive at the task level, with an assumption that the lamps produce their rated values. To incorporate the actual change in light output, the BF of the ballast must be considered in the calculations.

Thermal Factor. Although it has been long known that the fluorescent lamp is the only major light source that is sensitive to ambient temperatures, this effect on the operational characteristics of lamps used in different types of luminaires has remained grossly neglected in lighting-design calculations. Conventionally, designers refer to the respective manufacturers' catalogs for the lumen output data of different lamps and input power data of different lamp-ballast combinations, which are actually the results of a series of bench tests performed according to predetermined, fixed parameters. Although these bench-tests are performed in accordance with accepted national standards, their results are only true to the extent of the surrounding parameters at which they were tested. When lamps are inserted in luminaires and luminaires are installed in the ceiling, the performance of the lamps and ballasts differ substantially from their bench-test values and thus affect the overall efficiency. With a fixed room ambient temperature, the lamp cavity temperature and the bulb operating temperature are going to be different for different types of luminaires, and hence there will result a significant variation in the lumen output and input power values among the types. Until and unless these variations in lumen output- and input-power consumption are properly recognized for each type of luminaire in their mounting conditions, a true analysis of energy effectiveness of the system cannot be made. This means that designers now must consider not only catalog and nameplate ratings, but also the systems' thermal factors that can substantially affect the overall result.

In determining the TF, the first thing to recognize is the fact that lumen output and input power of a lamp-ballast combination will vary with temperature.

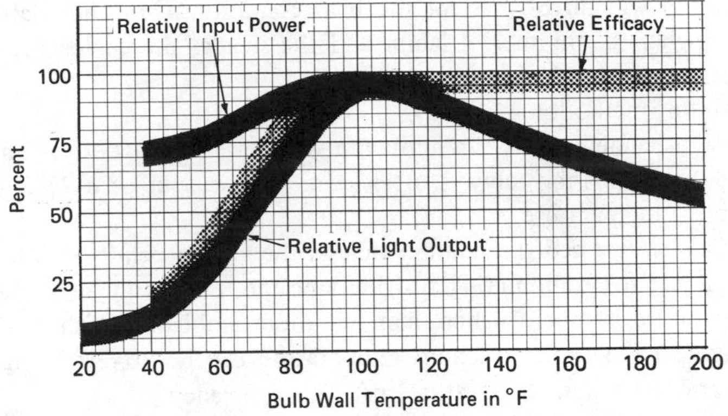

Fig. 4-8. Fluorescent lamp output and input watts versus bulb-wall temperature. The shaded portion indicates varience in readings for different types of fluorescent lamps. Light output and input power decreases simultaneously almost by same proportion at bulb-wall temperatures higher than 100°F. The efficacy at this range is almost constant. As temperature drops under 100°F, relative drop of light output is much steeper than the input power, resulting in a fast decrease in efficacy values.

Figure 4-8 shows curves representing these characteristics of a typical indoor-type fluorescent lamp. The most critical area of a fluorescent lamp is the coldest spot on the bulb surface (there may be more than one spot). As we have seen in Chapter 1, the amount of light produced by a fluorescent lamp is directly proportional to the ultraviolet energy available in the lamp. When the temperature in some spot or spots on the bulb reduces, the mercury tends to condense at these areas, resulting in a low overall pressure. This causes less ultraviolet energy and hence low light output. Most fluorescent lamps are designed to peak in light output around 100°F bulb wall temperature. If the temperature exceeds this value, too much mercury vapor is present in the lamp; causing a shift in wavelength produced that is nearer to the visual spectrum, and hence less effect on the phosphor. This results in a subsequent reduction in both light output and input power. As can be seen in the figure, at bulb-wall temperatures higher than 100°F the lumen output and input power reduces by the same proportion; however, in lower temperatures lumen output decreases much more drastically compared to the input power, resulting in a much lower efficacy at these temperatures. The power consumption in the lower temperatures remains relatively high compared to the lumen output primarily because power is still being used to heat the lamp at these temperatures.

The factors affecting bulb-wall operating temperature, can be generalized into three groups as follows:

1. Luminaire type (TF_A). The number and types of lamps and ballasts, the volume inside the luminaire, ventilating conditions, and material of con-

struction will directly affect the heat condition of the luminaire interior. The higher the number of lamps and the smaller the enclosure, the higher the luminaire inside-temperature.

2. Mounting methodology (TF_B).
   a. A suspended open-strip luminaire will have the least effect on light output and possibly always offer peak light output at a 77°F stable ambient temperature.
   b. Ceiling recessed luminaires, such as the grid-suspended troffers, will be subject to accumulating a higher amount of heat, since the plenum temperature is much higher than the room ambient. With a 77°F room ambient, a typical plenum temperature can be 10-20°F higher. This raises bulb-wall temperature, affecting efficiency.
   c. Wraparound or other types of surface-mounted, enclosed luminaires will collect maximum heat because of the insulating effect of ceiling material.
3. Room ambient temperature (TF_C). In general, the room ambient temperature is controlled thermostatically and kept at a predetermined constant reading for a season. However, a number of different factors may change its value and thus affect the bulb-wall temperature.
   a. During winter season, it is a common practice to lower the temperature during nonoperating hours such as weekends, holidays, and weekday nights. This will result in lower room ambient temperature, at least when the luminaires are first turned on.
   b. Luminaires near heating elements, such as unit heaters or radiant panels, will be subject to higher room ambient temperature.
   c. During the summer season, luminaires near low-temperature air diffusers will be subject to lower room ambient temperature.

The TF that will encounter all these effects can be expressed as follows:

$$TF = TF_A \times TF_B \times TF_C$$

Any change in lumen output due to internal thermal conditions in a luminaire is automatically encountered in the CU factors. A good example of this can be seen when we examine the CU data of a 2 × 4-foot luminaire with four lamps with those for one with two lamps. The fewer the lamps, the higher the CU. Based on these facts, the TF_A of a luminaire can be assumed to be unity if the luminaires used in reality and those used in calculations are the same. The designer should always insist on the photometrics of a luminaire with the exact type of lamps and ballasts that he intends to use. Manufacturers will submit such a report, usually free of cost to the designer, as an investment towards getting the order.

A TF due to mounting methodology (TF_B) must be taken into account, since the changes in operating characteristics of some luminaires can be substantial, and not reflected in the available photometric report. It is important to remember that all photometric tests of luminaires are carried out in open air at 77°F ambient and do not take into account how they are mounted. This factor will vary substantially from luminaire to luminaire with different mounting characteristics. While showing each instance is beyond the scope of this book, Table 4-6 is a useful table of such thermal factors that are related to most frequently used luminaires.

A change in operating characteristics as a result of a change in room ambient temperature can be substantial. However, to establish a fixed number of account for all the fluctuating reasons that vary from one luminaire to another and from season to season is almost impossible. For this reason, this factor, TF_C, may be assumed to be unity for most practical purposes.

All factors shown in Table 4-6 are determined from practical experiments and readings taken after the light output has stabilized. In all three cases, maintained ambient temperature is 77°F. For recessed luminaire, plenum temperature is approximately 95°F.

Table 4-6. Thermal Factor, TF_B, for Suspended, Recessed, and Surface-Mounted Luminaires.

				Thermal Factor, TF_B		
Key	Ballast	Lamps	Ballast Factor, BF	Suspended Luminaire	Recessed Luminaire	Surface-Mounted Luminaire
A	Standard	Standard	0.95	1.0	0.92	0.85
B	Standard	Lite-white	0.895	1.0	0.95	0.90
C	Low-loss	Standard	0.935	1.0	0.92	0.87
D	Low-loss	Krypton	0.87	1.0	0.96	0.92
E	Low-loss	Lite-white	0.875	1.0	0.98	0.95
F	Wave-modified low-loss	Lite-white	0.95	1.0	0.97	0.92

Suspended Luminaire: Ambient temperature = 77°F

Recessed Luminaire: Ambient temperature = 77°F; Plenum temperature = 95°F

Surface Mounted Luminaire: Ambient temperature = 77°F

Note that in determining the effect of TF on light output, the net TF must be introduced in lighting calculations, either in terms of one of the nonrecoverable factors towards constructing the LLF or as a factor to modify the CU value. Although, according to the definition of LLF, it is categorized as a part of the LLF, the TF will be automatically encountered in the CU value if the photometric tests were done with appropriate mounting conditions. In determining the effect on input power, these factors must be multiplied by the bench-test figures published by manufacturers. A detailed technique for implementing these items in calculations, is shown in the design procedure and example.

Design Procedure. In order to show basic techniques for lighting design, two separate examples will be given to generalizc about the two most common situations of commercial interiors: (1) areas where task locations are not known and (2) areas where task locations arc known.

Commercial Interiors, Task Locations Not Known. Speculation-type office buildings with large open spaces are the typical examples of this type of interior. For years, lacking advance knowledge of task type and locations, designers have been compelled to distribute even illuminance throughout the room. This, along with the spacing limitations set by a ceiling grid, resulted in a footcandle level exceeding requirements. Earlier attempts to avoid energy wastage by reducing the lighting level failed, since this resulted in a sharp decline in productivity. It is important to remember that the main goal of energy saving lighting design is to provide adequate lighting with least consumption of power.

In the midst of ever-changing tenants' requirements, the most practical solution for an open space is still to provide an even but reasonable amount of lighting throughout the room for the minimum amount of power. With the wide variety of low-power-consuming products available today, the design can now be fine-tuned to a specific requirement, particularly for a lower level of footcandles–this was not possible in the past. Good judgment and proper selection of such products will provide adequate lighting, maintain productivity, and yet consume the least amount of power.

A wide range of light output with a wide variance in input-power consumption is accomplished with various combinations of low-power-consuming and standard lamps and ballasts. In general, their outcome can be loosely divided into three groups:

1. Reduces light output, consuming less power
2. Maintains same light output with less power
3. Increases light output, with less power

Note that while all three above are suitable for retrofits, the third is particularly suitable for newer applications, where the adequacy of lighting is to be accomplished with least consumption of power. Figures 4-9 and 4-10 illustrate the comparative operating characteristics of six combinations of these products when equipped in a grid-suspended troffer and a surface mounted wraparound respectively. To simulate the average commercial interiors, the troffer selected is a 2 X 4-foot, static type, equipped with four lamps and two ballasts and an

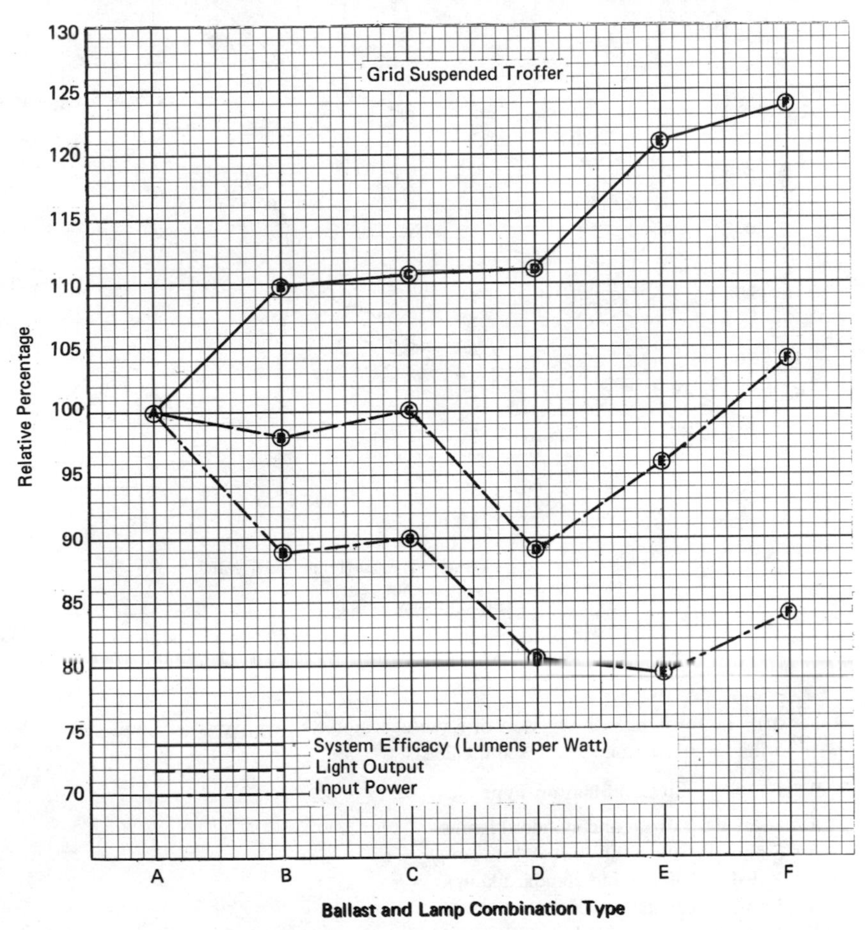

Fig. 4-9. Performance comparison of a grid suspended troffer equipped with various combination of lamps and ballasts.

Ballast and Lamp Combination Type.

A = Standard ballasts and standard lamps
B = Standard ballasts and Lite-white lamps
C = Low-loss ballasts and standard lamps
D = Low-loss ballasts and Krypton lamps
E = Low-loss ballasts and Lite-white lamps
F = Wave-modified low-loss ballasts and Lite-white lamps

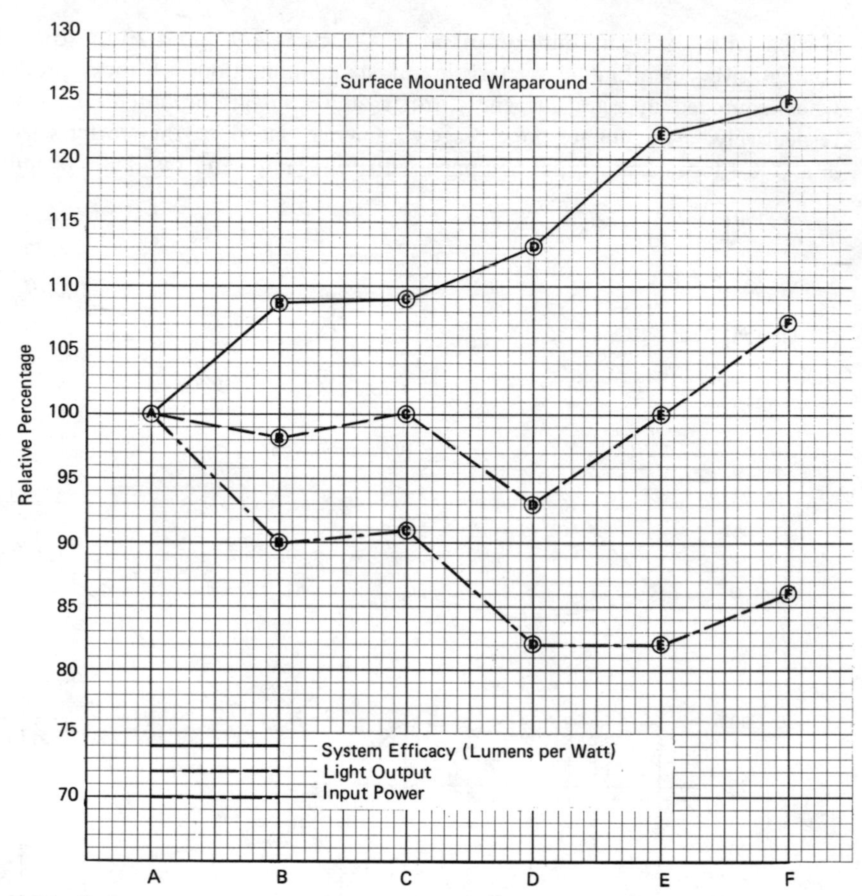

Fig. 4-10. Performance comparison of a surface mounted wraparound luminaire equipped with various combination of lamps and ballasts.

Ballast and Lamp Combination Type.

A = Standard ballasts and standard lamps
B = Standard ballasts and Lite-white lamps
C = Low-loss ballasts and standard lamps
D = Low-loss ballasts and Krypton lamps
E = Low-loss ballasts and Lite-white lamps
F = Wave-modified low-loss ballasts and Lite-white lamps

acrylic prismatic lens. Plenum is a non-air-return type, with an approximate temperature of 95°F. The surface mounted wraparound has four lamps, two ballasts, and acrylic-prismatic wrapped around lens mounted against a low-density ceiling with average insulation. In establishing the curves, all readings were taken in an ambient of 77°F after the light output stabilized. For the purpose of ease in reading and use, all readings are expressed in terms of relative percentage of a standard system, as shown by combination A.

Note that use of the curves in the two figures are for comparative study and preliminary selection only. A number of factors, such as the luminaire's design quality, the ambient conditions, draft conditions, and BF, will definitely alter the readings to some extent. In making a final selection, a calculation must be made and then compared with the preliminary selection. Let us take an example to illustrate this

Given: Room dimensions of an office: $L = 60$ ft.; $W = 60$ ft.; $H = 8.5$ ft.
Task height = 2.5 ft. above floor.
Illuminance required = 70 fc.
Ceiling type: 2 × 4-ft., grid-suspended.
Surface reflectances: 80/50/20 (ceiling/walls/floor).
RCR = 1.0

In grid-suspended ceilings, where luminaire spacing arrangement ultimately dictates the actual illuminance level, usually it is more convenient to refer to curves like those shown in Figures 4-11 through 4-14. These curves, having drawn against RCRs, directly offer the various maintained illuminance levels for various room dimensions and layout. Although a wide range of such curves is possible, usually it is sufficient to draw a few curves that represent the average

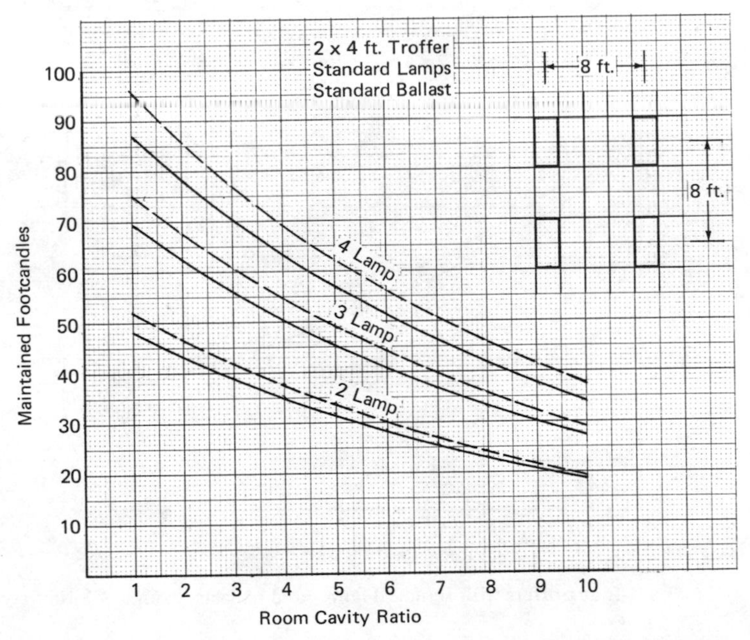

Fig. 4-11. 2 × 4-foot troffers with standard lamps and ballasts, in an 8 × 8-foot layout.

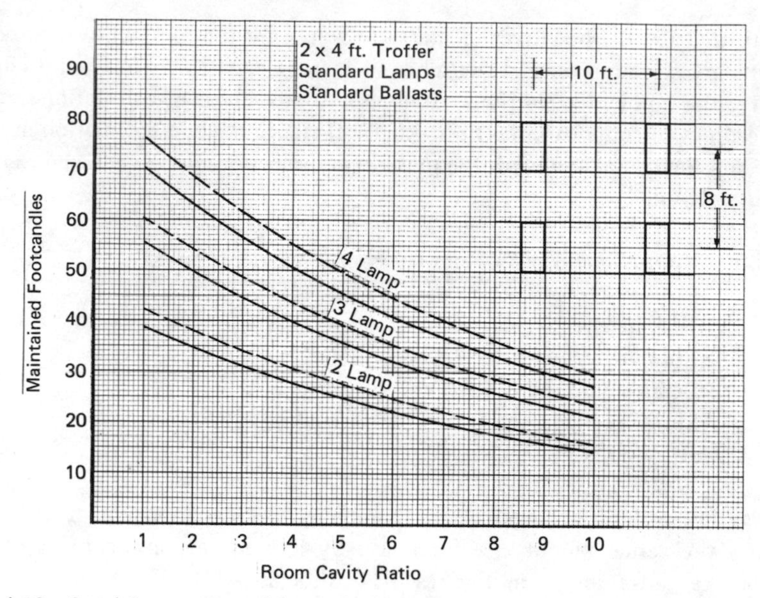

Fig. 4-12. 2 × 4-foot troffers with standard lamps and ballasts, in an 8 × 10-foot layout.

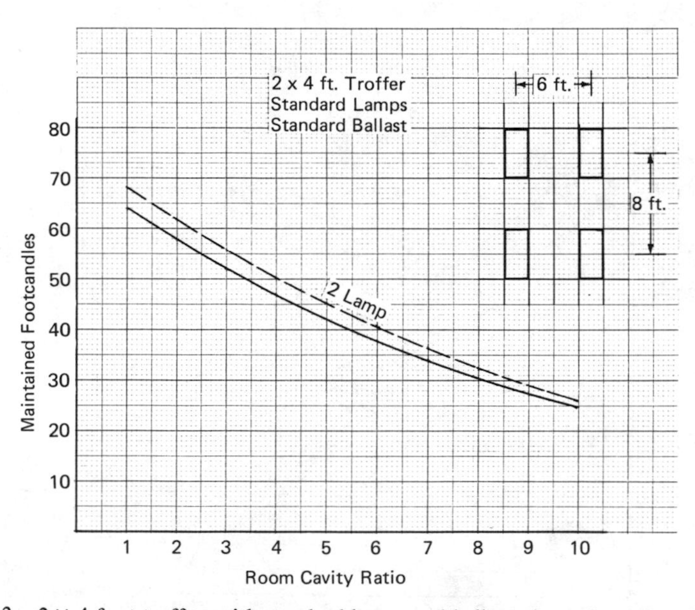

Fig. 4-13. 2 × 4-foot troffers with standard lamps and ballasts, in an 6 × 8-foot layout.

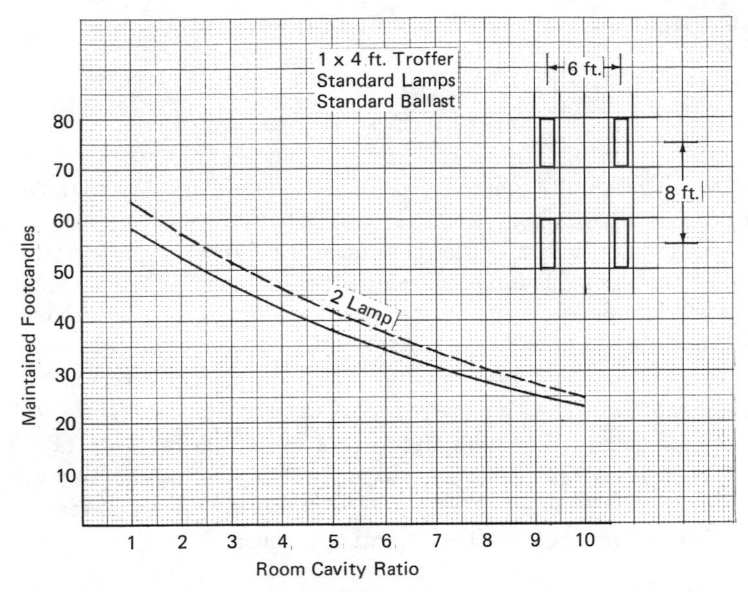

Fig. 4-14. 1 X 4-foot troffers with standard lamps and ballasts, in an 6 X 8-foot layout.

situations, e.g., 2 X 4-foot luminaires laid in 6 X 8-foot, 8 X 8-foot, 8 X 10-foot ceiling spacings. This will be pretty good start for a preliminary selection. A final selection can be made reviewing all necessary factors involved, in a step by step order. Note that all light outputs are expressed by broken lines and solid lines--representing the initial (before thermal effect) and stabilized (after thermal effect) and stabilized (after thermal effect) conditions respectively. For an accuracy in result, all calculations must be based on the solid line readings since this represents the stabilized light output which encompasses the effects of lamp cavity and ceiling space thermal conditions.

Working with the requirements of 70 fc, at RCR = 1 of our example, a review of Figures 4-11, 4-12, 4-13 and 4-14 shows the following possibilities.

Case 1: Use of 2 X 4-foot luminaires with four lamps in an 8 X 10-foot layout (Figure 4-12)

Case 2: Use of 2 X 4-foot luminaires with three lamps in a 8 X 8-foot layout (Figure 4-11)

Case 1. Observing Figure 4-12, note that starting with an Illuminance of 76 fc at initial condition, it settles at 70 fc when stabilized. Although a maintained 70 fc can be produced by this arrangement, the biggest obstacle is the uniformity, that requires a minimum S/MH ratio equal to:

$$\frac{10}{(8.5 - 2.5)} = 1.67$$

A satisfactory uniformity cannot be achieved with the standard lenses, since their S/MH ratio usually lies between 1.2 to 1.4. There are some specially designed lenses available whose S/MH ratio varies up to 1.8. These are, however, substantially more expensive and also produce direct glare, reducing VCP considerably.

The power consumption at the stabilized condition can be calculated as follows. From manufacturer's catalog data, power consumption for four standard lamps with two standard ballasts is 96 + 96 = 192 watts. However, this is the bench-test result only, and it must be modified to account for the thermal effect at a stabilized condition by multiplying by a TF. From Table 4-6 $TF = TF_A \times TF_B \times TF_C = 1 \times 0.92 \times 1 = 0.92$. So the net power consumption per luminaire in stabilized condition is $192 \times 0.92 = 176$ watts. Thus power density in an 8×10-foot layout is 176/80 = 2.2 watts per square foot.

Case 2. The minimum S/MH ratio in this system required is 8/(8.5 - 2.5) = 1.33. This is within the reach of the S/MH ratio provided by a standard acrylic prismatic lens. Note that as shown in Figure 4-11, starting with a footcandle level of 75 initially, the stabilized illuminance level settled at approximately 70 fc, which serves our purposes adequately. The power consumption with this arrangement at stabilized condition can be found by multiplying the bench-test data by the TF, as shown earlier.

The published data of a twin-lamp ballast and a single-lamp ballast is 96 and 53 watts, respectively. So the net power consumption at stabilized condition is $[(96 + 53) \times 0.92]/64 = 2.14$ watts per square foot. Between the two choices, from an energy-saving standpoint, obviously Case 2 is preferable to Case 1. Initial cost, however, will be more in Case 2, since it will require an approximate 80/64 = 25% more luminaires than Case 1, but some of this cost will be offset due to the fact that special lenses must be used in Case 1 for uniformity.

We must investigate the use of lower-watt-consuming lamps and ballasts in achieving the same illuminance level with possibly, the least amount of power. To make a preliminary selection, we refer to Figure 4-9, which must be verified by calculations later. The first selection, obviously, will be combination F, followed by E. Combination F produces approximately 4% more light and improves system efficacy by almost 20% over the standard system (combination A). With combination F as our choice, we must select the luminaire type and spacing that will produce the required 70 fc with minimum power consumption overall. In other words, to investigate the right luminaire we will review Figures 4-11 through 4-14 again, looking for a maintained illuminance of 70/1.04 = 67.3 fc.

A quick review shows two possibilities:

Case 3: Use of 2 X 4 ft. luminaire with two lamps in a 6 X 8-foot layout, (Fig. 4-13), and

Case 4: Use of 2 X 4 ft. luminaire with three lamps in an 8 X 8-foot layout, (Fig. 4-11).

Case 3: As can be seen in Figure 4-13, starting from a value of 70 fc the illuminance level finally settles at 66 fc at stabilized condition–which is quite close to the required 67.3 fc mentioned earlier. With the use of combination F, which consists of lite-white lamps and wave-modified low-loss ballasts, the expected illuminance level will increase to 66 X 1.04 = 68.64 fc.

Expected power consumption for this system at stable condition can be found from Figure 4-9. Consumption = 84% of power consumed by standard system at stable condition = 0.84 X (96 X 0.92) = 74.18 watts. Power density with this system for a 6 X 8-foot layout = 74.18/48 = 1.54 watts per square foot.

Case 4. Reading directly from Figure 4-11, the stable illuminance level of a standard three-lamp is 69 fc. With wave-modified low-loss ballasts and lite-white lamps, this value is expected to increase to 69 X 1.04 = 71.75 fc, which is slightly higher than the required 70 fc. The power consumption under this arrangement will be lower and can be calculated directly from Figure 4-9. Consumption = 84% of power consumer by standard system at stable condition = 0.84 X [(96 + 53) X 0.92] = 115.14 watts, or 115/64 = 1.8 watts per square foot. The advantage of the use of lower-watt-consuming products is obvious. Both cases 3 and 4 offer the required 70 fc, with considerably less power than that consumed by Cases 1 and 2 using standard lamps and ballasts.

Discussion: Although from an adequacy in lighting, uniformity, and energy-saving standpoint, Case 3 shows a favorable result, it may require a substantially increased initial cost because of the higher number of luminaires involved. Compared to Case 4, the total number of luminaires increases by 64/48 = 34%. The cost difference between a 2 X 4-foot luminaire is equipped with three lamps and the same with two lamps is as low as the price of one lamp and one ballast (which is about $6), and this will usually discourage the owner. All points considered, Case 4 is the preferred solution, since any higher initial costs due to special lamps and ballasts, can be recovered within a short period of time by the amount of energy saved.

In order to cut initial cost and yet comply with the two-lamp units in a 6 X 8-foot layout, one possibility is to use 1 X 4-foot luminaires instead of 2 X 4-foot ones. This is shown in Figure 4-14. Note that although cost is about 75% of that of 2 X 4-foot ones, these luminaires do not offer as good efficiency as the

2 × 4-foot ones, primarily because of the relatively higher thermal effect. Although both luminaires have two lamps each, because of the volume difference, the 1 × 4-foot luminaires retain more heat inside. Manufactured by the same company, with same type of material and surface reflectances, these luminaires are approximately 92% as efficient as the 2 × 4-foot counterparts. Note that, corresponding to RCR = 1, the maximum stabilized illuminance with this system is about 60 fc as seen in Figure 4-14. What this means is, even with the use of wave-modified low-loss ballasts and lite-white lamps, the maximum light output will be around 60 × 1.04 = 62.4 fc, which is much too low for our purpose. Needless to mention, if the illuminance requirement was, say 60 fc, the total outlook in final luminaire selection may be different.

Calculations and Layout. With Case 4 as our solution, the calculations for determining the number of luminaires and power consumption can be done as follows:

Given: Room dimensions: $L = 60$ ft; $W = 60$ ft; $H = 8.5$ ft
Surface reflectances = 80/50/20%
Luminaire type = 2 × 4-foot troffer, acrylic prismatic lens, three lite-white lamps and two wave-modified low-loss ballasts
Spacing layout = 8 × 8-foot center to center
CU table is shown in Table 4-7, which is developed with the required lamps and ballasts.

Corresponding to RCR = 1, and surface reflectances = 80/50/20, from Table 4-7, the required CU is equal to 0.78.

In determining LLF, thermal factor TF and ballast factor BF must be introduced as the non-recoverable factors. From Table 4-6, these values are 0.97 and 0.95 respectively. Assuming the recovering factors for the depicted example equal to 0.7,

$$\text{LLF} = 0.7 \times 0.95 \times 0.97$$
$$= 0.645$$

Using these values in the lumen formula,

$$\text{Number of Luminaires} = \frac{\text{A} \times \text{E}}{\text{L} \times \text{CU} \times \text{LLF}}$$
$$= \frac{60 \times 60 \times 70}{(3050 \times 3) \times 0.78 \times 0.645}$$
$$= 55 \text{ luminaires.}$$

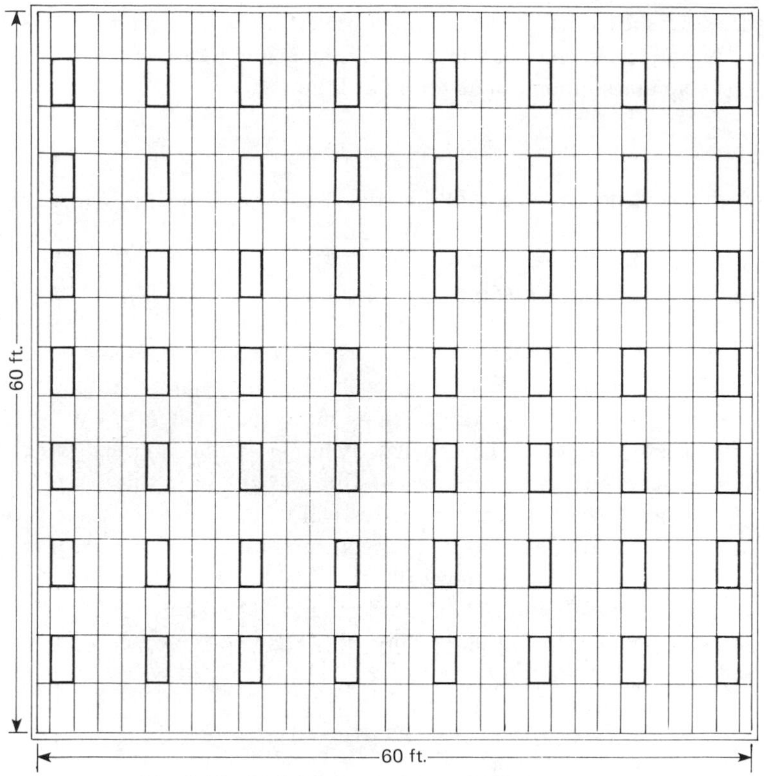

Fig. 4-15. Lighting layout of the depicted example.

Table 4-7. CU Table of a Troffer with Three Lite-White Lamps and Two Wave-Modified Low-Loss Ballasts.

% Effective Ceiling Cavity Reflectance	80				70				50			30			10			
% Wall Reflectance	70	50	30	10	70	50	30	10	50	30	10	50	30	10	50	30	10	0
1	81	78	76	73	79	77	75	72	73	71	70	70	69	68	68	66	65	64
2	74	70	66	63	73	69	65	61	65	62	60	63	61	59	61	59	68	56
3	69	62	58	54	68	62	57	54	59	56	52	58	54	51	56	52	50	49
4	64	57	51	46	63	56	51	46	54	49	46	52	48	45	50	47	45	43
5	59	51	45	41	58	50	44	39	48	44	39	47	43	39	46	42	38	37
6	55	47	40	36	54	45	40	35	44	38	35	43	38	35	42	37	34	33
7	50	43	36	31	50	41	36	31	39	34	31	38	34	31	37	33	31	29
8	46	38	31	28	45	37	31	27	35	31	27	35	30	27	34	30	27	25
9	43	34	28	24	42	33	28	23	32	29	23	31	27	23	31	27	23	22
10	39	31	25	22	39	30	24	21	29	24	21	29	23	20	28	23	20	10

% Effective Floor Cavity Reflectance = 20.

Figure 4-15 shows the layout of the complete system. Note that to maintain symmetry, the total number of luminaires is increased to 56.

The power consumption can be found as follows:

Total power consumption at stabilized condition:

= bench-test values (catalog published data) $\times$ TF $\times$ number of luminaires
= (78 + 45) $\times$ 0.97 (from Table 4-6) $\times$ 56
= 6681.3 watts
Power density = 6681.3/3600 = 1.85 watt per square foot

It is important to remember that two single-lamp ballasts consume more power than one two-lamp ballast. For instance, the ballast operating the single lamp in our example consumes 45 watts which is more than half of 78 watts for the two-lamp ballast. This can be avoided by using all two-lamp ballasts only arranged in a master and slave or master-satellite service. What this basically means is all six lamps of two adjacent luminaires will be connected by three two-lamp ballasts instead of the conventional arrangement with a combination of two two-lamp ballasts, and two single-lamp ballasts. The master unit will accommodate two ballasts and the satellite will have the other. This is shown in Figure 4-16.

Since each two-ballast consumes only 78 watts and the total number of ballasts will be 56 $\times$ 3/2 = 84, the net power will be:

$$84 \times (78 \times 0.97) = 6355.4 \text{ watts, or}$$

$$6355.4/3600 = 1.76 \text{ watt per square foot}$$

Factors like VCP and ESI must be considered on an individual application basis. Luminaires with parabolic louvers or with special lenses producing high ESI are substantially more expensive and should be used for specific applications when task types and locations are known.

The techniques of lighting design discussed earlier have been limited to the use of conventional types of luminaires with prismatic lens since these are the most economically feasible and frequently used units found in most commercial interiors. Although quality aspects of lighting design are important in most commercial applications, selection of products based on statements like "X-luminaire produces maximum ESI or Y-luminaire produces maximum VCP" should be avoided. Even if a product does produce what is claimed, a blind application of the product may be a sheer waste of power and dollars since it may not be suitable for the user's type of task and sitting locations. For most applications, obtaining a glare-free, good-quality light may not be any more trouble than holding the task at a different angle or moving the desk or luminaires around by 90°.

The increasing need for closer corelations between anticipated and actual il-

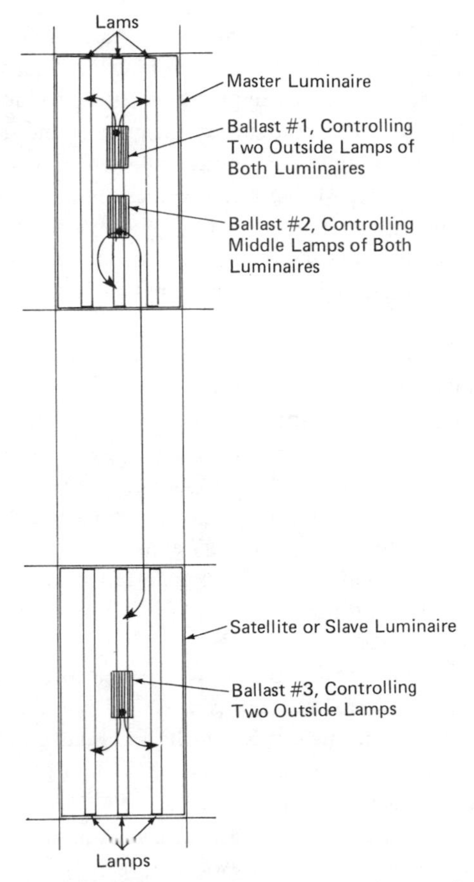

Fig. 4-16. Two 3-lamp luminaires connected by three two-lamp series ballasts. This eliminates the use of two single-lamp ballasts and saves energy.

luminance levels does not mean that assumptions will not continue to be made. As discussed earlier, fluorescent luminaires can be subjected to widely varying ambient temperatures, depending upon type and application. On the average, for static luminaires, the ceiling-cavity temperature is usually between 10 and 20°F higher than the room ambient. Temperatures in return air plenums sustain a closer gap since much of the lamp heat is carried away, resulting in a significant change of light output and input power. This is discussed in detail in Chapter 8. Restrictions on heating- and cooling-maintained temperatures, type of luminaire, room-surface reflectances, ambient dirt conditions, the number of hours of operation, and cleaning intervals are all subject to change and have an impact on overall luminaire performance. Assumptions are unavoidable, but well-communicated and carefully considered assumptions will bring calculated results closer to reality.

Commercial Interiors, Task Locations Known. A task-oriented lighting system will be energy effective if it aids in accomplishing productive results. As discussed earlier, for most commercial applications, the productive results are influenced by the visibility of the task and by the glare in the field of view, which may cause discomfort and slow visual-task assimilation. Three important factors of consideration for an energy-effective, task-oriented lighting system are (1) comparing the amount of consumed power, (2) comparing the relative visibility of the task in terms of ESI, and (3) comparing visual comfort characteristics in terms of VCP. Task-oriented lighting system can be accomplished by two general means:

1. A ceiling-mounted, task-oriented lighting system.
2. A task/ambient lighting system.

Ceiling-Mounted, Task-Oriented Lighting System. This system has luminaires precisely located on the ceiling to provide adequate lighting on the task in terms of good visual comfort (heads-up position) and good visibility of task (heads-down position). The different types of task-location arrangement in a room can be loosely divided into three groups:

1. Single location. Small office rooms with single work stations are typical examples of this kind. These rooms are usually within 16 feet in dimensions, lengthwise or widthwise. Since the luminaires are located above or away from the field of view (heads-up position), a high VCP is not as important as having high ESI values in these conditions. A high ESI can be accomplished by considering a minimum of two luminaires in these rooms. The luminaires must be oriented away from the offending zones, so that there will be minimum veiling reflection and increased task contrast. Placing one luminaire at each side of the work station will cancel each other's shadow and will accomplish this (see Figure 4-17).

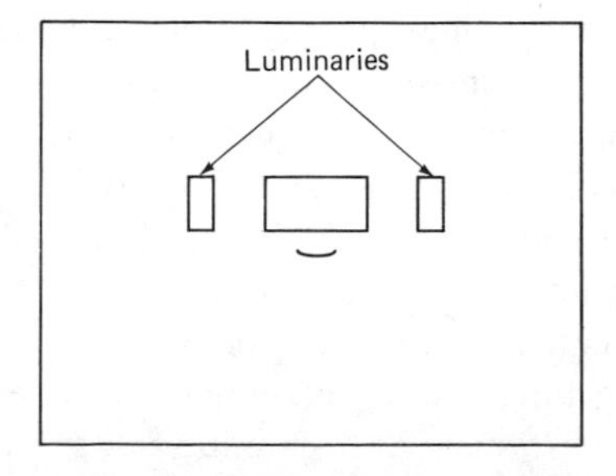

Fig. 4-17. A minimum of two luminaires should be considered. Placement of a luminaire in each side of the work station will cancel each other's shadow and increase task contrast.

2. Multilocation in symmetrical rows. Large rooms with several task locations placed in a predetermined symmetrical order fall under this category, e.g., institutional areas such as classrooms and libraries. If task locations are in a row, as shown in Figure 4-18, the best way to illuminate these areas is to place luminaires in between rows. Lamp length should be always parallel to the field of view, since fluorescent luminaires usually produce less glare lengthwise than crosswise. Placement of luminaires in each side of rows will produce good contrast in task. Luminaires of high VCP will be desirable in such applications, since the direct glare of luminaires is in the field of view. A high S/MH (SC) will make it possible to place luminaires far apart and possibly increase ESI, but this will increase glare sideways. It is thus important that the field of view be parallel to luminaires lengthwise.
3. Multilocation with no symmetry. These are the areas where a number of work stations exist but there is no specific symmetrical pattern. These are the most difficult areas to light, especially if the quality of light is the main concern. In order to get high ESI values for each location, a mirror may be used to position luminaires away from offending zones. However, this may create a haphazard unsymmetrical pattern on the ceiling, causing an unpleasing esthetic effect. The only practical means is to locate luminaires uniformly on the ceiling, which will produce an even illuminance level all over the room.

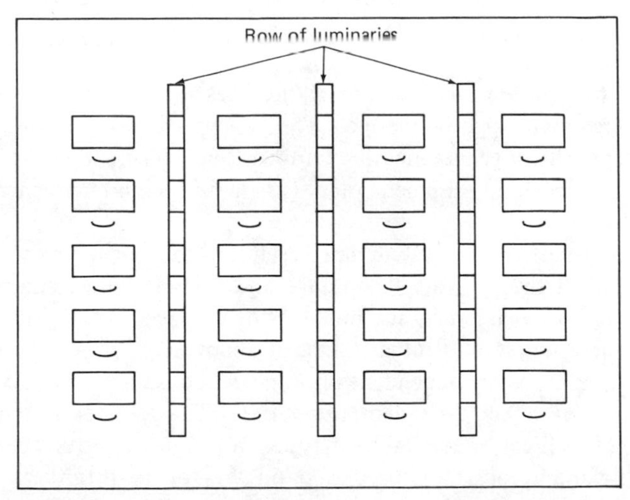

Fig. 4-18. If task locations are in rows, the best way to illuminate these areas is by placing luminaires in betwen the rows.

In order to receive maximum amount of light for less power, the guideline discussed earlier in this chapter should be observed. The main intent is to select a lamp–ballast combination that will serve this purpose best.

A precise amount of ESI value can be obtained only by trial and error. A preliminary luminaire layout is to be made from experience. The main consideration in this respect will be to locate luminaires away from the offending zones. After a preliminary layout has been made, the system should be analyzed with a minimum of two or three different types of luminaires, to determine which produces the closest amount of required ESI. If the readings are too low, a new layout must be made and tried again. Luminaires producing batwing-type distribution usually offer high ESI values, although this may not be true for all applications. Every application should be treated on an individual basis. A comparison of various types of luminaires for a particular layout is about the only means of selecting the highest-ESI-producing luminaires.

VCP is an important consideration, although its importance will vary on room dimensions. VCP values of some luminaires will improve as the room gets larger; for other luminaires, the relationship is just the opposite. For all practical considerations, the importance of high VCP usually increases with the size of a room and is the least important consideration for smaller rooms.

The basic problem of obtaining simultaneously high ESI, VCP, and efficiency of a lighting system lies in their opposing principles of operation. A high ESI would usually mean low VCP and vice versa, and any effort made to improve both these factors results in lower efficiency. In order to receive high values of VCP, ESI, and efficiency simultaneously, the lighting industry is now looking at other systems that are considerably different from the conventional ceiling lighting system. A task/ambient lighting system is one such consideration.

Task/Ambient Lighting. This is one of the latest developments and considerations in energy-saving lighting design. The concept basically relies upon providing lower amounts of general lighting (ambient) all over the room for noncritical purposes; and localized, supplementary adequate amount of lighting for task purposes.

The ambient part of this system can be either direct or indirect. For a direct type of ambient lighting, the techniques and methods of calculations can be done by conventional lumen method. Widely spaced, ceiling-mounted luminaires provide a low level of illuminance of about 20 to 30 fc. The selection of these luminaires is mainly dependent upon power consumption and VCP values—and least or not at all dependent upon ESI. The indirect system, however, makes use of indirect or semiindirect types of luminaires, either pendant from ceiling, self-standing on floor, or placed on the top of furnitures. Lamps are either fluorescent or HID (mercury vapor or metal halide). Available illuminance on the task is always less than that from direct lighting, since light bounces from the ceiling before arriving on the task. The net result is an evenly spread, diffused light with a lower illuminance level. From this sense, the efficiency of an indirect system is lower than that of a direct system.

The fluorescent indirect luminaires are usually installed in a tray of 3 to 5 inches depth, incorporated in the top of cabinets. The placement of the luminaires should be above the eye level to avoid direct glare. If the cabinet height is low and/or the source is visible, it is necessary to use low-brightness shielding to avoid the direct glare. This, however, may reduce luminaire efficiency to as low as 50%, depending upon the type of shielding used.

Indirect ambient lighting with HID sources is usually equipped with either mercury or metal-halide lamps. These may be built into the furniture or may be free-standing. For most applications, the lamps are no larger than 250 watts and fit into a space of approximately 15 × 15 × 10 inches. Some typical examples are shown in Figure 4-19.

There are various design considerations with indirect lighting units:

1. There should be sufficient distance between the ceiling and luminaires to avoid a concentration of light above the luminaires. Too close a reflecting surface will make the ceiling the dominating feature and may be esthetically quite unpleasing. Distribution of light should be well spread in a radial batwing fashion so that concentration of light at a confined area can be avoided. Such a distributing pattern can be accomplished by using either proper refractor or reflector or by using both. There should also be sufficient distance between the luminaires and the floor, so as to avoid direct glare. If the mounting height is low, louvers may be considered.
2. The ceiling surface should be made of diffusing matte material that is highly reflective. A matte, highly-reflective surface with slight texture will help in diffusing light. If an exposed, grid tee-suspended ceiling is used, the tees should be matte finish, preferably of the same color as the ceiling panels.
3. Select a luminaire with tempered glass top. A flat horizontal surface will aid in maintenance. A properly enclosed luminaire with tempered-glass top will avoid exposure to ultraviolet rays should a lamp shatter. The use of such a glass top however will reduce a luminaires' efficiency by approximately 10%. Use of clear Teflon may be considered; a 5 mil Teflon will reduce efficiency by approximately 5%. A sealed and gasketed luminaire with Teflon will avoid dirt accumulation inside, yet allow luminaire to "breathe."

The main purpose of using an indirect system is to produce uniform luminance across the ceiling area and uniform, diffused, shadow-free light throughout the room. This will increase visual efficiency. From the standpoint of esthetics and psychology, however, the effect may be boring and unpleasing. Selection of

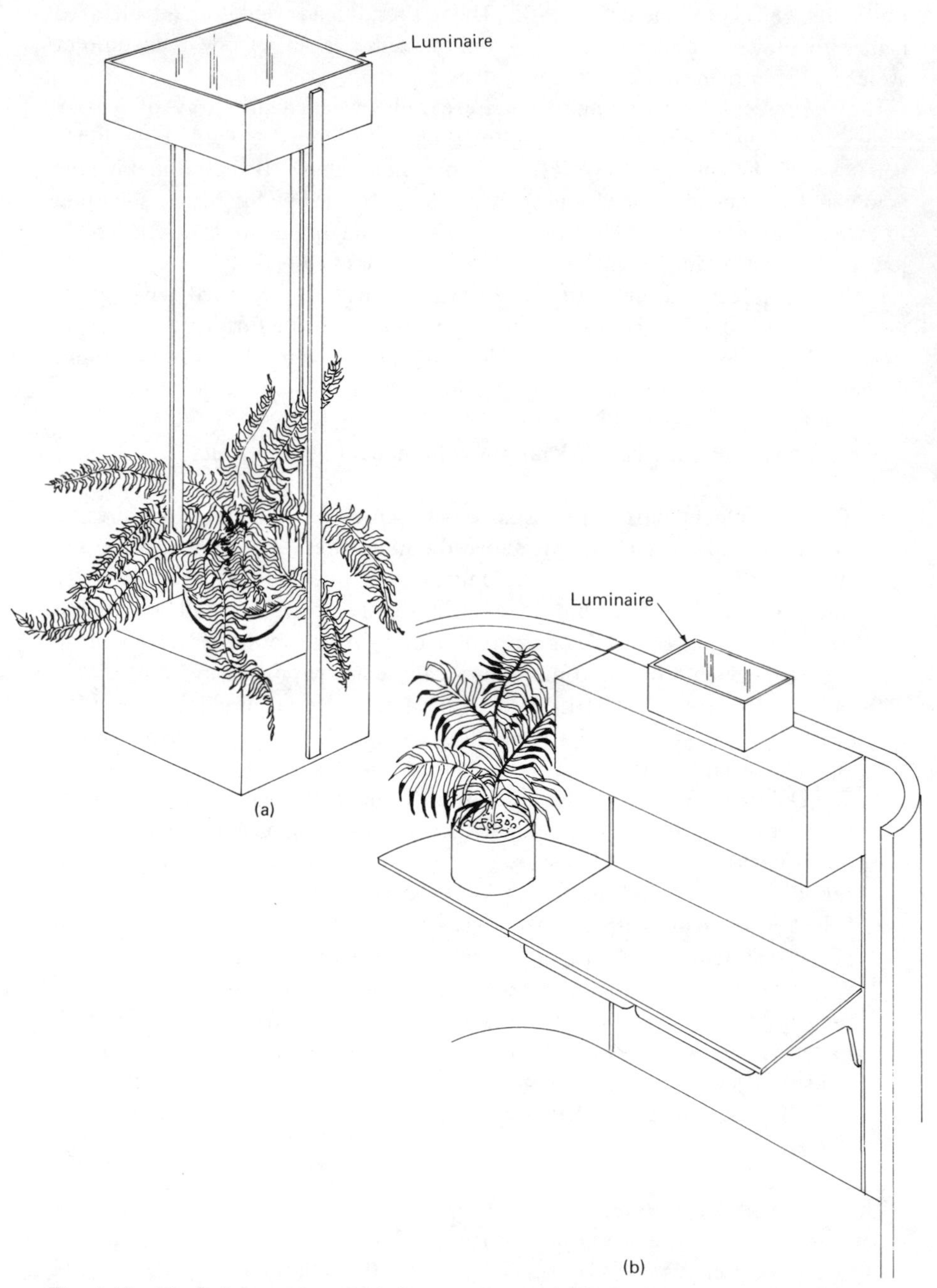

Fig. 4.19. Typical examples of (a) free-standing and (b) furniture-mounted, indirect ambient lighting luminaires.

furniture with pleasing colors, with accent lights highlighting pictures or decorative items, will provide visual relief from the boredom of the uniform diffused illuminance caused by indirect lighting systems.

The task-lighting part of a task/ambient system mostly uses fluorescent lamps that are mounted in the furniture. The enclosure usually has a conventional prismatic, plastic refractor or an open bottom. Others have specially designed refractors that produce batwing distribution. While the location of most lamps is directly above the user and parallel to the task, others are side mounted, or there may be a combination of the two. Figure 4-20 shows some typical examples.

Furniture-mounted task lighting should be properly designed, with equal importance given to esthetic and technical aspects. The following are some considerations for a good technical performance:

1. The task area should have even and uniform light. If the lamp is located towards the rear, it is important that proper refractors and/or reflectors are implemented to produce an equal amount of light towards the front and rear; this must be done, however, with minimum veiling reflection. The use of open-bottom enclosures will inevitably produce a large amount of light directly underneath, making the rear the dominant feature of the task area. The use of conventional prismatic plastic refractor will diffuse light, but it still may produce high veiling reflections. A better result can be accomplished with the use of batwing-type distributing refractors. A proper placement of such refractor will tend to produce even lighting with a minimum of veiling reflections. Side-mounting task lighting usually causes a large amount of light at the two sides and provides a higher overall illuminance level than is required. For most single-person task systems, use of one lamp is enough.
2. High luminance contrast between the task and the immediate furniture-surface area should be avoided and limited to a 3:1 ratio. Most tasks are white papers of approximate 85% reflectance, so the reflectance of furniture color should not be any lower than 25%. If the luminaire is located at the rear and light spills on the back, vertical surface of the work station, it is important that this surface be uniformly lighted and produce luminance within the 3:1 ratio.
3. From the standpoint of ease in maintenance, there should be easy access to lens, louvers, or lamp removal. The plastic selected should be acrylic as opposed to polystyrene to avoid yellowing. The luminaire finish is important in that if it is glossy, it will be highly reflective; if too porous, it will hold fingerprints. A semigloss finish is usually the best solution.
4. The luminaire should be carefully concealed so that it will not produce direct glare.

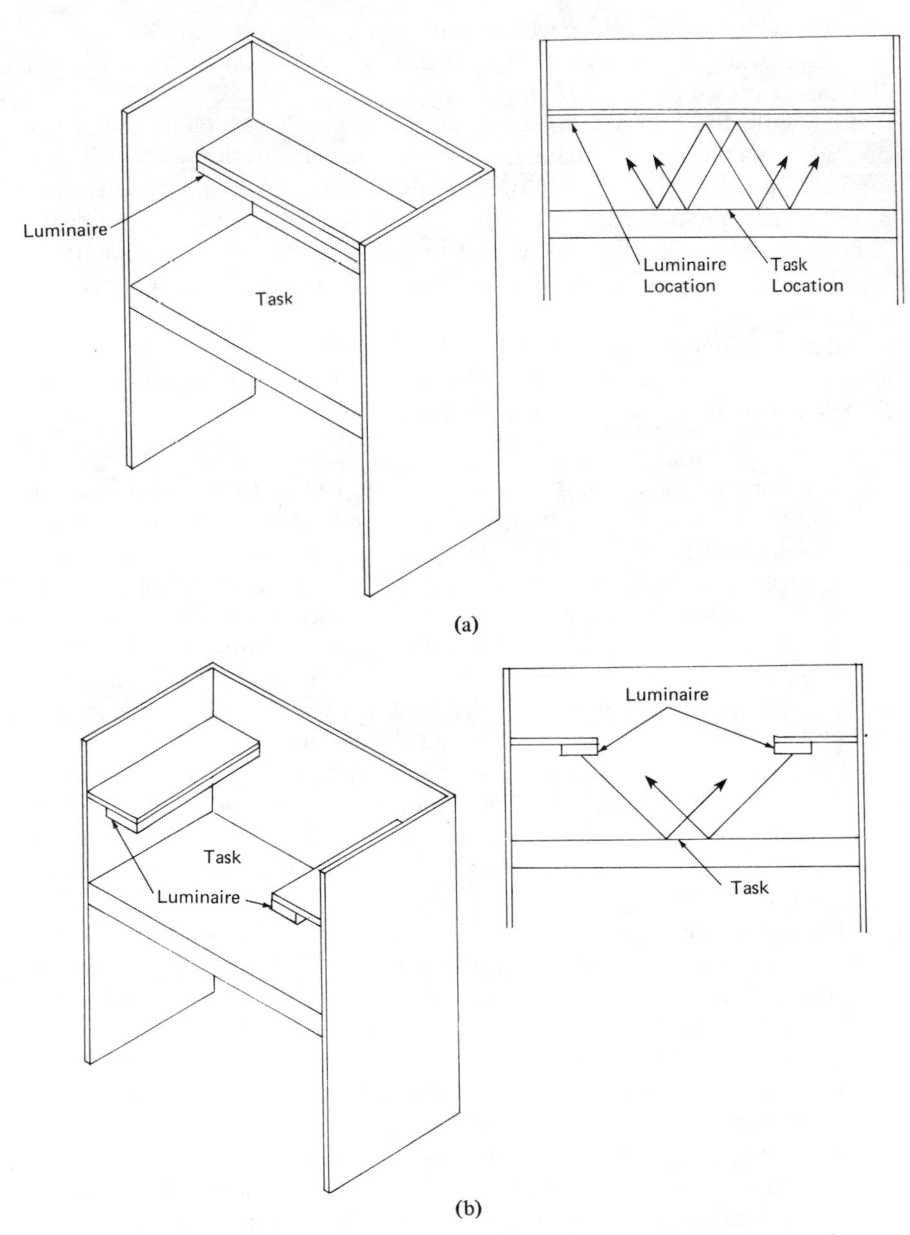

Fig. 4-20. (a) A batwing distributing will redirect reflected rays away from offending zones. (b) Light rays arriving from two lamps, one at each side, will provide a large amount of light on the sides and less in the middle. Lighting on desk must be uniform and no more than required.

5. Ballasts that produce least noise should be selected. Furniture-integrated luminaires have ballasts that are too close to user. Any ballast noise under these circumstances will be louder than with the ceiling mounted system. All wiring must be concealed and must comply with applicable electrical and other standards. Total assembly should bear a U.L. label. The overall system that is selected should not have exposed sharp edges or corners that can damage the electrical cord and cause a hazard.

REFERENCES

Barnes, J. T. "Plastics–Pros and Cons for Outdoor Lighting." *Lighting Design & Application*, December 1972, pp. 29–32.

Burkhardt, W. C. "High-impact Acrylic for Lenses and Diffusers." *Lighting Design & Application*, April 1977, pp. 12–20.

Helms, R. N. "Luminaire Selection and Design–An Alternative to Manufacturer's Catalogs." *Lighting Design & Application*, August 1974, p. 26.

IES. "Light Control and Luminaire Design." *IES Lighting Handbook, 1981 Reference Volume*, Sect. 6. pp. 6-1 to 6-33.

Luminaire Selection and Specification, Electrical Design Library, NECA, June 1980.

McGowan, T. K., and A. L. Hart. "The Application of Reduced Wattage Fluorescent Lamps and High Efficiency Ballasts to General Lighting Systems." *IEEE*, 1979.

Sherwood, M. "Aluminum Reflectors: Improved Characteristics." *Lighting Design & Application*, December 1972, p. 34.

Shmitz, Sylvan R. "Evaluating the Quality of Task/Ambient Lighting." *Lighting Design & Application,* January 1979, p. 25.

Sorcar, Prafulla C. *Rapid Lighting Design and Cost Estimating.* New York: McGraw-Hill, 1979, Chs. B and C.

Chapter Five
Industrial Lighting

The techniques and basic considerations behind an industrial lighting design are considerably different from that of a commercial area. Industrial areas mainly deal with manufacturing, production, or storage types of operations, so more attention is given to the luminaires' functional values than to their esthetics. Most industrial areas usually have high ceilings and relatively dirtier environmental conditions. The level of lighting also varies substantially. A warehouse is basically a huge storage area, so a lighting level of 5 fc may suffice; the level in manufacturing plant may exceed 100 fc, on the other hand, and additional task lighting may still be needed, depending on the degree of fine work to be done. High illuminance level requirements, compounded with high ceilings and relatively lower-reflecting room surfaces, require more luminaires for the same number of footcandles, when compared to a typical commercial application. Consequently, energy consumption also rises by the same proportion.

The main difference between an industrial lighting design and a commercial application is probably the ambient environment condition. Depending on the type of work performed, an industrial area can be fire hazardous, corrosive, humid, hot, wet, etc., and its air can carry a range of pollution, from plain dirt to electrostatically charged contaminents, or simply dirt mixed with vapor of oil or water. Each of these conditions would have a serious impact on the selection of the luminaires. The luminaires selected, thus, will have to have special design features that will resist foreign attacks and yet provide all the light required.

Energy saving is a relative term. How much energy a lighting system will save can only be evaluated when comparison is made to another system with the same amount of light. For a retrofit application, such comparison can be made very easily. However, for a new installation an energy-saving industrial lighting system would be one that would resist the hostile environmental conditions the best and provide the required amount of light with least consumption of energy.

A light source that satisfies all other requirements of an industrial application and has the best efficacy obviously has the most energy-saving potential. However, when the source is enclosed inside a luminaire, the net result may not be the most energy saving anymore. A lamp can be only as efficient as the luminaire, and the luminaire is only as good as the quality of its components and the kind of engineering know-how that has gone into its design. In this chapter we will analyze the different design aspects, physical properties of components, and operational characteristics that affect the energy efficiency of an industrial luminaire. We will also show a quick and reasonably accurate method for selecting a source and type of luminaire that may be most energy saving for an application.

LUMINAIRES

Different types of luminaires are available for various types of industrial application. However, they are all created to provide electrical connections to the lamps and to provide a controlled lighting distribution suitable for specific applications. The basic laws of light control include that of reflection, refraction, and diffusion (see Chapter 4).

Reflector Function and Design

The light distribution of a bare lamp is seldom suitable for direct industrial application. A reflector is used around the lamp to redirect the light to necessary areas. The reflector, in essence, acts as another light source. This can be seen in a polished reflector, where the image of the source is somewhat visible. The design of the reflector thus should be such that the luminaire can make best use of the lighting distributions of the bare lamp and the reflector image at the same time. Some industrial luminaires are designed so that the lamp can be installed in different socket positions in the same lamp axis, under the same reflector. Changing lamp position provides a change in lighting distribution characteristics. A lower position of the lamp would usually indicate wider beam spread, whereas, a higher position would tend to narrow the beam. Obviously, the range of such beam spread is limited by the reflector dimensions, and a wider range of beam spread would require a new luminaire with a different reflector size. Each industrial luminaire manufacturer thus has a series of different dimension luminaires to provide a complete range of beam spread for different types of lamps and wattages.

The reflector can be smooth, specular finished, or rough. The candlepower and hence the amount of light distributed from the luminaire, is much dependent on the reflector design, material, and the type of finishes that have gone into it. A specular reflector can be highly efficient if it is precisely designed for a particular lamp position and operated with a clear or phosphor-coated lamp. If not precisely designed, use of clear arc source in a highly polished reflector can result in streaks and uneven rings of light, especially when lamp position is changed. Important to note that a highly polished or specular-finished surface can reflect the image of the light source back on to the arc tube, causing adverse effects on its operating characteristics. A high-pressure sodium lamp, for instance, would have reduced life under such circumstances.

When, because of its reflector position, the source reflects its image back on the arc tube, energy is redirected to the arc tube. With an HPS, this accelerates the increase in voltage drop across the arc tube even faster than by its normal aging process. Luminaire manufacturers normally do not mention how much redirected energy their units may induce this way; however, a concerned designer would investigate the loss expected. This artificial aging reduces the lamp life substantially. The end of the lamp is apparent when it shows the "cycling" characteristics. The lamp starts but it does not stay lit since the voltage needed to sustain the arc is higher than the amount ballast can supply. The lamp lights up each time the voltage matches that supplied by the ballast and then goes off as it exceeds. The frequency of the cycle generally starts every hour or so and gradually shortens to only a minute or less.

A lighting distribution from a specular-finished or highly polished reflector would usually mean a higher utilization factor. However, because it may cause brightness, streaks of light rings, and other disadvantages mentioned above, some degree of diffusion is usually added to all reflectors. Treating, etching, or coating the surface in some manner would bring diffusion. Some reflectors have two materials bonded together; some have a sheet of glass with infinitesimal prisms; others are painted. Flat paint and other matte finishes such as white plaster or terra-cotta are also typical examples of such diffuse reflecting surfaces. Enamels are organic pigmented coatings that are applied for protection, decoration, and reflectance. They cure by oxidation, by means of air or forced drying, or by polymerization by means of baking or catalytic action. This results in a very tough finish that protects the reflecting surface from scratches and other physical damage. Some luminaires have ceramic coatings, such as porcelain enamels that are fired onto the metallic surface at temperatures in excess of 1000°F. This results primarily in high resistance to corrosion, good diffused reflectance, and easy maintenance, but there is poor impact resistance. A list of reflecting materials and their reflectances is given in Table 4-1 in Chapter 4. Obviously, each of these reflecting surfaces and finishes would result in a different pattern of light output and affect the efficiency to a certain degree.

If painted, luminaire efficiency and utilization factors vary with the degree of its reflectance. A highly reflective painted surface usually means higher utilization. On the average, a 1% rise in painted reflectance will result in 0.9% rise in CU factors. Large variations in the quality of products from one manufacturer to another may result in as much as 15% variance in luminare efficiency with the same reflector, design, and material. In selecting a luminaire, the designer

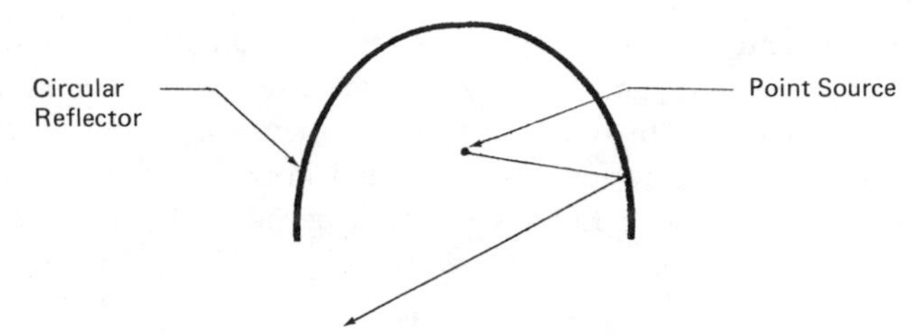

Fig. 5-1. A point source is easier to control.

should always request specific product information from several manufacturers and then compare.

Circular reflectors are ideal for small, compact sources. All high-intensity discharge sources have arc tubes that are between 2 and 6 inches in length, and their intensity varies throughout the length. A long arc tube, as can be imagined, would spread the light much wider than a point source. This is because each point in the arc source would act as a point source and reflect the light rays as shown in Figures 5-1 and 5-2. The larger the size of the arc tube, the wider the light pattern. A phosphor-coated lamp hides the arc tube, and the entire length of the lamp acts as the light source. With this, the beam spread tends to be even wider, making it more difficult to obtain a precise narrow distribution. Light emitted in wider pattern will not affect the efficiency of the luminaire; however, a drop in the CU values would be expected.

Industrial environments are often corrosive, damp, dusty, or even volatile. Each such condition would tend to decrease the quality of performance and reduce the longevity of the material and finishes. Special precautions, suitable for the specific environmental condition, thus should be taken for optimum performance. For a corrosive environment, all exposed rivets, screws, or die-cast parts either should be made of corrosion-resistant material or should be coated with corrosion-resistant paint. For use in web, damp, or dusty environments, special gasketing should be used to seal all areas where moisture and/or

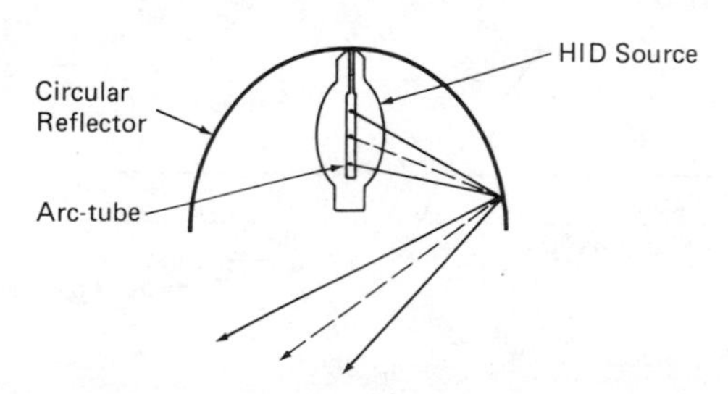

Fig. 5-2. A long arc tube will spread light much wider than a point source, since the arc tube behaves as a combination of several point sources.

dust is likely to enter. Use of specially treated aluminum, glass-bonded-to-aluminum, ceramic, or organic coatings on metal are quite common in such conditions. The character of reflected light, the percentage of reflectance, and the resistivity to heat, corrosion, abrasion, and impact vary in each case. A comparative view of all such finishes has been given in Table 4-2 in Chapter Four.

Lensed Luminaires

An industrial luminaire can be open or enclosed. The enclosed luminaires are usually equipped with a "lens" that is either a flat, clear glass or a special transmitting medium that either refracts or diffuses the light output. Flat, clear glass usually has two purposes: first, to protect the inside of the luminaire from outside dirt, and second, to act as a lamp guard. An impact-resistant glass protects the lamp from external forces and prevents ultraviolet rays from spreading when a lamp shatters. In any case, a clear, flat piece of glass attached to the bottom of a luminaire does not improve the lighting distribution in any respect. In fact, depending on the quality and thickness, a glass can be responsible for reducing light output by as much as 10%. When a ray of light passes through a piece of flat glass, approximately 4–6% of the incident ray is reflected off the glass surface at the first strike and the same amount is lost inside the glass when the ray emerges. This is shown in Figure 5-3. This results in lower efficiency and utilization factor of the luminaire.

To minimize the loss, some manufacturers prefer to use Teflon instead of glass in the luminaire. Teflon is a special plastic product that is clear like glass but shatter-proof, highly resistant to chemicals and solvents, and excellent for high-temperature use. A 5 mil thickness of Teflon, which is commonly used, has approximately 5% better transmission capabilities than glass. This improves the efficiency as well as the utilization factors of a luminaire. Figure 5-4 shows a

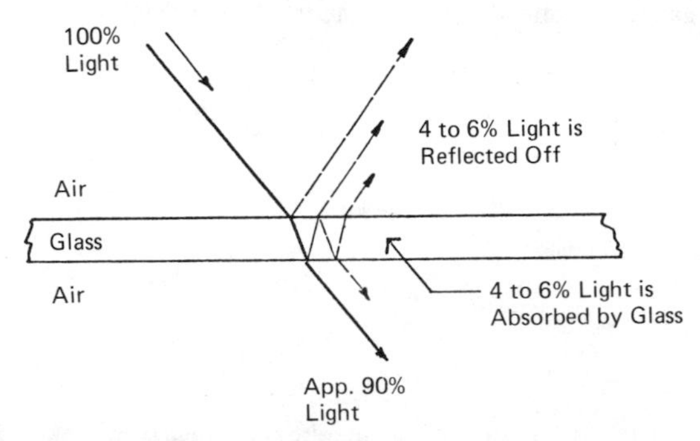

Fig. 5-3. Transmission of light through glass. Depending on its quality and thickness, a glass can be responsible for reducing light output by as much as 10%.

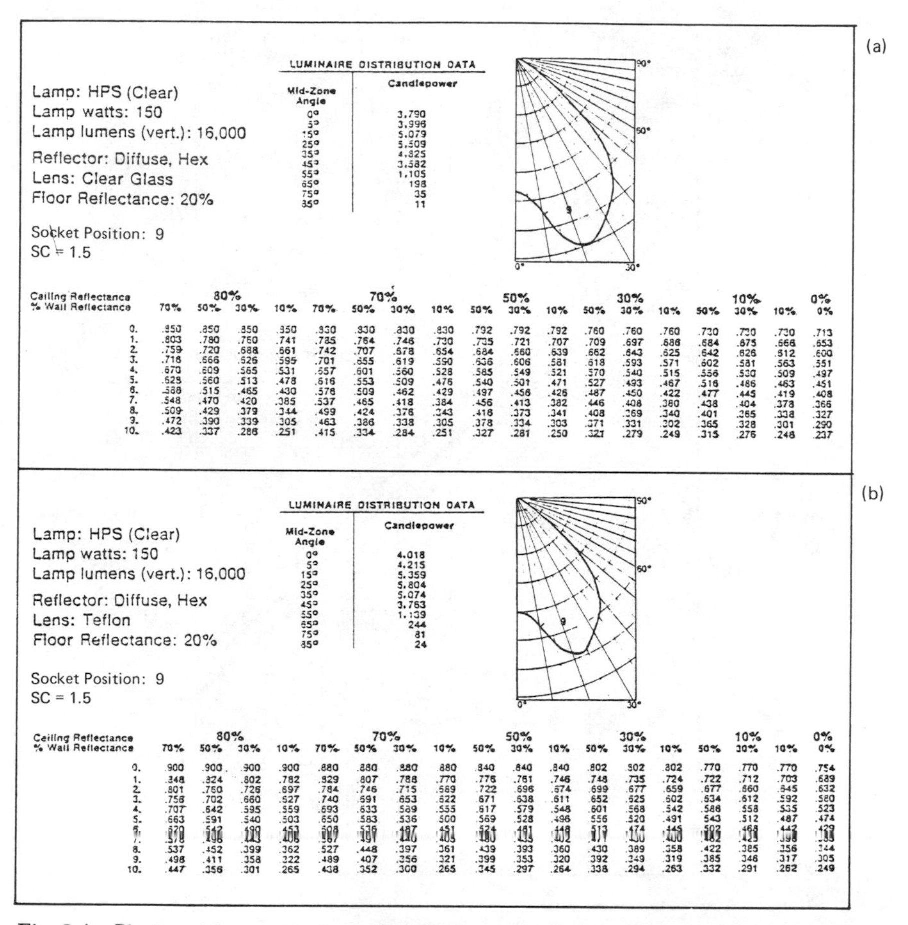

(a)

Lamp: HPS (Clear)
Lamp watts: 150
Lamp lumens (vert.): 16,000
Reflector: Diffuse, Hex
Lens: Clear Glass
Floor Reflectance: 20%

Socket Position: 9
SC = 1.5

LUMINAIRE DISTRIBUTION DATA

Mid-Zone Angle	Candlepower
0°	3,790
5°	3,996
15°	5,079
25°	5,509
35°	4,825
45°	3,582
55°	1,105
65°	198
75°	35
85°	11

Ceiling Reflectance	80%				70%				50%			30%			10%			0%
% Wall Reflectance	70%	50%	30%	10%	70%	50%	30%	10%	50%	30%	10%	50%	30%	10%	50%	30%	10%	0%
0.	.850	.850	.850	.850	.830	.830	.830	.830	.792	.792	.792	.760	.760	.760	.730	.730	.730	.713
1.	.803	.780	.760	.741	.785	.764	.746	.730	.735	.721	.707	.709	.697	.686	.684	.675	.666	.653
2.	.759	.720	.688	.661	.742	.707	.678	.654	.684	.660	.639	.662	.643	.625	.642	.626	.612	.600
3.	.718	.666	.626	.595	.701	.655	.619	.590	.636	.606	.581	.618	.593	.571	.602	.581	.563	.551
4.	.670	.609	.565	.531	.657	.601	.560	.528	.585	.549	.521	.570	.540	.515	.556	.530	.509	.497
5.	.628	.560	.513	.478	.616	.553	.509	.476	.540	.501	.471	.527	.493	.467	.516	.486	.463	.451
6.	.588	.515	.465	.430	.576	.509	.462	.429	.497	.456	.426	.487	.450	.422	.477	.445	.419	.408
7.	.548	.470	.420	.385	.537	.465	.418	.384	.456	.413	.382	.446	.408	.380	.438	.404	.378	.366
8.	.509	.429	.379	.344	.499	.424	.376	.343	.418	.373	.341	.408	.369	.340	.401	.365	.338	.327
9.	.472	.390	.339	.305	.463	.386	.338	.305	.378	.334	.303	.371	.331	.302	.365	.328	.301	.290
10.	.423	.337	.286	.251	.415	.334	.284	.251	.327	.281	.250	.321	.279	.249	.315	.276	.248	.237

(b)

Lamp: HPS (Clear)
Lamp watts: 150
Lamp lumens (vert.): 16,000
Reflector: Diffuse, Hex
Lens: Teflon
Floor Reflectance: 20%

Socket Position: 9
SC = 1.5

LUMINAIRE DISTRIBUTION DATA

Mid-Zone Angle	Candlepower
0°	4,018
5°	4,215
15°	5,359
25°	5,804
35°	5,074
45°	3,763
55°	1,139
65°	244
75°	81
85°	24

Ceiling Reflectance	80%				70%				50%			30%			10%			0%
% Wall Reflectance	70%	50%	30%	10%	70%	50%	30%	10%	50%	30%	10%	50%	30%	10%	50%	30%	10%	0%
0.	.900	.900	.900	.900	.880	.880	.880	.880	.840	.840	.840	.802	.802	.802	.770	.770	.770	.754
1.	.848	.824	.802	.782	.829	.807	.788	.770	.776	.761	.746	.748	.735	.724	.722	.712	.703	.689
2.	.801	.760	.726	.697	.784	.746	.715	.689	.722	.696	.674	.699	.677	.659	.677	.660	.645	.632
3.	.756	.702	.660	.627	.740	.691	.653	.622	.671	.638	.611	.652	.625	.602	.634	.612	.592	.580
4.	.707	.642	.595	.559	.693	.633	.589	.555	.617	.579	.548	.601	.568	.542	.586	.558	.535	.523
5.	.663	.591	.540	.503	.650	.583	.536	.500	.569	.528	.496	.556	.520	.491	.543	.512	.487	.474
6.	.620	.542	.490	.453	.608	.536	.487	.451	.524	.481	.448	.513	.474	.445	.502	.468	.442	.429
7.	.578	.496	.443	.406	.567	.491	.440	.405	.480	.435	.402	.471	.430	.400	.462	.425	.398	.385
8.	.537	.452	.399	.362	.527	.448	.397	.361	.439	.393	.360	.430	.389	.358	.422	.385	.356	.344
9.	.498	.411	.358	.322	.489	.407	.356	.321	.399	.353	.320	.392	.349	.319	.385	.348	.317	.305
10.	.447	.356	.301	.265	.438	.352	.300	.265	.345	.297	.264	.338	.294	.263	.332	.291	.262	.249

Fig. 5-4. Photometric report comparing the same luminaire equipped with (a) clear glass lens and (b) Teflon. Overall efficiency improves by approximately 5% with Teflon.

comparative study of the effect of using Teflon over clear glass in the same luminaire. The luminaire is a dust-proof, sealed, and gasketed, equipped with a 150-watt high-pressure sodium, clear lamp. The socket position in either case is the same. Note that although the SC in either case is the same (1.5), the candlepower values at different vertical angles and the CU factors at all room cavity ratios and reflectances have improved substantially with the Teflon lens. The centerline or the nadir candlepower has changed from 3790 to 4018 candela, offering about 6% improvement. Figure 5-5 shows the improvement in the CU values for RCRs ranging from 1 through 10 at 50/30/20 effective surface reflectances. The overall improvement is regarded as between 5 and 6%.

The other advantage of the use of a Teflon lens is that, in a sealed luminaire,

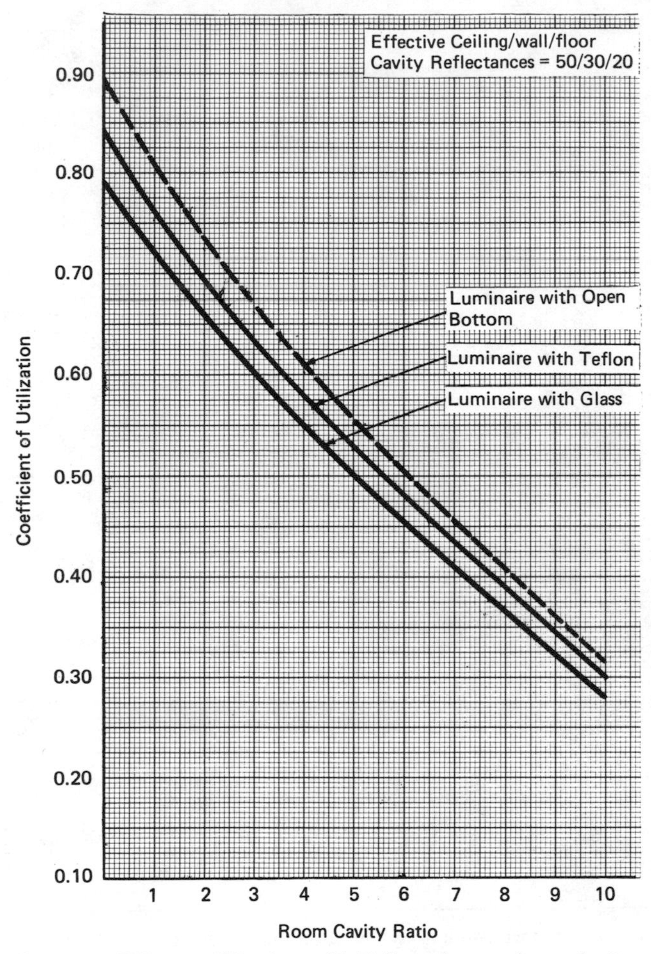

Fig. 5-5. Effect on CU values with Teflon, glass, and open bottom.

as the pressure increases with temperature build-up in a lighted luminaire, it expands to accommodate the extra pressure. This somewhat minimizes the possibility of "breathing in" dirt from outside (this phenomenon is explained later in this chapter).

Teflon has some severe disadvantage as well. The biggest disadvantage is that although a 5 mil Teflon offers better efficiency, it does not protect against lamp shattering and the spreading of ultraviolet rays, making it suitable for only certain types of lamps and wattages. A 10 mil Teflon does not have these disadvantages, but it offers transmission as low as that of glass. In addition, although highly impact resistant, an external pressure exerted on a Teflon surface deforms its shape and causes a sag very easily; this can distort the lighting characteristics.

However, as the temperature rises inside the luminaire, with the heat build-up, the sag disappears slowly and the lens regains its normal appearance. Stretched Teflon can be penetrated or cut very easily with a sharp pointed object or knife, making it a target of vandalism.

Refractors

The refractor luminaires are of two types: (1) luminaires with a refractor bottom or (2) all-refractor luminaires. Figure 5-6 shows an example of each. The basic purpose of the refractor is to bend the light into a specific pattern with the help

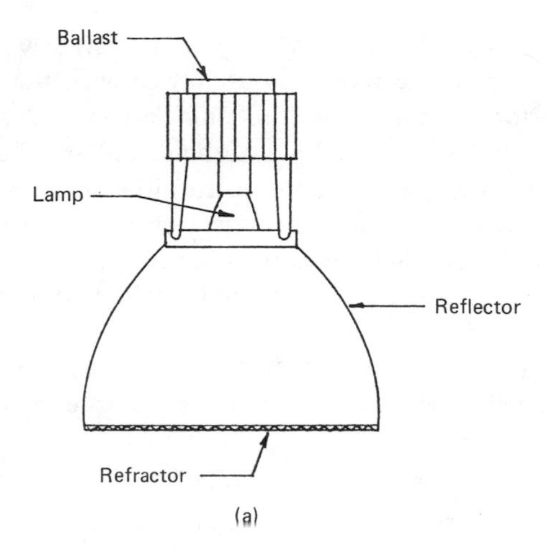

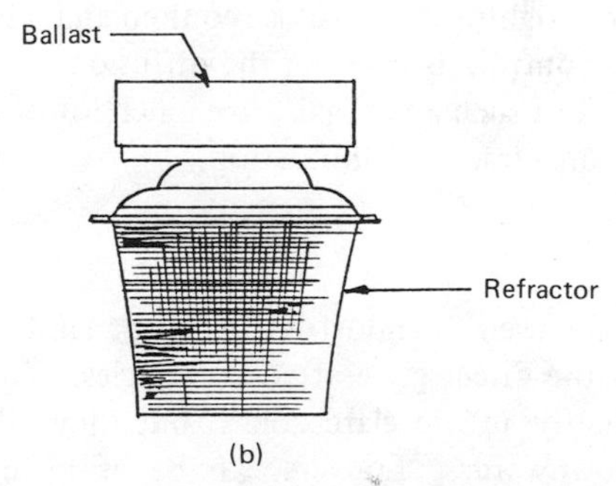

Fig. 5-6. (a) Luminaire with refractor bottom. The basic purpose of the refractor is to bend the light into a specific pattern with the help of tiny prisms. (b) All-refractor luminaire. This spreads light at much higher angles than are possible with refractor bottom luminaires.

of tiny prisms. Each ray of light falling on the prisms is refracted and bent at a higher angle than would be possible by the reflector alone. A controlled high-angle distribution thus can be achieved, with protection against lamp damage and dirt accumulation.

An all-refractor luminaire, as shown in Figure 5-6b, is basically used to spread light at much higher angles than are possible with either a reflector or reflector-refractor combinations. In fact, with these luminaires, the entire control of light output can be accomplished without a reflector. Because of high-angle spread and glare possibility, these are available in lower wattages only. They are used in low-bay areas, such as in garages, where prisms on the side of the refractor elevate light rays by 5–10° at high angles and spread the light to the farthest distance.

Refractors are usually made of glass, acrylic, polystyrene, or polycarbonate plastics. The glass or acrylic refractors usually provide the best results, since their quality of performance in terms of retaining original transmitting capability is much superior to polystyrene and polycarbonate plastic, both of which tend to turn yellow with continuous exposure to ultraviolet rays. Polycarbonate plastics are used because of their high resistance to external impact, but with prolonged exposure to ultraviolet rays they become brittle as they turn yellow. The LLF in each of these cases would be lower than that of units made of glass or acrylic.

All-refractor luminaires can be quite high in efficiency, but their CU is not as high as that of luminaires with a reflector or a reflector-refractor combination.

Diffusers

The main purpose of a diffuser is to hide the lamp and spread the brightness of the source over a larger area so that the peak brightness is minimized. It can be made of any translucent material that will hide the lamp. Diffusers are mostly used for areas where brightness control is required and when there is no need for precise photometric control. Because of the diffusion quality, the efficiency and the utilization factors of such a luminaire are much lower than those with a plain reflector or a reflector-refractor combination.

Louvers and Visors

Louvers or visors are used in industrial lighting mainly to improve the visual comfort, by hiding the direct glare at certain angles. They are physical barriers to light output chiefly in the glare zones, but they allow the light to travel relatively freely downwards. Louvers can be of different shapes, sizes, and colors. Because much of the light is directly obstructed by louvers or visors, luminaires equipped with such devices suffer the poorest efficiency. The degree of efficiency loss depends on their physical shape and size and can be as low as 40%. For a lighting design where louvered luminaires are to be used, a designer should always use the photometric data that have been developed with the louvers in place.

BEAM SPREAD, SC, AND CU

There is an inherent relationship among beam spread, spacing criteria, and the coefficient of utilization. In the past, when beam spread used to be classified by names, the range of S/MH ratio and corresponding beam spread classification was as follows: 0–0.5, highly concentrated; 0.5–0.7, concentrated; 0.7–1.0, medium spread; 1.0–1.5, spread; 1.5–2.0, wide spread.

With the new concept of SC, beam spreads are no longer classified by names. However, the fact still remains that a wider beam spread would mean higher SC. As a luminaire tends to have higher SC, it starts producing more light at the higher vertical angles than it does at or towards the nadir. A quick way to observe this is as the ratio of the candlepowers at higher angles (35-degree or higher) to that at nadir increases, beam spread gets wider and SC increases. Higher SC luminaires are desirable for uniformity in lighting level; however, a very high SC value may cause extreme visual discomfort.

A luminaire with concentrated down-light would obviously throw a majority of its light on the work plane. Compared to this situation, a luminaire with high beam spread would result in a relatively lower amount of light on the work plane, since the light emitted at higher angles would have to bounce back and forth on reflector and room surfaces before falling on the work plane. This results in relatively lower CU values for wider beam spreads than for concentrated one. Thus, theoretically, although a luminaire may have the same efficiency with different lamp-socket positions, its CU values can vary considerably.

Figure 5-7 illustrates this phenomenon. The luminaire is a dust-tight, sealed, and gasketed, with a 400-watt clear high-pressure sodium lamp. Three lamp-socket positions offer 1.0, 1.5, and 1.9 SCs. Note that as the beam spreads from SC = 1.0 to SC = 1.9, significant changes take place in candlepower distribution and CU values. The nadir candlepower at SC = 1.0 is 21,743; that at SC = 1.9 is 7290, showing a reduction of as much as one-third variation. There is also a gradual drop of the CU values as the SC increases from 1.0 through 1.9. Note that the drop in CU values is more prominent at higher RCR values and lower effective surface reflectances. This is mainly because with lower RCR values (which usually means smaller rooms), CU values are increasingly dependant upon room-surface reflectances—particularly the walls. Figure 5-7b shows the effect of CU values for all SCs in the example, at various RCRs. The effective cavity reflectances are 50/30/20.

A lighting designer must be careful, therefore, in selecting a luminaire from its SC, CU, and visual comfort standpoints. Luminaires from different manufacturers should be studied before a decision is made. If several luminaires are found to have the same required SC, select the one with highest CU value.

Lamp: HPS (Clear)
Lamp watts: 400
Lamp lumens (vert.): 50,000

Reflector: Diffuse, Hex
Lens: Clear Glass
Floor Reflenctance: 20%

Socket Position	SC
#4	1.0
#7	1.5
#9	1.9

LUMINAIRE DISTRIBUTION DATA

Mid-Zone Angle	Candlepower		
0°	21.743	12.824	7.290
5°	21.637	12.924	7.602
15°	21.209	14.789	10.197
25°	17.850	15.159	12.469
35°	13.162	13.676	12.760
45°	8 854	11.063	11.313
55°	2 998	5.376	7.150
65°	547	988	1.611
75°	131	193	271
35°	32	38	51

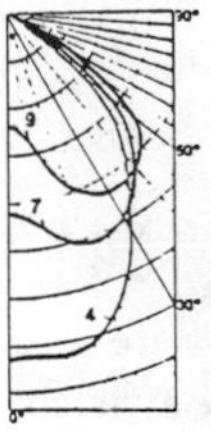

Ceiling Reflectance	80%				70%				50%			30%			10%			0%
% Wall Reflectance	70%	50%	30%	10%	70%	50%	30%	10%	50%	30%	10%	50%	30%	10%	50%	30%	10%	0%
0.	.330	.830	.330	.830	.310	.810	.810	.310	.771	.771	.771	.740	.740	.740	.710	.710	.710	.694
1.	.785	.764	.746	.729	.768	.749	.733	.718	.721	.708	.598	.695	.685	.675	.671	.663	.555	.543
2.	.746	.711	.682	.658	.730	.599	.673	.651	.678	.655	.636	.656	.638	.623	.638	.622	.610	.598
3.	.708	.633	.628	.600	.694	.653	.621	.595	.635	.608	.588	.518	.595	.576	.602	.584	.567	.556
4.	.668	.614	.574	.544	.655	.606	.569	.540	.591	.559	.534	.577	.550	.528	.564	.541	.521	.511
5.	.631	.571	.529	.497	.620	.564	.525	.495	.552	.517	.491	.540	.510	.486	.529	.503	.482	.471
6.	.596	.531	.488	.456	.586	.526	.485	.455	.515	.479	.452	.505	.473	.448	.496	.468	.445	.435
7.	.561	.493	.449	.417	.552	.488	.446	.416	.479	.441	.414	.471	.437	.412	.463	.433	.410	.399
8.	.528	.457	.412	.381	.519	.453	.410	.381	.445	.408	.379	.437	.403	.377	.431	.399	.376	.365
9.	.495	.422	.378	.348	.487	.419	.376	.347	.412	.373	.346	.406	.370	.345	.400	.367	.343	.333
10.	.451	.374	.329	.298	.443	.371	.327	.298	.365	.325	.297	.359	.322	.298	.354	.319	.295	.285
0.	.850	.850	.850	.350	.830	.830	.830	.830	.790	.790	.790	.760	.760	.760	.730	.730	.730	.711
1.	.797	.774	.752	.733	.779	.758	.739	.721	.729	.713	.599	.702	.690	.678	.577	.668	.658	.645
2.	.750	.710	.676	.648	.734	.697	.666	.640	.673	.648	.626	.652	.631	.612	.631	.614	.599	.586
3.	.705	.651	.610	.577	.690	.541	.603	.572	.622	.590	.563	.504	.577	.554	.587	.565	.545	.533
4.	.657	.592	.545	.509	.643	.584	.540	.508	.568	.530	.500	.553	.520	.494	.539	.511	.488	.475
5.	.612	.541	.491	.454	.599	.534	.487	.452	.520	.479	.447	.507	.471	.443	.495	.464	.439	.427
6.	.570	.493	.442	.404	.558	.487	.438	.403	.475	.432	.400	.465	.426	.397	.455	.421	.394	.382
7.	.529	.448	.395	.358	.518	.442	.393	.357	.432	.388	.355	.423	.383	.353	.414	.378	.351	.339
8.	.490	.406	.353	.317	.480	.401	.351	.316	.393	.347	.315	.385	.343	.313	.377	.340	.312	.300
3.	.454	.368	.316	.281	.444	.364	.315	.280	.357	.311	.279	.350	.308	.278	.343	.305	.277	.265
10.	.408	.319	266	.231	.399	.316	.265	.231	.309	.262	.230	.303	.259	.229	.297	.257	.228	.216
0.	.331	.831	.831	.931	.812	.812	.812	.812	.780	.780	.780	.743	.743	.743	.712	.712	.712	.700
1.	.773	.752	.729	.708	.760	.737	.716	.697	.708	.691	.675	.581	.668	.655	.557	.646	.536	.522
2.	.726	.682	.645	.615	.709	.669	.536	.608	.646	.618	.594	.624	.601	.581	.504	.585	.568	.555
3.	.678	.618	.573	.537	.660	.608	.566	.533	.588	.553	.524	.570	.540	.515	.553	.528	.507	.494
4.	.623	.554	.503	.464	.609	.545	.498	.461	.529	.488	.455	.513	.478	.449	.499	.469	.444	.431
5.	.575	.498	.444	.404	.562	.490	.440	.402	.477	.432	.398	.464	.424	.394	.451	.417	.390	.377
6.	.530	.447	.391	.351	.518	.441	.388	.350	.429	.382	.347	.418	.376	.344	.407	.370	.342	.328
7.	.486	.399	.342	.302	.474	.393	.340	.301	.383	.335	.299	.373	.330	.297	.364	.325	.295	.283
8.	.445	.356	.299	.260	.435	.351	.297	.250	.342	.293	.258	.334	.289	.257	.326	.285	.255	.242
9.	.409	.318	.262	.225	.400	.314	.251	.224	.306	.257	.223	.299	.254	.222	.292	.251	.221	.208
10.	.365	.272	.216	.179	.356	.268	.214	.178	.261	.211	.177	.254	.209	.176	.248	.206	.178	.163

(a)

Fig. 5-7. (a) Photometric data of a 400-watt HPS luminaire in three-socket positions. As SC changes from 1.0 to 1.9, significant changes occur in candlepower and CU values. (b) Effect on CU values with three SC values.

Maintenance and Light Loss Factor

As seen in Chapter 3, one of the key factors that determines the total number of luminaires and hence the power consumption of a lighting system, is the light loss factor. No lighting system will be fully effective for long unless a scheduled and appropriate maintenance is provided to retain the initial light output of the luminaires. This is particularly true with industrial lighting systems since they are subject to so many different types of environment conditions. A good, regular, and scheduled maintenance will insure clean and functional luminaire and room surfaces, to retain lighting level as close as possible to the initial values. However, despite adequate maintenance, a luminaire may suffer a loss of 25-35% of its initial lumens at the end of a maintenance period because of its inherent, unavoidable light loss characteristics. Infrequent and insufficient maintenance may pull this value down to 50% or less, depending on the ambient condition and the types of luminaire selected.

Different factors that construct an LLF consist of the nonrecoverable and the recoverable factors. This subject has been discussed in Chapter 3; the discussion there applies for industrial lighting systems as well. In this section we will mainly stress those factors that have significant impact on industrial lighting systems.

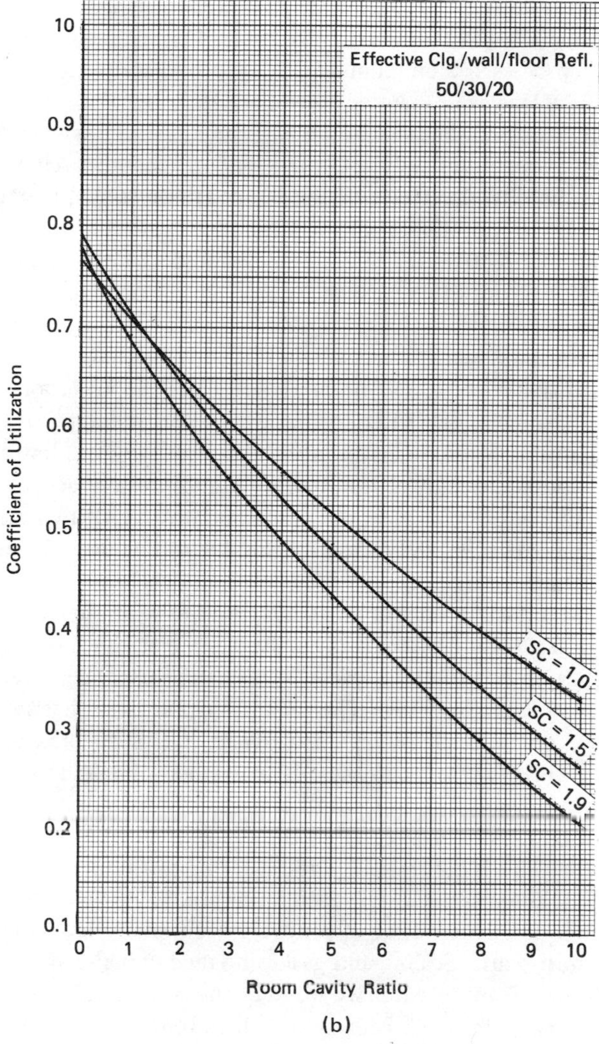

(b)

Fig. 5-7. *(Continued)*.

The most effective way to minimize nonrecoverable light losses is to select the appropriate types of luminaire in the beginning. All materials used in luminaire construction should be suitable for the environmental condition. The expected environment ambient condition must be studied first; then manufacturers should be consulted for a luminaire selection. For HID units any normal change in luminaire ambient temperature has insignificant or no effect in light output. Ballast factors are important and may vary significantly for fluorescent lighting systems. For HID units the BF is usually unity. Luminaire surface depreciation is about the only item in nonrecoverable factors that may have a significant impact in light output. It results from adverse changes in metal, paint, and plastic components that reduce light output. Luminaires using glass, porcelain, or specially processed aluminum are the most tolerant and have insignificant depreciation. However, painted surfaces, including baked enamel, suffer a constant depreciation because of their inherent porous characteristics. In plastics, polystyrene and polycarbonate tend to become yellow with constant exposure to the ultraviolet rays of the lamps (fluorescents, mercury vapor, and metal halide). HPS produces almost nil, and LPS does not produce ultraviolet rays. The most effective way to minimize nonrecoverable light losses is to select the luminaire by consulting with the manufacturer or with a luminaire maintenance company.

Two of the major items in recoverable factors that have significant impact in industrial lighting systems are the luminaire direct depreciation (LDD) and the lamp lumen depreciation (LLD). For most industrial applications LDD alone can be responsible for depreciating total light output to as low as 50% or even lower. This loss is mainly a result of the ambient dirt accumulation on and inside the luminaire surface. In order to minimize LDD, two factors that must be understood are the environmental dirt condition and the luminaire's dirt-protective means.

The typical industrial environment dirt condition can be loosely divided into three categories. The first consists of the plain dirt and dust that floats in the air and settles on horizontal surfaces. This can be easily removed by wiping the luminaires with a plain cloth. The second type is the adhesive or sticky type, the result of dirt carried by vapors of water, oil, mist, etc., or simply by the fumes. The only way to clean these is to wash the luminaires with a detergent, rinse and wipe clean. The third kind consists of dirt like hair, fibers, lints, etc. that is electrostatically charged from the machine operation and clings to luminaire surfaces. This is prevented by destaticizing the luminaire lenses.

An enclosed luminaire, e.g., one with a flat piece of glass at the bottom, suffers the maximum dirt depreciation if proper precautionary measures are not taken to make it dust-tight. Sealing and gasketing the luminaire at places where moisture or dust are likely to enter would help the situation; however, they will not stop the luminaire from "breathing in" dirt from outside. When an enclosed and gasketed luminaire is turned on, with the build-up of heat inside, the pressure also increases. As the pressure rises, the only means of its release takes place through the gasketing or the spring latching device—wherever the least resistance exists. When the luminaire is turned off and temperature is reduced, the process is reversed and outside air is drawn into the luminaire through similar openings. The process is known as "luminaire breathing." If the ambient air is dirty, the contaminants will settle on the lamp, the reflector and the inside surface of the bottom enclosure. Unfortunately, expanded air driven out of the luminaire does not carry all the dirt back, and as a consequence, in the course of time the accumulation of dirt gets even worse than if the luminaire were open at the bottom.

Industry offers two solutions to this problem. The first is to make the luminaire totally dust-tight, allowing no air to come in or go out; the second, to allow the luminaire to "breath" through a low-pass filter.

The advantage of a dust-tight luminaire can be observed by the illustration shown in Figure 5-8. A ray of light passes through a dust film two times: first, as it approaches the reflector, and second, as it leaves the reflector. Each time light is lost (see Figure 5-8a.) For an open-bottom and vented-type luminaire, a ray of light produced by the source would pass through the dust layers at least three times: first, the ray will pass through the film of dust on the lamp; second, while approaching the reflecting surface; third, leaving the reflecting surface (see Figure 5-8b.)

An enclosed, but non–dust-tight luminaire would compound the problem even further. As shown in Figure 5-8c, the same ray of light would now have to pass through two additional layers of dust, accumulated on the inside and outside surfaces of the bottom enclosure.

Comparing the same phenomenon for a totally dust-tight luminaire, the ray of light will have to pass only one time through the dust layer, which is on the outer surface of the bottom enclosure. This is shown in Figure 5-8d. To get an idea of how much light each of these examples would finally reflect, let us assume a loss of 14% (0.86 transmission) for each dust layer. Let us assume, too, that 15% of the light is directly emitted downward and the remaining 85% is emitted towards the reflector. With these assumptions, we can tabulate the effective net reflectances as follows:

	Luminaire		
	A	B	C
	Open-Vented	Enclosed, Non–Dust-Tight	Dust Tight
Dust layers through which light must pass	3	5	1
Reflected light	0.85(0.86 × 0.86 × 0.86) = 0.54	0.85(0.86 × 0.86 × 0.86 × 0.86 × 0.86) = 0.39	0.85(0.86) = 0.73
Direct light	0.15(0.86) = 0.12	0.15(0.86 × 0.86 × 0.86) = 0.095	0.15(0.86) = 0.12
Total	0.54 + 0.12 = 0.66	0.39 + 0.095 = 0.485	0.73 = 0.12 = 0.85

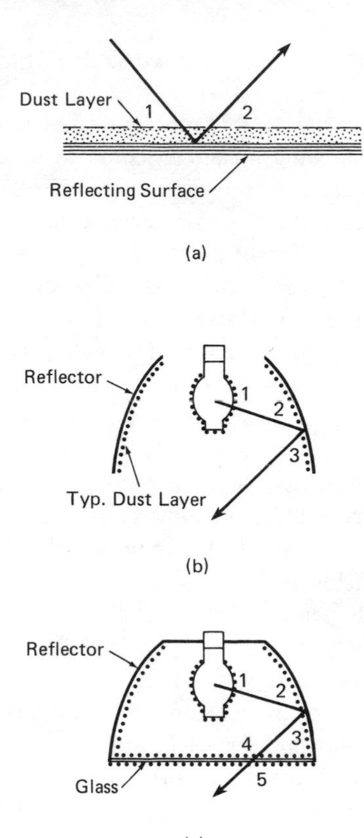

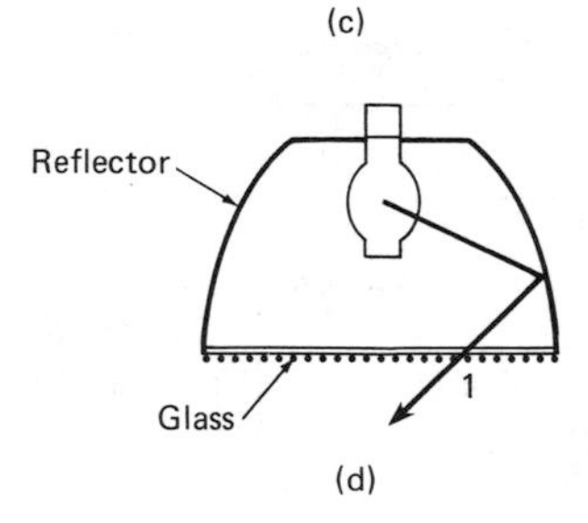

Fig. 5-8. (a) A ray of light passes through a dust film two times: first, as it approaches the reflector; second, as it leaves the reflector. Each time, light is lost. (b) Open-vented luminaire. In these types of luminaire, a ray of light passes through three dust layers: (1) leaving the lamp, (2) approaching the reflector, and (3) leaving the reflector. (c) Enclosed, non-dust-tight luminaire. A ray of light passes through five dust layers: (1) leaving the lamp, (2) approaching the reflector, (3) leaving the reflector, (4) approaching the glass, and (5) leaving the glass. (d) Enclosed, sealed, gasketed, dust-tight luminaire. A ray of light passes through only one layer of dust, as it leaves the glass.

The advantage of a dust-tight luminaire is obvious. The figures shown are for comparison only. While the actual values may be different, the proportion would be about the same.

The concept of total dust-tight luminaires is good from a maintenance standpoint in that dirt is prevented from accumulating inside. But any leakage in the assembly caused during manufacturing or during lamp replacement would defeat the purpose and allow the luminaire to breathe-in dirt from outside. Some manufacturers recommend the use of Teflon in place of glass for a dust-tight type of luminaire. As the temperature inside the luminaire increases, it expands and inflates outwards slightly to accommodate the built-up additional pressure.

Besides dirt particles of various sizes, the other contaminants that significantly affect light control and efficiency of a non-dust-tight luminaire are vapors and gases that either corrode the optical control surfaces or deposit films that are subsequently baked on by lamp heat. Chemical analysis of typical contaminants removed from reflector surfaces usually include nitrogen dioxide, unburned hydrocarbons, and sulphur dioxide, all of which tend to bond to the surface of a specular aluminum reflector. A light, thin film of such accumulation on the reflector surface may be the largest contributing factor in light loss. In most cases, even under ideal cleaning conditions and using the best cleaning solutions known, it is almost impossible to regain the original light reflectance once these contaminants are baked onto the reflecting surfaces because of lamp heat. The use of appropriate absorbtive filters is an effective means to minimize these problems. Using a filter necessitates sealing the luminaire totally except for the filtered opening through which the luminaire can breathe.

There are three means of filtering now being used: activated charcoal, Dacron felt, and fiberglass. While all three offer good resistance to dirt accumulation, the activated charcoal filter is the one that produces the most effective result by absorbing the molecular species of hydrocarbons, nitrogen oxide, nitrogen dioxide, and sulfur dioxide. This reduces their concentration on reflecting surfaces as the luminaires breathe. Statistically, a well-designed luminaire with an activated charcoal filter helps keep the light losses because of internal contaminants to an average of 1% per year, as compared to 4–5% for a well-designed, average nonfiltered and non-dust-tight luminaire for the same application.

LDD values recommended by the manufacturers are usually too optimistic, since these values are developed based on the assumption that proper, scheduled maintenance work is done by owners. However, such is seldom the case. Besides, even if the industry is careful about maintenance, it is still hard to estimate a constant figure for LDD since the environment dirt condition may change significantly with any change in manufacturing procedure. The value of LDD is best determined from experience with similar luminaires or by consulting a luminaire maintenance company. If such a figure is unknown, manufacturers' suggestions or the IES handbook may be referred to for a suitable factor.

SOURCE SELECTION

There are five major types of light sources that are commonly used for industrial lighting: (1) fluorescent, (2) mercury vapor, (3) metal halide, (4) HPS, and (5) LPS. The use of high-wattage, mogul-base incandescents has been quite popular in the past; however, with the development of other efficient light sources, their use in industry has been almost eliminated. Incandescents are sometimes used for supplementary task lighting for certain types of work, where color rendering is of prime importance.

All types of source mentioned above and their ballasts have been discussed in detail in Chapter 1. They have been analyzed from their physical and operational characteristics to determine what individual condition and lamp-ballast combination will produce the most stable and energy-saving condition. All these discussions still hold true in selecting a lamp-ballast combination for industrial lighting. In this section, we will review some of those factors from a comparison standpoint.

In making a source selection, five of the major considerations are (1) efficacy, (2) color, (3) lumen maintenance, (4) life, and (5) cost. In selecting a source, all these points almost have to be considered together. However, since energy saving is the main criterion here, it will be logical to select sources from their efficacy aspect first. LPS is an excellent choice from this standpoint. It produces more than 180 lumens per watt, which is the maximum amount of light produced by any commercially available light source. But it has some strong disadvantages. The first is its unique and unusual color rendering, monochromatic yellow, which makes it suitable only for applications where color rendering is of the least or of no importance; unfortunately, such applications are extremely limited. In addition, LPS tends to gray almost all colors except for yellow. Seeing things other than their normal color has an adverse psychological effect on workers', decreasing their efficiency. Its use is thus limited only to areas where no color matching or reading work is done, when the room is basically used for storage purposes. Some warehouses are good examples for such application; other good examples are cold-storage areas where a lot of light is required with sources that produce the least heat. Controlling light distribution from a linear source like the LPS is also a difficult task. A linear lighting source basically operating on low-pressure gaseous discharge has its total candlepower distributed all over its length. This makes it more suitable for low-height ceiling areas.

Another disadvantage of LPS is the gradual decreasing of its system efficacy (lamp + ballast) concomitant with an increase in power consumption as the lamp ages. This is one factor of which a designer should always be careful. An LPS lamp is expected to live about 18,000 hours. According to its characteristics, its lumen output for some lamps actually increases with age. However, as the lamp ages, the power consumption also increases at an even faster rate, reducing the system efficacy to as low as 79% of its original value at the end of rated life. A designer must include these facts in the lighting calculations.

The second greatest energy-saving source is the HPS of the HID sources. An HPS source has a maximum of 140 lumens per watt efficacy, which is the highest among all "white" sources. It is available in 70, 100, 150, 250, 400, and 1000 watts, which gives it a wide range of ceiling-height and illumination-level applications. The color rendering is quite acceptable for most industrial purposes, excluding except, probably, close color matching. High-intensity capability, compact source, and ease in photometric control make it ideally suitable for medium-to-high bay areas. It has excellent life (24,000 hours) and lumen maintenance. With all these factors combined, an HPS today has become the most-used industrial lighting source.

The metal halide lamp is the next below the HPS in efficacy and is usually the second economic choice. It has the best color rendering among all HID sources and is suitable for medium to high bay areas. The life rating of metal halide lamps, however, is quite poor, and better than incandescents only. It is also position-sensitive. A metal halide lamp can lose much of its light if burned in any position other than the one for which it was specifically designed. The relatively lower initial output, poorer lumen maintenance and shorter life usually results in higher cost of light when compared to the HPS.

Fluorescent lamps, while slightly lower in efficacy than metal halide, may be preferred in some instances where, by virtue of large surface area and diffusion, they produce softer reflections in shiny material and/or equipment. Like the LPS, being low-pressure gaseous-discharge linear sources, they are more suitable for low-height ceilings. The lamps are available in different lengths and produce a wide variety of acceptable colors. High efficacy, wide color choice, instant starting ability, and extremely low cost make fluorescent ideally suitable for many industrial areas where the ceiling is low.

Mercury vapor lamps, because of their superb life (24,000+) and moderate color rendering, were a favorite source for years, until the development of other efficient HID sources.

Incandescents are seldom used for industrial general lighting because they have the lowest efficacy and the shortest life. As indicated earlier, they are useful in seldom-occupied areas and/or where instant turn-on is required. They are also suitable, particularly in some self-contained reflector types, for some inspection and accent lighting and for additional tool-point illuminance. Low-cost, instant-lighting incandescents are also suitable for battery-operated emergency lighting.

Quartz-iodine is a form of incandescent light that is sometimes used in conjunction with HID sources for emergency lighting purposes. These special luminaires accommodate the incandescent and HID sources under the same reflector. When the switch is turned on, the incandescent provides instant lighting and stays lit until the HID source comes to about 60% of its full brightness. This helps the workers to get by with some degree of instant light during the HID warm-up period.

Figure 5-9 has been developed to allow comparison of the gaseous discharge sources of different wattages to their initial lumens. The numbers on the curves are the rated nominal lamp wattages, and the readings directly below on the horizontal axis indicate the total power consumed by the lamp and ballast. The initial lumens are to be read on the vertical axis. For instance, to find the total system power consumption and the initial lumens produced by a 250-watt HPS lamp, draw a vertical and horizontal line from the number 250 on the HPS curve and read 310 watts and 30,000 lumens respectively. This lumen value divided by the system wattage will give the initial system efficacy, e.g., in this case, 30,000/310 = 97 lumens per watt. The wattage and lumen values found from these curves are approximate and to be used for comparison purposes only. The actual values will differ from manufacturer to manufacturer and from one type of ballast to another. Note that the relative locations of the curves automatically shows their overall efficacy differences. LPS shows the maximum and mercury vapor the minimum; note, too, that metal halide basically continues the ended curve of fluorescent, indicating that its efficacy is almost equal to that of fluorescents at lower wattages and gradually improves but always less than that of HPS.

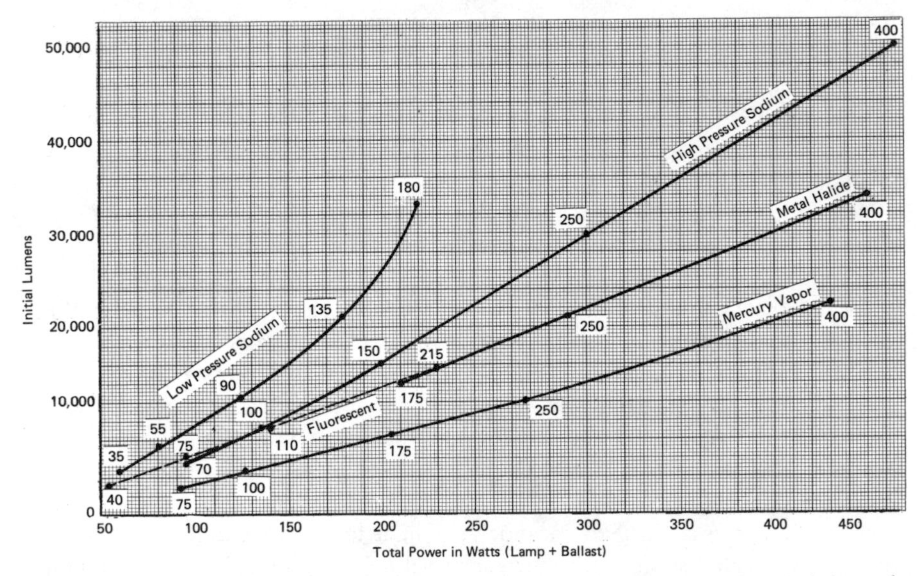

Fig. 5-9. Curves representing initial lumens output of various light sources and their total input watts.

DESIGN GUIDE AND SAMPLE PROBLEM

A typical industrial lighting design includes evenly distributed illuminance throughout the room. Luminaire spacing is the main criterion in obtaining uniformity. Each luminaire is specified with a spacing criterion that, when multiplied by the mounting height, aids in determining a maximum allowable spacing between luminaires for uniformity.

For most industrial applications, as far as selection of sources is concerned, the debate is ultimately between fluorescent and HID. A fluorescent luminaire, typically, is more suitable for low-ceiling applications, since its brightness is evenly distributed over the entire surface of the luminaire. Occupying a larger area of a room with uniformly distributed brightness, it spreads the light and brings uniformity. HID luminaires, however, because of their tremendous amount of energy output from a compact source, are more suitable for higher ceilings. If the HID sources are used in low-height ceilings and installed within spacing limitations, uniformity in illuminance will probably be achieved, but at the expense of higher footcandle level and glare. By the same context, use of fluorescents in high-ceiling areas would bring uniformity but would be inefficient in producing high illuminance levels. As a general rule, for uniform lighting, fluorescent should be considered for low ceilings—and HID for high ceilings only.

A low but uniform illuminance level in low-ceiling areas can also be achieved by lower-wattage HID sources. However, there are some strong limitations. First, there is not a wide range of low-wattage lamps in HID sources; second, their efficacy sharply declines in lower wattages. Mercury vapor sources have lower efficacy than fluorescents, so they are of least interest to us. Metal halide lamps are available at 175, 250, 400, 1000, and 1500 watts, and the HPS lamps are available at 70, 100, 150, 175, 250, 400, 1000, and 1500 watts. Although the HPS sources are available at lower wattages, their efficacy is quite low at the low wattages, and it is not until the use of a 150-watt lamp that any energy saving can be accomplished over the fluorescent units.

Figures 5-10 and 5-11 have been drawn to aid the designer in making a preliminary selection between HID sources and fluorescents, with a consideration of luminaire mounting height and SC, for uniformity and the least power consumption. Figure 5-10 serves to make the selection between metal-halide and fluorescents; Figure 5-11 is drawn for HPS and fluorescents. Before using any of these figures, a preliminary decision has to be made between HPS and metal halide sources. From an efficacy standpoint, we have already seen the superiority of HPS over metal-halide, so the decision has to be based on other factors, such as color, lamp longevity, and economy. Once a decision has been made, the designer can use the specific figure to determine a preliminary wattage of the source and the minimum SCs of the luminaire that would maintain uniformity

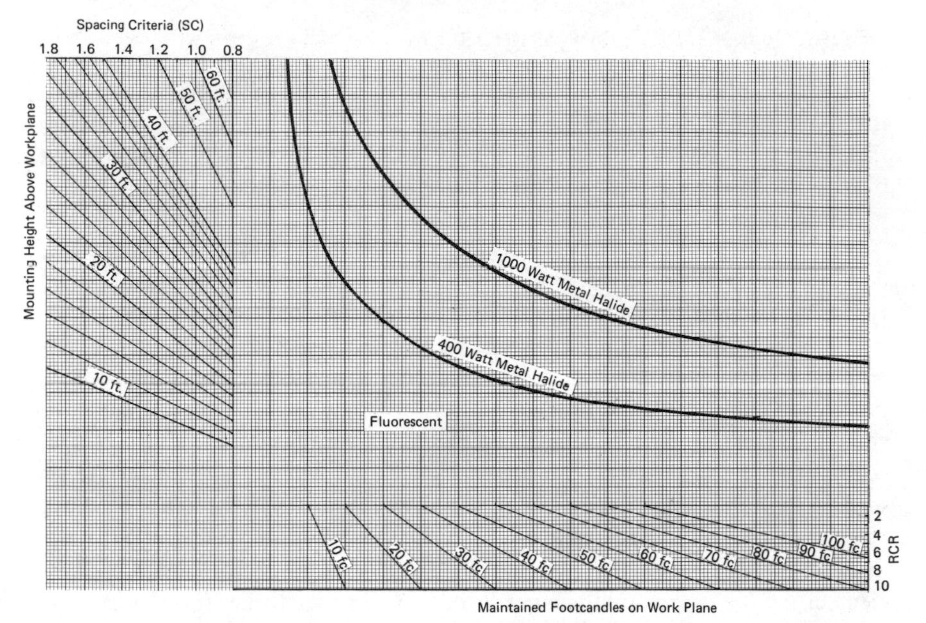

Fig. 5-10. Curves representing an empirical method for quickly determining which light source between metal-halide and fluorescent is more appropriate for energy saving.

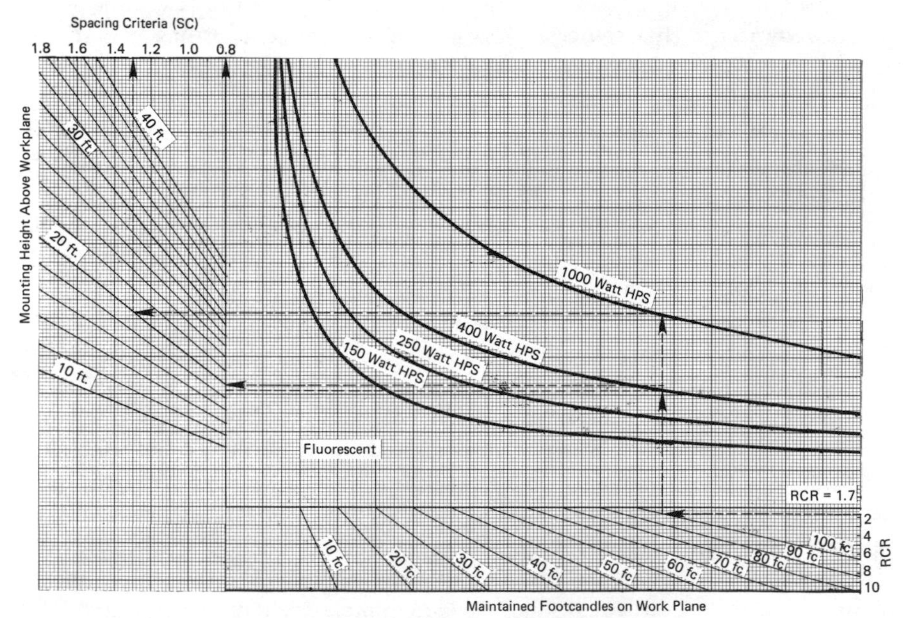

Fig. 5-11. Curves representing an empirical method for quickly determining which light source between HPS and fluorescent is more appropriate for saving energy.

and yet provide maximum energy saving for a given mounting height. The curves have been developed from the basic definition of footcandle, i.e., the lumens per area, and has been precisely drawn to separate lamps of different wattages for uniformity and saving energy. The horizontal axis represents the maintained footcandles above work plane for RCRs ranging from 1 to 10, and the vertical axis represents luminaire mounting heights above the work plane for different SCs. The technique of using the figure is as follows:

1. Determine the RCR of the room and draw a horizontal line from the RCR to the required footcandle level.
2. Now draw a vertical line from the footcandle level and go across all the curves above. Draw horizontal lines from each of these points of intersection and extend them to the required mounting height.
3. Draw vertical lines from these intersections to the SC scale.
4. The largest-wattage size of the group and the corresponding SCs represent the source type, wattage, and the possible minimum SC of the luminaire that will be most energy saving with uniform illuminance for the application.

Further discussion on the use of these figures has been done in the sample problem.

Once the source selection has been made, the next step is to decide on a luminaire. Luminaires of different lamp wattage have different candlepower distribution. It varies from lamp to lamp, from wattage to wattage, and from luminaire to luminaire in design qualities. Figure 5-12 has been developed to aid the designer in selecting a luminaire by knowing the centerline candlepower (nadir candlepower) for certain specific footcandle levels and mounting-height requirements. Note that since curves do not reflect contribution from room surface reflectances, the actual candlepower selection should be somewhat less than what is found in the curves. Also note that the technique applies only for luminaires that produce direct light, and the illuminance level under consideration is on a plane directly underneath the luminaire.

SAMPLE PROBLEM

There is no typical industrial environment that can be used for all calculations. Each application has its own environmental characteristics that would dictate the type of luminaire to be used. The sample problem that has been selected enables us to discuss most of the typical factors involved in energy-saving industrial lighting. Let us suppose the model problem is as follows:

Given: Room dimensions: $L = 150$ ft.; $W = 100$ ft.; $H = 25$ ft.
Luminaires are to be mounted under metal joists, 2′6″ below the

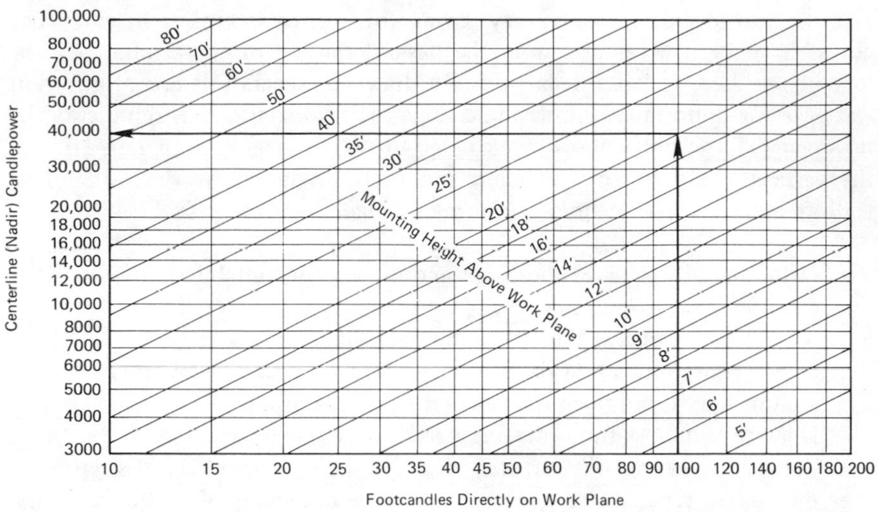

Fig. 5-12. Curves representing a quick method for determining the intial footcandles directly on the work plane knowing centerline (nadir) candlepower and luminaire MH above the work plane.

> ceiling. The type of work is some type of manufacturing that produces fumes and adhesive contaminants. The dirt condition is considered to be medium. The work plane is approximately 2′6″ above floor. The ceiling/wall/floor reflectances are 50/30/20. Work goes on 10 hours a day, 6 days a week. Relamping interval is not decided. Minimum maintained illuminance level requirement is 100 fc.

We will base our calculations on the given reflectances: 50/30/20%. For an existing building the actual reflectances should be measured. If a surface has different reflecting materials, each occupying a substantial portion of the surface, a weighted average value should be determined. The method of determining a weighted average reflectance has been shown in Chapter 2; a handy method was given for measuring any surface reflectance, along with a list of the reflectance values of typical industrial room finishes.

The main purpose of knowing the room surface reflectances is to determine the effective ceiling- and floor-cavity reflectances. The values of the effective cavity reflectances can be obtained by the formula shown below. However, in using this formula, it is assumed that the zonal cavities or areas are open and free of obstruction. In reality, such is not always the case. Beam locations, joists, mechanical ducts, are frequently present in typical industrial areas, and these can alter the calculated effective cavity reflectance values. There are no methods available yet to take these factors into account.

As shown in Chapter 2, for a rectangular room the general formula of cavity ratio is given by

$$\text{Cavity ratio} = \frac{5 \times \text{cavity height} \times \text{perimeter of room}}{\text{Area of the room}}$$

So

$$\text{RCR} = \frac{5 \times 20 \times (150 + 100)}{150 \times 100} = 1.7.$$

Note that the room-cavity height is same as the mounting height of the luminaire above the work plane: 25 - (2.5 + 2.5) = 20 ft. Similarly,

$$\text{CCR} = \frac{5 \times 2.5 \times (150 + 100)}{150 \times 100} = 0.20,$$

and

$$\text{FCR} = \frac{5 \times 2.5 \times (150 + 100)}{150 \times 100} = 0.20.$$

Note that these values of the cavity ratio also can be found directly by using Table 2-1 in Chapter 2.

From here we go to Table 2-3 to determine the effective ceiling- and floor-cavity reflectances. Corresponding to ceiling and wall reflectances of 50% and 30%, respectively, and CCR = 0.2, the effective ceiling-cavity reflectance is found to be 47%. Similarly, with floor and wall reflectances of 20% and 30%, respectively, and FCR = 0.2, the effective floor cavity reflectance is found to be 19%. The numbers may be rounded off to 50% and 20%, respectively, since this change will bring insignificant difference in lighting output.

A proper analysis of the room is important before making any calculation. If the room has large machines or storage racks that create aisles and do not allow light to pass through one aisle to another, they should be treated as individual rooms. The cavity ratios in such cases obviously would be different. If there is an assigned area that needs a different illuminance level from that of the rest of the room, it should be treated separately. Formulas for determining cavity ratios of irregular-shape rooms are given in Chapter 2. In our example, the room is treated as a single big rectangular room since it is dealing with smaller machines (work plane is 2′-6″), and for every practical purpose it is obstruction free.

Source and Luminaire Selection

The designer, at this point needs to make a tentative choice of the source. Since energy saving is the main criterion, the choice should start from the sources with highest efficacies. LPS is a good selection for efficacy but an unsuitable one for our purposes since color rendition is important here. The choice is then HPS, since it is second best in efficacy and its color rendering is quite acceptable.

We now refer to Figure 5-11 to make a preliminary selection of the type of wattage of the source that may be most energy saving for our example. The method is very simple and reasonably accurate; however, a further verification by calculations must be made. To make a selection, we proceed as follows:

1. Draw a horizontal line from RCR = 1.7 up to the required 100 fc maintained-illuminance level.
2. Now draw a vertical line from 100 fc and go across all curves above. Draw horizontal lines from each of the intersections and extend up to the required mounting height of 20 ft., as shown in the figure.
3. Draw vertical lines from these intersections and extend to the SC scale above.
4. The 1000-watt HPS cuts the SC scale at 1.3, and the 400-watt HPS at 0.8.

This means that, for our example, the tentative most energy-saving selection is a HPS source of 1000 watts and a luminaire with a minimum of 1.3 SC. The second best choice is a 400-watt HPS luminaire with a minimum of 0.8 SC. Note that the horizontal line from the 400-watt HPS curve does not intersect the 20 ft. mounting height directly, indicating that the minimum SC value may be somewhat lower than 0.8. In practice, an SC lower than 0.8 is undesirable, since it increases luminaire quantity substantially. With 0.8 as minimum, the SC of the required 400-watt HPS may be as high as 1.3 for the example. This is shown by the two arrowheads intersecting the 20 ft. mounting height line.

Now we refer to Figure 5-12, which has been developed to give a preliminary idea of the maximum candlepower a luminiare should have at the nadir to produce certain footcandle levels at a given distance, directly below. The figure is developed from the basic relationship among candlepower, mounting height, and footcandles, given by $E = I/D^2$. Note that footcandle values obtained by this method assume that there is no contribution of light from any other neighboring luminaires or room surfaces, and the total illuminance at that location is produced by the luminaire alone. In reality, this is seldom the case, since a typical industrial area would involve several luminaires and an overlapping of light from different sources is encouraged for uniformity. Thus the candlepower requirement directly under a luminaire, at the mounting height, should be less than what would be found in the Figure 5-12. How much reduction in the nadir candlepower will be accepted depends upon the type of work performed and the equipment in the room. If there are tall machines in the area and a shadow-free environment is desirable, as much as 50% of the candlepower beneath a luminaire may be borrowed from the neighboring sources. For general purposes, 25-50% of the light directly below a luminaire may be borrowed from neighboring luminaires. This way, if a luminaire burns out, at least 25% of the illumination level would be available directly underneath, and a dark spot will be avoided. Let us suppose for our purposes that this value is 40%; hence, each luminaire should produce 60% of the light directly underneath.

From Figure 5-12 we see that for 100 fc and 20-foot mounting height the maximum nadir candlepower should be 40,000 candela. For our design purposes, this value should be 40,000 × 0.6 = 24,000 candela.

The next move is to look for a luminaire that has these qualifications. In addition, the luminaire should be totally dust-tight or properly filtered, since there are fume- and adhesive-type contaminants in the air. Figures 5-13 and 5-14 show separate photometrics of 1000-watt and 400-watt sealed and dust-tight luminaires with glass and Teflon lenses. According to Figure 5-11, our first choice should be a 1000-watt HPS luminaire with 1.3 SC and a nadir candlepower of 24,000 candela. The closest one we can get is the luminaire with 1.2 SC and 40,637 candlepower. This does not suit our purpose. Now we refer to Figure 5-14 (400-watt HPS) and make a similar analysis. The luminaire with Teflon lens, with SC = 1.0 and nadir candlepower of 22,826, is closest to our requirement. So this is our choice.

Determining the CU

Corresponding to RCR = 1.7 and effective reflectances of 50/30/20, we find the CU = 0.705. The value is interpolated as follows. Given at RCR = 1 the CU is 0.745 and at RCR = 2 the CU is 0.688, the CU value for 1.7 is 0.745 - [(0.745 - 0.688) × 0.7] = 0.705.*

Determining the LLF

As discussed earlier, two of the major factors in determining an LLF of an HID unit are the LLD and the LDD. All other factors can be assumed to be unity. Given that lamps will burn 10 hours a day, 6 days a week, for the whole year, let us analyze the effect of LLF for two group-relamping duration cycles, 12,000 hours, and 16,800 hours, representing 50% and 70% of the rated life (24,000 hours), respectively. Since per year 6 × 10 × 52 = 3120 hours of work is done, the group relamping will occur at approximately every 4 years and 5.5 years,

*The CU for 1.7 falls between the two values given and can be interpolated by deducting 7/10 of the value from 1 to 2 from RCR = 1.

Lamp: HPS (Clear) | Reflector: Diffuse, Hex
Lamp watts: 1000 | Lens: Clear Glass
Lamp lumens (vert): 140,000 | Floor Reflectance: 20%

Socket Position	
	SC
#5	1.0
#9	1.2

LUMINAIRE DISTRIBUTION DATA

Mid-Zone Angle	Candlepower	
0°	48,859	40,637
5°	48,173	41,030
15°	44,232	40,146
25°	37,059	35,525
35°	28,817	29,473
45°	23,122	25,855
55°	16,094	19,109
65°	7,440	9,399
75°	1,725	2,069
85°	171	191

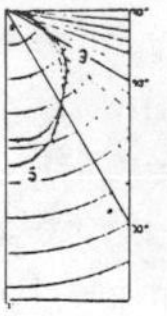

Ceiling Reflectance	80%				70%				50%			30%			10%			0%
% Wall Reflectance	70%	50%	30%	10%	70%	50%	30%	10%	50%	30%	10%	50%	30%	10%	50%	30%	10%	0%
0.	.800	.800	.800	.800	.730	.730	.730	.730	.744	.744	.744	.713	.713	.713	.684	.684	.684	.670
1.	.748	.725	.702	.684	.731	.709	.689	.671	.681	.665	.651	.655	.648	.631	.632	.622	.612	.600
2.	.699	.659	.625	.596	.684	.647	.615	.590	.624	.608	.577	.603	.582	.562	.584	.567	.551	.539
3.	.653	.599	.558	.525	.639	.591	.552	.522	.572	.540	.513	.555	.528	.504	.539	.516	.496	.484
4.	.608	.544	.497	.462	.595	.535	.493	.458	.520	.483	.453	.506	.474	.448	.493	.465	.442	.430
5.	.567	.496	.447	.410	.552	.491	.442	.408	.476	.435	.405	.463	.428	.401	.453	.422	.396	.385
6.	.530	.454	.404	.369	.517	.448	.402	.366	.437	.396	.364	.427	.390	.361	.417	.383	.358	.347
7.	.493	.416	.366	.322	.483	.411	.362	.329	.400	.357	.327	.391	.353	.324	.384	.349	.322	.311
8.	.460	.380	.331	.297	.450	.376	.328	.296	.367	.325	.294	.360	.321	.293	.352	.317	.290	.279
9.	.430	.351	.302	.268	.421	.347	.300	.258	.339	.296	.266	.333	.293	.255	.326	.290	.263	.253
10.	.391	.309	.260	.229	.393	.306	.260	.229	.300	.257	.227	.294	.254	.225	.289	.251	.224	.214
0.	.840	.840	.840	.840	.821	.821	.821	.821	.734	.734	.754	.751	.751	.751	.720	.720	.720	.710
1.	.795	.760	.734	.714	.787	.742	.720	.701	.713	.695	.679	.686	.671	.658	.661	.650	.639	.626
2.	.730	.686	.647	.615	.714	.672	.637	.609	.648	.619	.595	.626	.603	.580	.606	.587	.568	.555
3.	.679	.618	.573	.536	.664	.609	.566	.532	.590	.554	.523	.571	.540	.515	.555	.528	.506	.493
4.	.628	.557	.505	.465	.614	.549	.500	.462	.531	.490	.456	.516	.480	.451	.502	.471	.445	.431
5.	.582	.504	.449	.408	.567	.498	.444	.406	.482	.436	.402	.468	.428	.398	.457	.423	.394	.381
6.	.541	.458	.402	.363	.528	.451	.400	.360	.439	.393	.358	.428	.386	.355	.417	.380	.351	.339
7.	.501	.415	.360	.323	.490	.410	.356	.319	.399	.350	.317	.388	.346	.314	.381	.342	.312	.300
8.	.465	.377	.322	.234	.454	.372	.320	.283	.363	.316	.281	.355	.311	.280	.348	.307	.273	.266
9.	.433	.346	.292	.254	.424	.341	.239	.254	.333	.296	.253	.325	.292	.251	.319	.279	.249	.238
10.	.392	.303	.249	.214	.384	.299	.249	.214	.292	.245	.212	.286	.242	.210	.280	.239	.209	.198

(a)

Lamp: HPS (Clear) | Reflector: Diffuse, Hex
Lamp watts: 1000 | Lens: Teflon
Lamp lumens (vert.): 140,000 | Floor Reflectance: 20%

Socket Position	
	SC
#5	1.0
#9	1.2

LUMINAIRE DISTRIBUTION DATA

Mid-Zone Angle	Candlepower	
0°	57,550	41,694
5°	55,924	41,383
15°	50,167	40,979
25°	40,628	36,164
35°	30,384	29,959
45°	24,536	26,256
55°	16,599	19,312
65°	6,605	9,481
75°	1,377	2,430
85°	333	396

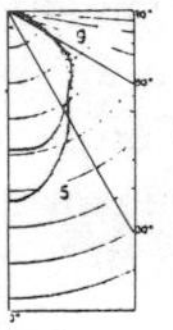

Ceiling Reflectance	80%				70%				50%			30%			10%			0%
% Wall Reflectance	70%	50%	30%	10%	70%	50%	30%	10%	50%	30%	10%	50%	30%	10%	50%	30%	10%	0%
0.	.844	.844	.844	.844	.824	.824	.824	.824	.790	.790	.790	.754	.754	.754	.723	.723	.723	.710
1.	.793	.770	.746	.727	.775	.752	.732	.714	.723	.707	.692	.696	.683	.671	.672	.661	.651	.638
2.	.743	.702	.667	.637	.729	.689	.656	.632	.665	.639	.618	.643	.622	.602	.623	.606	.589	.577
3.	.697	.641	.599	.566	.682	.632	.592	.561	.613	.580	.552	.595	.567	.543	.579	.555	.534	.522
4.	.650	.585	.537	.500	.637	.577	.532	.497	.560	.522	.491	.545	.512	.485	.531	.502	.479	.466
5.	.607	.535	.484	.447	.593	.530	.480	.445	.514	.472	.441	.501	.464	.435	.490	.459	.432	.420
6.	.569	.492	.440	.404	.556	.483	.438	.401	.473	.431	.398	.463	.425	.395	.455	.418	.392	.380
7.	.530	.451	.400	.365	.521	.446	.396	.362	.435	.390	.359	.425	.388	.356	.417	.382	.354	.343
8.	.497	.414	.363	.327	.486	.409	.360	.325	.400	.356	.324	.393	.352	.323	.384	.348	.321	.309
9.	.465	.383	.333	.297	.455	.379	.330	.297	.371	.325	.295	.364	.322	.294	.357	.319	.292	.281
10.	.423	.338	.288	.255	.415	.335	.288	.255	.329	.234	.253	.322	.281	.251	.317	.278	.250	.239
0.	.860	.860	.860	.860	.840	.840	.840	.840	.802	.802	.802	.770	.770	.770	.740	.740	.740	.722
1.	.802	.775	.749	.728	.783	.758	.735	.715	.723	.709	.693	.700	.685	.671	.675	.663	.652	.638
2.	.745	.699	.660	.627	.729	.686	.649	.621	.661	.631	.607	.638	.614	.591	.618	.598	.579	.565
3.	.693	.630	.584	.546	.678	.621	.577	.542	.601	.564	.533	.582	.551	.524	.565	.538	.515	.502
4.	.641	.563	.515	.474	.627	.560	.510	.470	.542	.499	.465	.525	.489	.459	.512	.480	.454	.439
5.	.594	.514	.458	.416	.579	.509	.453	.414	.492	.445	.410	.478	.437	.406	.466	.431	.401	.388
6.	.553	.467	.410	.370	.539	.460	.407	.367	.448	.401	.364	.437	.394	.362	.425	.387	.358	.345
7.	.511	.424	.357	.329	.501	.419	.363	.328	.407	.357	.323	.396	.353	.320	.388	.349	.318	.309
8.	.475	.385	.329	.290	.464	.380	.325	.289	.370	.322	.287	.362	.318	.286	.354	.285	.254	.242
9.	.443	.353	.298	.259	.433	.349	.295	.259	.340	.292	.258	.333	.288	.256	.325	.295	.254	.242
10.	.401	.309	.254	.219	.392	.305	.254	.219	.299	.250	.217	.292	.247	.215	.298	.244	.213	.202

(b)

Fig. 5-13. Photometric reports of 1000-watt luminaires with clear glass and Teflon.

respectively. Reviewing Figure 1-40 of Chapter 1, we find the LLD for 50 and 70% rated life are 0.90 and 0.85, respectively. Also reviewing Figure 1-41, we find the lamp burn out (LBO) values for the same period to be 0.96 and 0.85, respectively. In order to determine LDD, let us suppose that the luminaires will be wipe-cleaned every 24 months in either case. Referring to the LDD curves

Lamp: HPS (Clear)
Lamp watts: 400
Lamp lumens (vert.): 50,000

Reflector: Diffuse, Hex
Lens: Clear Glass
Floor Reflenctance: 20%

Socket Position	SC
#4	1.0
#7	1.5
#9	1.9

LUMINAIRE DISTRIBUTION DATA

Mid-Zone Angle	Candlepower		
0°	21.743	12.324	7.290
5°	21.637	12.924	7.602
15°	21.209	14.789	10.197
25°	17.850	15.159	12.469
35°	13.162	13.676	12.760
45°	8 854	11.063	11.313
55°	2 998	5.378	7.150
65°	547	988	1.611
75°	131	193	271
35°	32	38	51

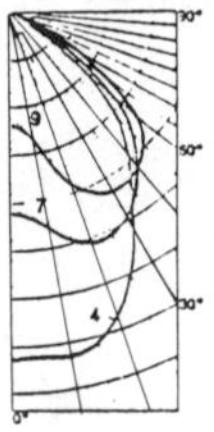

Ceiling Reflectance	80%				70%				50%			30%			10%			0%
% Wall Reflectance	70%	50%	30%	10%	70%	50%	30%	10%	50%	30%	10%	50%	30%	10%	50%	30%	10%	0%
0.	.330	.830	.830	.830	.810	.810	.810	.310	.771	.771	.771	.740	.740	.740	.710	.710	.710	.694
1.	.785	.764	.746	.729	.768	.749	.733	.718	.721	.708	.596	.695	.685	.675	.571	.663	.555	.543
2.	.746	.711	.682	.658	.730	.599	.673	.651	.678	.655	.636	.656	.638	.623	.636	.622	.510	.598
3.	.708	.633	.628.	.600	.694	.653	.621	.595	.635	.608	.586	.518	.595	.576	.602	.584	.587	.556
4.	.668	.614	.574	.544	.655	.606	.569	.540	.591	.559	.534	.577	.550	.528	.564	.541	.521	.511
5.	.631	.571	.529	.497	.620	.564	.525	.495	.552	.517	.491	.540	.510	.486	.529	.503	.482	.471
6.	.596	.531	.468	.456	.586	.526	.485	.455	.515	.479	.452	.505	.473	.448	.496	.468	.445	.435
7.	.561	.493	.449	.417	.552	.488	.446	.416	.479	.441	.414	.471	.437	.412	.463	.433	.410	.399
8.	.528	.457	.412	.381	.519	.453	.410	.381	.445	.406	.379	.437	.403	.377	.431	.399	.376	.365
9.	.495	.422	.378	.348	.487	.419	.376	.347	.412	.373	.346	.406	.370	.345	.400	.367	.343	.333
10.	.451	.374	.329	.298	.443	.371	.327	.298	.365	.325	.297	.359	.322	.296	.354	.319	.295	.285
0.	.850	.850	.850	.350	.830	.830	.830	.330	.790	.790	.790	.760	.760	.760	.730	.730	.730	.711
1.	.797	.774	.752	.733	.779	.758	.739	.721	.729	.713	.699	.702	.690	.678	.577	.668	.658	.645
2.	.750	.710	.676	.648	.734	.697	.666	.640	.673	.648	.625	.652	.631	.612	.631	.614	.599	.586
3.	.705	.551	.610	.577	.690	.541	.603	.572	.622	.590	.563	.604	.577	.554	.587.	.565	.545	.533
4.	.657	.592	.545	.509	.643	.584	.540	.508	.568	.530	.500	.553	.520	.494	.539	.511	.488	.475
5.	.612	.541	.491	.454	.599	.534	.487	.452	.520	.479	.447	.507	.471	.443	.495	.464	.439	.427
6.	.570	.493	.442	.404	.553	.487	.438	.403	.475	.432	.400	.465	.426	.397	.455	.421	.394	.382
7.	.529	.448	.395	.358	.518	.442	.393	.357	.432	.388	.355	.423	.383	.353	.414	.378	.351	.339
8.	.490	.406	.353	.317	.480	.401	.351	.316	.393	.347	.315	.385	.343	.313	.377	.340	.312	.300
9.	.454	.368	.316	.281	.444	.364	.315	.280	.357	.311	.279	.350	.308	.278	.343	.305	.277	.265
10.	.408	.319	266	.231	.399	.316	.265	.231	.309	.262	.230	.303	.259	.229	.297	.257	.228	.216
0.	.331	.831	.831	.831	.812	.812	.812	.812	.780	.780	.780	.743	.743	.743	.712	.712	.712	.700
1.	.773	.752	.729	.708	.760	.737	.716	.697	.708	.691	.675	.681	.668	.655	.557	.646	.536	.622
2.	.726	.682	.645	.615	.709	.669	.536	.608	.648	.618	.594	.624	.601	.581	.604	.585	.568	.555
3.	.676	.618	.573	.537	.660	.606	.566	.533	.588	.553	.524	.570	.540	.515	.553	.528	.507	.494
4.	.623	.554	.503	.464	.609	.545	.498	.461	.529	.488	.455	.513	.478	.449	.499	.469	.444	.431
5.	.575	.498	.444	.404	.562	.490	.440	.402	.477	.432	.398	.464	.424	.394	.451	.417	.390	.377
6.	.530	.447	.391	.351	.518	.441	.388	.350	.429	.382	.347	.418	.376	.344	.407	.370	.342	.328
7.	.486	.399	.342	.302	.474	.393	.340	.301	.383	.335	.299	.373	.330	.297	.364	.325	.295	.283
8.	.445	.356	.299	.260	.435	.351	.297	.260	.342	.293	.258	.334	.289	.257	.326	.295	.255	.242
9.	.409	.318	.252	.225	.400	.314	.251	.224	.306	.257	.223	.299	.254	.222	.292	.251	.221	.208
10.	.365	.272	.216	.179	.356	.268	.214	.178	.261	.211	.177	.254	.209	.176	.248	.206	.178	.163

Lamp: HPS (Clear)
Lamp watts: 400
Lamp lumens (vert.): 50,000

Reflector: Diffuse, Hex
Lens: Teflon
Floor Reflectance: 20%

Socket Position	SC
#4	1.0
#7	1.5
#9	1.9

LUMINAIRE DISTRIBUTION DATA

Mid-Zone Angle	Candlepower		
0°	22.826	13.593	7.723
5°	22.699	13.648	3.104
15°	22.272	15.580	10.808
25°	18.781	15.986	13.032
35°	13.784	14.377	13.384
45°	3.223	11.557	12.412
55°	3.116	5.485	7.348
65°	684	1.121	1.774
75°	241	317	423
85°	71	88	105

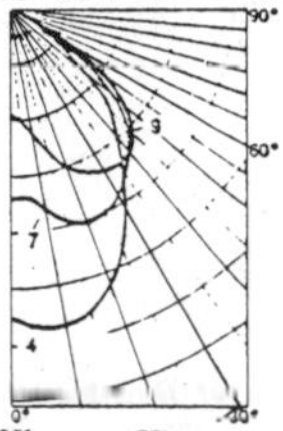

Ceiling Reflectance	80%				70%				50%			30%			10%			0%
% Wall Reflectance	70%	50%	30%	10%	70%	50%	30%	10%	50%	30%	10%	50%	30%	10%	50%	30%	10%	0%
0.	.871	.871	.871	.871	.851	.851	.851	.851	.313	.813	.813	.780	.780	.780	.750	.750	.750	.732
1.	.827	.805	.785	.767	.809	.739	.771	.755	.759	.745	.732	.732	.721	.710	.707	.598	.590	.676
2.	.785	.748	.717	.692	.769	.735	.707	.684	.712	.688	.569	.690	.671	.654	.669	.654	.540	.628
3.	.746	.697	.660	.630	.731	.687	.652	.625	.567	.638	.615	.650	.625	.605	.633	.613	.596	[illegible]
4.	.703	.645	.603	.570	.690	.637	.597	.587	.621	.587	.560	.506	.577	.554	.592	.587	.547	.536
5.	.664	.600	.555	.522	.652	.593	.551	.519	.580	.543	.515	.567	.535	.510	.556	.528	.506	.494
6.	.627	.558	.512	.479	.616	.552	.509	.477	.541	.503	.474	.531	.497	.471	.521	.491	.467	.456
7.	.591	.518	.471	.438	.581	.513	.469	.437	.504	.464	.435	.495	.459	.432	.486	.454	.430	.419
8.	.555	.480	.433	.401	.546	.476	.431	.400	.468	.427	.398	.460	.423	.396	.453	.419	.394	.384
9.	.521	.444	.397	.365	.513	.441	.396	.365	.433	.392	.363	.427	.389	.362	.420	.386	.361	.350
10.	.475	.394	.345	.313	.466	.390	.344	.313	.384	.341	.312	.378	.338	.311	.372	.336	.310	.299
0.	.892	.892	.892	.892	.871	.871	.871	.871	.832	.832	.832	.800	.800	.800	.764	.764	.764	.750
1.	.839	.814	.791	.771	.820	.797	.777	.758	.767	.750	.735	.728	.725	.713	.712	.702	.692	.678
2.	.789	.746	.710	.680	.772	.733	.700	.672	.708	.681	.657	.685	.662	.643	.663	.645	.629	.616
3.	.741	.685	.641	.606	.725	.674	.633	.601	.653	.619	.591	.634	.606	.581	.617	.593	.572	.559
4.	.691	.623	.573	.535	.678	.613	.567	.531	.598	.556	.525	.581	.546	.518	.566	.538	.512	.499
5.	.644	.568	.516	.476	.630	.561	.511	.474	.546	.503	.470	.533	.495	.465	.521	.487	.461	.448
6.	.600	.519	.464	.425	.587	.512	.461	.423	.500	.454	.420	.489	.448	.417	.478	.442	.414	.401
7.	.557	.471	.416	.377	.545	.466	.413	.376	.455	.408	.373	.445	.403	.371	.436	.398	.369	.356
8.	.516	.428	.372	.334	.505	.423	.370	.333	.414	.366	.331	.405	.362	.330	.397	.358	.328	.316
9.	.478	.388	.333	.296	.468	.384	.331	.295	.376	.328	.294	.368	.324	.293	.361	.321	.291	.279
10.	.430	.337	.281	.244	.420	.333	.279	.243	.326	.276	.242	.319	.273	.241	.312	.271	.240	.228
0.	.876	.876	.876	.876	.856	.856	.856	.856	.818	.818	.818	.783	.783	.783	.751	.751	.751	.736
1.	.819	.792	.767	.745	.800	.775	.753	.733	.745	.727	.710	.717	.702	.689	.691	.679	.669	.654
2.	.764	.717	.678	.645	.746	.704	.668	.638	.679	.649	.623	.656	.631	.610	.634	.614	.597	.583
3.	.711	.650	.602	.564	.695	.639	.595	.559	.618	.581	.550	.599	.567	.542	.581	.555	.532	.518
4.	.558	.582	.528	.487	.640	.573	.522	.484	.555	.512	.477	.539	.502	.471	.524	.492	.465	.452
5.	.505	.523	.466	.424	.591	.516	.462	.422	.501	.454	.417	.487	.446	.413	.474	.438	.409	.395
6.	.558	.470	.411	.369	.545	.464	.408	.367	.451	.401	.364	.439	.395	.362	.428	.389	.359	.345
7.	.512	.420	.360	.318	.500	.414	.357	.317	.403	.352	.315	.393	.347	.313	.383	.342	.311	.297
8.	.469	.375	.315	.274	.458	.370	.313	.273	.360	.308	.272	.351	.304	.270	.343	.300	.269	.255
9.	.431	.335	.276	.237	.421	.331	.275	.236	.323	.271	.235	.315	.268	.234	.308	.264	.233	.220
10.	.385	.286	.227	.188	.375	.283	.226	.188	.275	.223	.187	.268	.220	.186	.262	.217	.185	.172

Fig. 5-14. Photometric report of 400-watt luminaires with clear glass and Teflon.

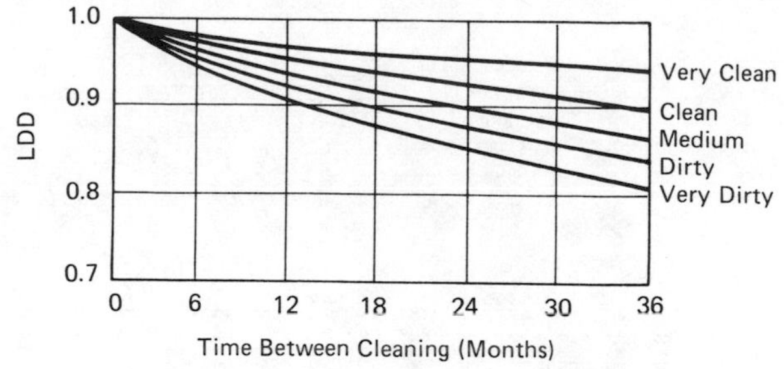

Fig. 5-15. LDD of the depicted example.

furnished by manufacturer (Figure 5-15), the LDD is found to be 0.89, corresponding to a cleaning interval of 24 months and the given "medium" environmental condition. With these figures, the LLF values can be found as follows:

Group Relamping Every 12,000 Hours (50% of rated life):

$$\begin{aligned} \text{LLF} &= 0.96 \times 0.90 \times 0.89 \\ &= 0.77 \end{aligned}$$

and, Group Relamping Every 16,800 Hours (70% of rated life):

$$\begin{aligned} \text{LLF} &= 0.85 \times 0.85 \times 0.89 \\ &= 0.64 \end{aligned}$$

Determine the Number of Luminaires

The number of luminaires can be found by the lumen formula:

$$\text{NOL} = \frac{\text{A} \times \text{E}}{\text{L} \times \text{LLF} \times \text{CU}}.$$

So with LLF = 0.64 and CU = 0.705,

$$\text{NOL} = \frac{150 \times 100 \times 100}{50{,}000 \times 0.64 \times 0.705} = 66,$$

and with LLF = 0.77 and CU = 0.705,

$$\text{NOL} = \frac{150 \times 100 \times 100}{50{,}000 \times 0.77 \times 0.705} = 55.$$

The benefit of cleaning and group relamping at 12,000 hour intervals is obvious.

Area occupied per luminaire = (150 X 100)/55 = 272.72 sq. ft.; spacing between luminaires (square pattern) = $\sqrt{272.72}$ = 16.5 ft. So the SC is 16.5/20 = 0.825, which is close to what we found in Figure 5-11. This is the best selection. The power consumption is 55 X 475 watts = 26,125 watts.

We rejected the 1000-watt HPS luminaire based on the conclusion we made from Figure 5-12. Let us see, however what the consequence would be if we did use this luminaire for our example. The CU for RCR = 1.7 is again interpolated and can be found to be = 0.706.

The LLF at 12,000 hour group relamping:

LBO = 0.96;
LLD = 0.90;
LDD = 0.89.

So, LLF = 0.96 X 0.90 X 0.89 = 0.76.

The number of luminaires = (150 X 100 X 100)/(140,000 X 0.76 X 0.706) = 20. The area per luminaire is (150 X 100)/20 = 750 sq. ft.; spacing between luminaires (square spacing) = $\sqrt{750}$ = 27.3 feet. So the SC is 27.3/20 = 1.37. Power consumption = 20 X 1100 watts = 22,000 watts.

Thus, although this would save a power consumption of 26,125 - 22,000 = 4,125 watts compared to the 400-watt system, the lighting level will be nonuniform, since the minimum SC required is 1.37 and the luminaire is designed for a maximum of 1.2. In addition, because the nadir candlepower is more than 40,000 candela, the illuminance directly underneath the luminaires will be much higher than 100 fc. This will result in discomforting hot and dark spots throughout the room.

Note, too, that if a luminaire could be found that had a minimum of 1.37 SC and centerline candlepower around the required 24,000 candela, it would be more energy saving than the second-best choice, the 400-watt luminaire.

Use of Fluorescents in Industrial Lighting

With the continuous development in HID sources, the use of fluorescents in industrial application is diminishing. Their use in industrial applications was dominant for a number of years until mercury vapor sources started replacing them, especially in high-bay areas. Lately, HPS is replacing almost all sources, primarily because of its long life and excellent energy-saving characteristics.

Fluorescent luminaires, however, are still the ideal solution for some unique industrial applications, e.g., long and narrow areas, such as between vertically stocked aisles in warehouses. These areas usually require equal importance in horizontal and vertical footcandles, which are difficult to provide by ceiling-mounted point sources. Continuous rows of fluorescent strips with reflectors can provide uniform horizontal and vertical illuminance more efficiently than any other type of luminaire under this condition. The mounting height of the continuous row of luminaires can be adjusted easily (suspended from ceiling by chain) to have sufficient light all along the area and yet waste no light above or beyond the aisles.

Calculations with Linear Sources

If the test distance is at least five times the larger dimension of the linear source, the inverse-square law can be used within 1% accuracy. This means that, in order to get a reasonably accurate result, a four-foot long fluorescent strip has to be a minimum of $4 \times 5 = 20$ feet away from the point of interest. The reader will recall the inverse-square law:

$$E = \frac{I}{D^2}$$

where

E = Illuminance in footcandles
I = Luminous intensity in candela
D = Distance in feet between the source and point of calculation

In order to determine the illuminance at a point that is not directly perpendicular to the source, an appropriate trigonometrical function must be considered to account for the change. This is shown in the following example.

Suppose that we are to determine the net illuminance at a point P on the floor that is obtaining lighting from two 4-foot double lamp sources, as shown in Figure 5-16. In order to find the appropriate candlepower, angle θ must be determined first and then the corresponding candlepower value can be looked up from the candlepower distribution diagram developed at a plane parallel to the lamp axis. This is shown in Figure 5-17. From Figure 5-16,

$$\cos\theta = \frac{11.5}{D} = \frac{11.5}{\sqrt{11.5^2 + 20^2}} = \frac{11.5}{23}$$

or

$$\cos\theta = 0.50$$

or

$$\theta = 60^\circ.$$

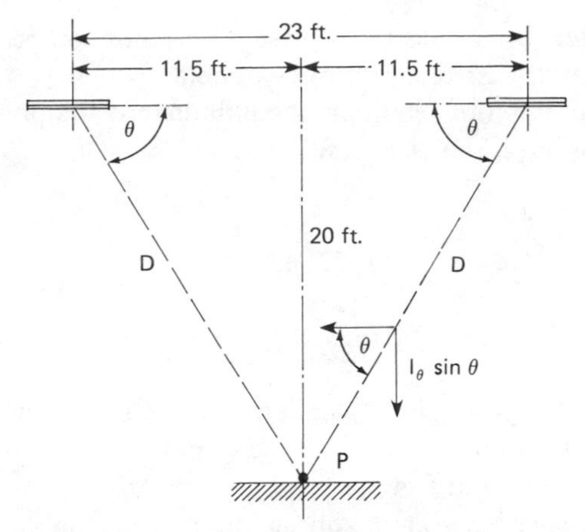

Fig. 5-16. Point P is located 20 feet below the luminaires, on the floor.

From Figure 5-17, candlepower at $\theta = 60°$ is 440 candela. The net illuminance at point P then, is;

$$E_P = 2\left[\frac{I_\theta \sin\theta}{D^2}\right] = 2\left[\frac{440 \times \sin 60°}{23^2}\right]$$

$$= 2\left[\frac{440 \times 0.866}{529}\right] = 1.44 \text{ fc.}$$

If the linear source is continuous, as is often necessary in warehouses, in between aisle spaces, the inverse-square law cannot be used. The illuminance from a line source of infinite length varies inversely as the distance to source and not

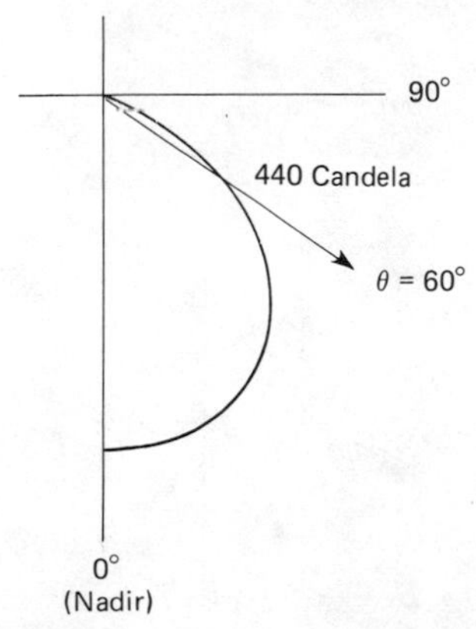

Fig. 5-17. Candlepower distribution diagram at a plane parallel to the axis of the lamps.

as the square of the distance as in the case with point sources, or special linear source cases such as discussed in previous example.

A simplified expression to determine the illuminance at a point directly underneath a continuous linear source is as follows:

$$E = \frac{L \times W}{2D}$$

where

E = Illuminance at a point directly underneath the source, in footcandles
L = Brightness of the linear source in footlamberts
W = Width of the source in feet
D = Shortest distance between the linear source to the point, in feet

This formula is exact if the source is linear and infinitely long. For other purposes, it will be accurate to within 10% if the distances d_1 and d_2 are greater than 1.5D, and to within 5% if both d_1 and d_2 are greater than 2D (see Figure 5-18).

Let us suppose we are to determine the illuminance at a point P on the floor that is 20 feet away and directly underneath a row of linear sources. Let us assume the luminaires used are 8-foot-long fluorescent strips with reflectors,

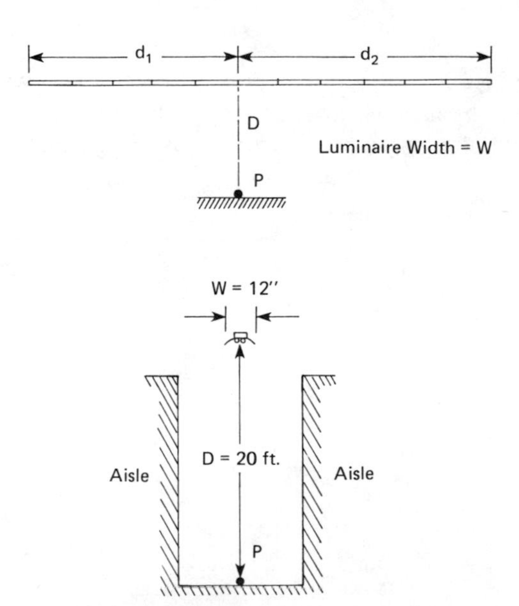

Fig. 5-18. Point P located directly underneath a row of lumen sources.

installed end to end, in between aisles, where length $d_1 = 40$ feet and $d_2 = 50$ feet (see Figure 5-18).

The first step is to determine L, the luminaire brightness. This can be found either by the formula

$$L = \frac{I_0 \times \pi}{A},$$

where

I_0 = Nadir candlepower
A = Area of the source

or directly from the luminaire's photometric data. Let us say, from the photometric report of the double-lamp, 8-foot-long fluorescent strip, the (average) brightness at 0° is 1196 footlamberts. Now, applying this value into the formula, $E = (L \times W)/2D$,

$$\text{Illuminance at point } P = \frac{1196 \times 1}{2 \times 20} = 29.9 \text{ fc.}$$

Of course, the results obtained in both the examples do not include any lighting contribution from room surface reflectances. In addition, these values represent the initial conditions only. A suitable LLF must be taken to account for any light loss as a result of environmental conditions or LLD.

Fluorescent-Strip Luminaires with Reflectors

Fluorescent-strip luminaires with reflectors have the advantage of providing somewhat higher candlepower towards the nadir. The main purpose of the reflectors is to minimize the upper-angle (70° and higher) light and reflect them towards more useful areas. This feature further makes these luminaires ideally suitable for long and narrow areas where uniform lighting is required on horizontal and vertical surfaces.

The reflectors are separately attached to the body of the luminaire. Removing a reflector from one side provides an asymmetrical lighting pattern that concentrates light at the side with no reflector. This provides a pool of light at one side with virtually no direct glare at the other side. The reflectors are of two types: those with and those without a vent (aperture). A vented reflector offers a better LLF by allowing a somewhat lower LDD.

In order to reduce energy consumption, cut operating costs, and make the overall system work more efficiently, lighting equipment manufacturers have introduced a variety of new "energy-saving" products in recent years; these were discussed in Chapters 1 and 4. Products that consume less power to produce an equal or a lower amount of light should be considered for retrofits only. Others producing higher amounts of light at lower amounts of power should be used for new applications. Careful analysis should be made before making a decision to use any combination of these products. None of these low-energy-consuming lamps is recommended for use with low-power-factor or dimming ballasts; these lamps should always be used at room temperatures of 60°F or higher. If they are operated below this temperature or if they are in the path of strong air drafts that blow directly on bare lamps, lamp flickering may occur. The krypton lamps operating at a slightly higher current cause a proportionate increase in ballast capacitor voltage. If these lamps are used in existing standard ballasts, they impose undesirable higher voltage across the existing capacitors and may cause some premature ballast failure. Typically, at a normal room ambient temperature of 70°F or higher or in enclosed luminaires that have a higher luminaire ambient temperature, the voltage rise is minimal (about 2-7%). At lower temperature, however, the capacitor voltage may rise up to 15%. When these lamps are equipped with suitable ballasts, the lower overall heat generation because of lower wattages will increase the ballast life as much as two-fold. The capacitor voltage rise in use with high-output or very-high-output, reduced-wattage lamps is negligible and has an insignificant effect on ballast failure.

Fluorescent light output and power consumption, as we have seen earlier, is significantly affected by its bulb-wall temperature. A fluorescent lamp is designed to peak in light output when the bulb-wall temperature is operating at 100°F. Any deviation from this value alters the lumen output and power consumption significantly. When a vented industrial fluorescent luminaire is operated at normal room temperaure (70°-77°F), the bulb wall operates at about 100°F, offering almost the peak light output. However, in a sealed and gasketed, enclosed luminaire, the effect is just the opposite. Enclosed trapped heat from continuous lamp burning increases the bulb-wall temperature, considerably reducing the light output as well as the power consumption. Use of lite-white lamps with low-loss ballasts would improve the situation substantially since the power consumption of this combination is inherently the lowest; as a result, the heat produced is also proportionally lower.

INTERCHANGING HID SOURCES

Interchanging an older, inefficient lamp with newer efficient HID source would bring more light for less operating cost and least initial expenses. However, the designer should be extremely careful in selecting the source. As shown in Chapter 1, each type of HID lamp operates on its own electrical characteristics, which include the special starting and operating voltages, operating current and its waveshapes, etc., which can be only provided by their appropriate ballasts.

Operating a source with a different type of ballast would not meet these requirements and result in unsatisfactory performance and possibly a premature lamp and ballast failure.

To make a lamp perform satisfactorily, it must be operated with the correct type of ballast for the designed wattage. If a correct type of ballast is used, an HID lamp of lower wattage probably would work in the high-wattage ballast. But, although the lamp will start, this will result in poorer lumen output, shorter life, and possibly a bulging of the arc tube that causes it to shatter. The use of a higher-wattage lamp in a lower-wattage ballast will cause the ballast to overheat and result in premature failure.

There are some specially designed lamps commercially available that can be interchanged with some existing ballasts. Most of these are designed to replace an older, inefficient source with more efficient ones. These are either metal halide or HPS to replace mercury lamps. There are also some special self-ballasted mercury vapor lamps, specifically designed to replace incandescent sources. In all applications, the manufacturer should be consulted to establish which of these products is suitable to work with the existing type of ballast.

Specially designed lamps like the ones mentioned above and in Chapter 1, in general have inherently low efficacy and shorter life when compared with their regular counterparts. The initial cost is also considerably higher. In addition, the use of some of these lamps to replace existing lamps inside the same reflector may cause undesirable photometric characteristics. With a different position of the new arc tube, the luminaire may produce light streaks or rings and even may reflect the energy back on the arc-tube. In HPS, this will bring further shortening of life of the lamp. Thus, the designer should evaluate the initial expenses saved in relation to the possible adverse outcome carefully and decide whether installing a brand-new luminaire system would be more beneficial. The consideration to interchange HID lamps should only be given where installation of a new system is impractical. In the majority of instances, a new system would certainly bring worthwhile, long-term benefits.

REFERENCES

American National Standard Practice for Industrial Lighting, ANSI/IES RP-7, 1979.

Baer, E. *Engineering Design for Plastics*. Huntington, N.Y.: Kreger, 1975.

Burkhardt, W. C. "High-impact Acrylic for Lenses and Diffusers." *Lighting Design & Application*, April 1977, pp. 12–20.

Clark, Francis. "Accurate Maintenance Factors–Part Two (Luminaire Dirt Depreciation)." *Illuminating Engineering*, January 1966, pp. 37–45.

Eagan, Connie. "Getting the Most Out of HID Lamps." *Electrical Construction and Maintenance*, March 1980, pp. 74–76.

Elmer, W. B. *The Optical Design of Reflectors*, 2nd ed. New York: Wiley, 1979.

IES. "Light Control and Luminaire Design, Lighting Calculations, and Light Sources."

IES Lighting Handbook, 1981 Reference Volume. XXX, Illuminating Engineering Society, 1981.
IES, Committee on Light Control and Equipment Design. "IES Guide to Design of Light Control–Parts I, II, III, and IV." *Illuminating Engineering,* 1959, 1967 and 1970.
McGowan, T. K., and A. L. Hart. "The Application of Reduced Wattage Fluorescent Lamps and High Efficiency Ballasts to General Lighting Systems." *IEEE,* 1979.
Sherwood, M. "Aluminum Reflectors: Improved Characteristics." *Lighting Design & Application,* December 1972, p. 34.

Chapter Six
Lighting and Energy Conservation Standards

In order to conserve energy and make efficient use of the country's resources, most states now have adopted energy-conservation standards. From the standpoint of conserving electrical energy, the following are some of the features that are common for most applications:

Selection of service voltage
Voltage drop in circuits or feeders
Individual metering systems
Power factor of electrical systems
Lighting switching
Lighting power budget

SELECTION OF SERVICE VOLTAGE

Where a choice of service voltages is available, a computation may be made to determine which service voltage would produce the least energy loss, and that voltage should be selected. A choice of service voltage is mostly available for commercial and industrial applications. For single family residences or small multifamily dwellings, the choice is usually limited to 120/240 volts, single-phase, three-wire systems. While almost all lighting sources are available in a wide range of voltages (except for incandescents, which are in general available for 120 volts only), the selection of service voltage does not have any influence on their energy consumption. A proper choice of service voltage is important in energy savings if the loads are mostly motors. A higher-horsepower motor works more efficiently with higher voltage. Voltage drop, wiring, circuit-breaker sizes, and maximum fault current availability are some of the other reasons for a proper voltage selection.

VOLTAGE DROP IN CIRCUITS OR FEEDERS

In buildings, the maximum voltage drop is not to exceed 3% in branch circuits or feeders, for a total of 5% to the farthest outlet based on steady-state design-load conditions. A drop in operating voltage will reduce light output in most cases, may shift color, or even extinguish the arc for HID units. For the purpose of efficiency, it is necessary that all lighting fixtures operate at rated voltage, or within the specified limits of voltage variation.

A quick method for determining the voltage drop at a point is by using the following formula:

$$VD = \frac{I \times L \times C}{V \times N}$$

where

VD = Voltage drop
I = Current of load
L = Length of conductor from source to load of calculations, in hundreds of feet
V = Applicable line voltage
N = Number of parallel runs
C = Conductor constant (voltage drop per ampere per 100 feet run

The values of C for different copper conductors are shown in Table 6-1. Let us take an example to illustrate its use. Suppose a group of luminaires is located 50 feet away from the panel board, which is 150 feet away from the utility company's transformer (see Figure 6-1). Find voltage drop at panel board (point A) and at the luminaires (point B). In the figure, the following information is given:

I = 400 amperes
V = 208 volts
L = 150 feet ÷ 100 feet = 1.5
N = 2
C = 0.0135 (from Table 6-1, for 208 volt three-phase and #3/0 conductors)

Applying the formula,

$$VD \text{ at point } A = \frac{400 \times 1.5 \times 0.0135}{208 \times 2}$$

$$= 0.0194, \text{ or } 1.94\%.$$

Table 6-1. Copper Conductor Constants for Determination of Voltage Drop.

Conductor Size (Copper)	Constant "C"	
	Single Phase	Three Phase
14 TW	0.6100	0.5280
12 TW	0.3828	0.3320
10 TW	0.2404	0.2080
8 TW	0.1502	0.1316
6 THW	0.0970	0.0840
4 THW	0.0614	0.0531
3 THW	0.0484	0.0420
2 THW	0.0382	0.0331
1 THW	0.306	0.0265
1/0 THW	0.0243	0.0210
2/0 THW	0.0194	0.0168
3/0 THW	0.0155	0.0135
4/0 THW	0.0123	0.0107
250 MCM	0.0105	0.0091
300 MCM	0.0088	0.0076
350 MCM	0.0076	0.0066
500 MCM	0.0054	0.0047
750 MCM	0.0037	0.0032
1000 MCM	0.0028	0.0025

The available voltage at this point is then 118/204 volts. Now for the voltage drop at point B from point A, consult Figure 6-1:

$I = 16$ amperes
$V = 118$ volts
$L = 50 \div 100 = 0.5$
$N = 1$
$C = 0.3828$ (from Table 6-1 for single-phase, and #12 conductors)

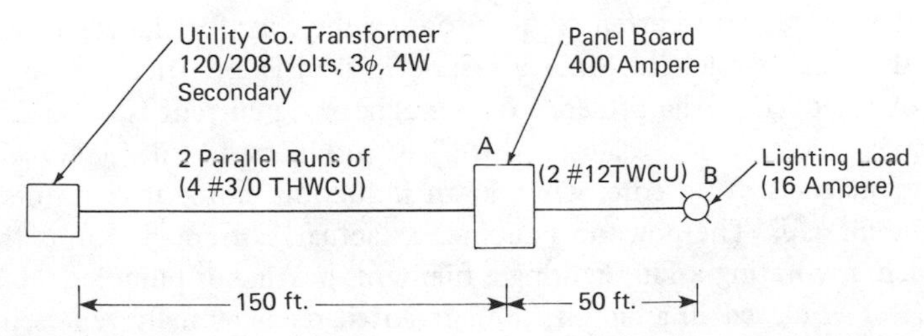

Fig. 6-1. One line diagram of the example to calculate voltage drop at points A and B.

Applying the formula,

$$VD \text{ at point B} = \frac{16 \times 0.5 \times 0.3828}{118 \times 1}$$

$$= 0.025, \text{ or } 2.5\%.$$

The voltage at this point is thus $114.9 \approx 115$ volts. The total voltage drop at this point is $(120 - 115)/120 \times 100 = 4.1\%$.

INDIVIDUAL METERING SYSTEM

This is a requirement for all multitenant residential buildings, e.g., condominium or apartments. A common metering system usually has little or no impact in energy savings since the tenants do not directly pay for their energy use; as a result, they usually do not care how long their lights are left on or how high the comfort heating level actually is. Individual metering systems make them energy conscious since the cost is directly reflected on their utility bills.

POWER FACTOR OF ELECTRICAL SYSTEMS

For most states, it is necessary that the overall power factor of a building or industry be at least 90%. In some states, the energy conservation standard additionally requires that all utilization equipment rated greater than 1000 watts and lighting equipment rated greater than 15 watts with an inductive reactance load component have a power factor of not less than 85% under rated load conditions. If the overall power factor is less than 85%, it is required to be improved to at least 90% under rated load conditions.

Before we discuss the reasons for improving power factor, or how it is related to lighting, it is necessary that we understand what power factor is. Operation of any inductive load is done by two types of current, namely, the magnetizing current (also known as the reactive current) and the power-producing current (also known as the actual current). Typical examples of inductive load are induction motors, welders, transformers, ballasts, etc. The main function of the magnetizing or reactive current is to produce the flux for the electromagnetic field of the inductive loads. Since a magnetic field is essential for proper functioning of the devices, the presence of a magnetizing current is essential. For a transformer, such as in a ballast, this flux produces the necessary electrical energy to magnetize the core, while in an induction motor it transmits energy across the air gap. The power-producing, or actual, current performs the main work, such as rotating a fan, lighting a filament, heating or pumping, etc. When an ammeter is placed in a circuit, the current it reads actually represents both

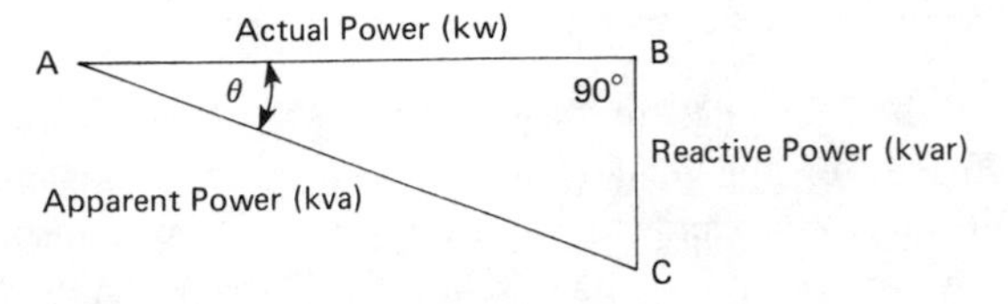

Fig. 6-2. Power triangle representing the Pythagorean Theorem.

the currents. This current is commonly known as the *apparent current*. Mathematically, the relation of the three types of current is given by

$$\text{Apparent current}^2 = \text{Actual current}^2 + \text{Reactive current}^2.$$

The relation can now be converted to power by multiplying the operating input voltage by each of the three types of current:

$$\text{Apparent power}^2 = \text{Actual power}^2 + \text{Reactive power}^2.$$

The relation representing the pythagorean theorem of the right-angle relation can be expressed as shown in Figure 6-2. The unit for the apparent power is the kilovolt-ampere (KVA); actual power is kilowatt (KW); reactive power is reactive-kilovolt-ampere (KVAR).

Note that if the angle between KW and KVA is θ, trigonometrically,

$$\text{KVA} \times \cos\theta = \text{KW} \quad \text{or} \quad \text{KW/KVA} = \cos\theta.$$

The factor $\cos\theta$ of this expression is known as the *power factor*, or simply pf.

Power factor is a numerical number expressed either as a decimal or in a percentage. The largest pf any load can have is unity, and this can happen only when the KW and KVA values of a load are equal, $\cos\theta = 1$, meaning angle $\theta = 0$. Similarly, the lowest pf any load can have is zero; and this can happen only when $\cos\theta = 0$, or $\theta = 90°$. What this basically means is that as the angle θ decreases from 90° to 0°, the pf increases from low to high.

It is important to note at this point that although the house meter (furnished by utility company) reads only the actual power or KW usage, the utility company furnishes the total power or the apparent power in KVA. A low pf will thus mean using a high KVA but paying only for KW. This will also mean a low pf will draw high apparent current, increasing the cost of conductors and disconnecting means for the customers, as well as for utility companies, and also in a small way, necessitating a larger generator. To discourage a low pf, many utility companies impose a penalty charge, which varies from one utility company to another. Usually a pf lower than 0.85 is considered low.

Leading- and Lagging-Type Power Factor

A pf can be of the "leading" or "lagging" type. These expressions indicate whether the current of the load is in phase, leading, or lagging the voltage. In an AC circuit, the voltage and current are represented by sinusoidal curves.

Figure 6-3 shows the current-voltage relationship for a system where the current is in phase with the voltage. This means that for the type of load, both the voltage and current sinusoidal curves start at the same instant and reach positive and negative peaks at the same angle and instant. The figure also shows the power development for each instant. The power curve is obtained from the product of volts times the current for any instant. Note that the power is always positive (+) in this type of volts-and-current characteristic, since at any instant, the product of the two is always positive.

Examples of loads producing in-phase, current-voltage sinusoidal curves are incandescent lights and resistance heaters. An in-phase relationship between voltage and current means that pf = 1 (unity), since angle $\theta = 0$, or $\cos \theta = 1$. This can only happen when KW = KVA of the load (see Figure 6-4a).

Figure 6-5 shows the typical current-voltage relationship of a lagging-type current load. Here current is found to start after an angle of θ has already elapsed. Note that the peak, positive or negative, and the zero current do not coincide with those of the voltage. The current at any instant, is lagging the voltage by an angle θ. Unlike the previous example, since the current and voltage are not always simultaneously of the same sign (positive or negative), there are instances where power is negative. The net usable power in this case is less than that found in Figure 6-3. The pf of this situation is always less than unity and is known as "lagging" (see Figure 6-4b). Examples of lagging-type loads are transformers, induction motors, and reactive ballasts.

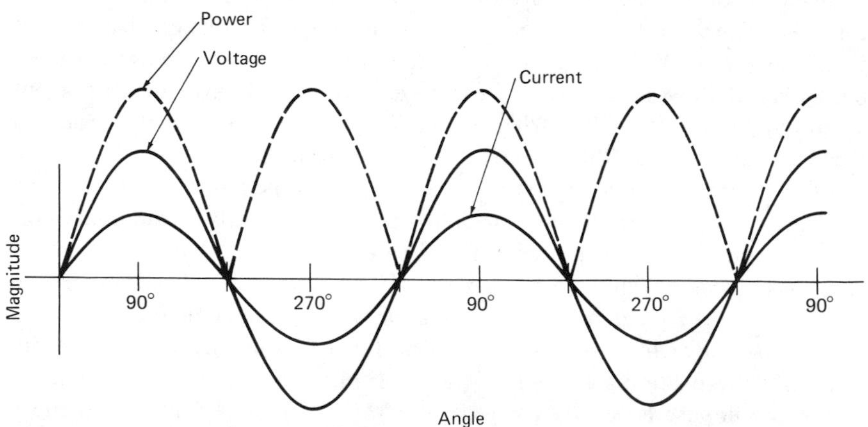

Fig. 6-3. Sinusoidal curves with current in phase with the voltage. Power curve is the product of the two and is positive at all instants. The pf of this type of load is unity.

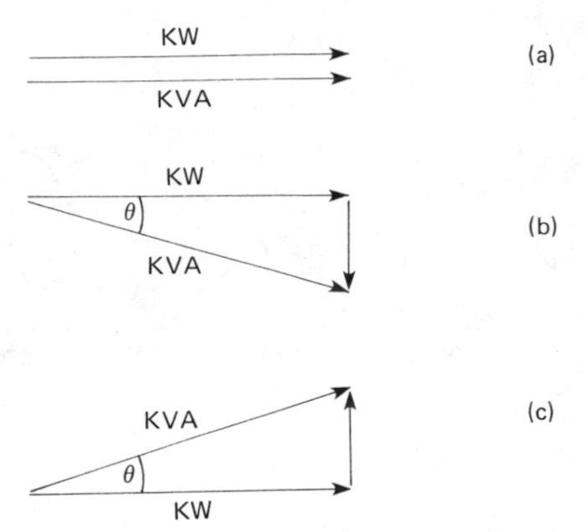

Fig. 6-4. (a) In phase, or unity pf. (b) Lagging pf. (c) Leading pf.

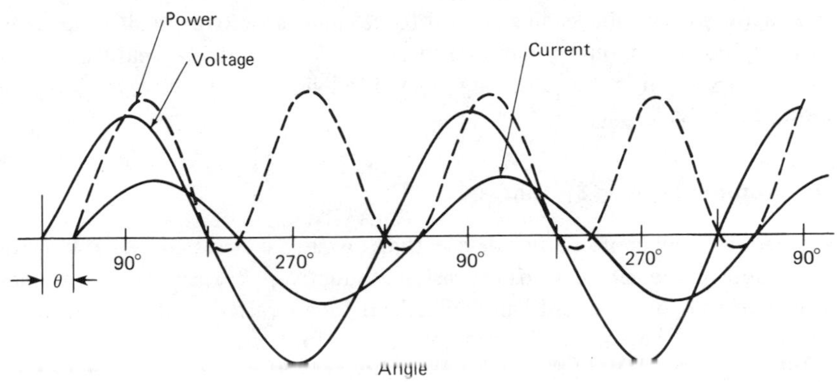

Fig. 6-5. Sinusoidal curves representing voltage, current, and power relationship of a load with lagging pf.

The concept of a leading pf can be developed similar to the example shown above, except that the current leads the voltage instead of lagging. This is shown in Figure 6-6. Note that here too the total net usable power is less than that found in Figure 6-3. The pf of this situation is also always less than unity and known as "leading" (see Figure 6-4c). Examples of leading-type loads are the "improved"-pf-type of ballasts, overexcited synchronous motors, and capacitors.

How to Improve Power Factor

There are several ways to improve a low pf. The main idea is to introduce a leading type of load to correct a lagging pf, and vice versa. Most industries utilizing large amounts of inductive loads are generally faced with the lagging-type pf. Use of capacitors is a good, economical solution to this problem. Capacitors

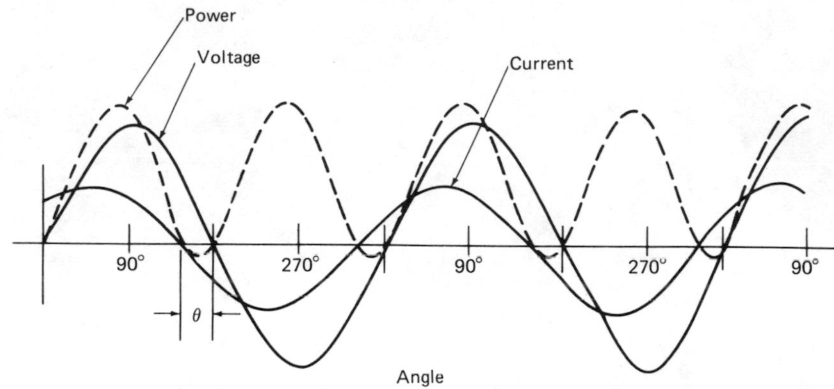

Fig. 6-6. Sinusoidal curves representing voltage, current, and power relationship of a load with leading pf.

correct a low pf since the leading current found in a capacitive circuit is opposite to the lagging current in an inductive circuit. When the two circuits are combined, the net result is the cancelling of each individual's influence, bringing the pf closer to unity.

Power Factor in Lighting Systems

Power factor is inherent in any device that consumes electricity. As we have seen previously, the nature of load (resistive, inductive, or capacitive) determines the type of pf. Incandescent lights, which include quartz-iodine and tungsten-halogen, being of the resistive filament type, offer unity pf.

In fluorescent and HID lighting systems, however, the pf may be high or low, and either lagging or leading. The gaseous discharge sources make use of ballasts that contain transformers and are thus inherently inductive in nature. If no correction is made, the pf of a ballast may be as low as 30–40% and lagging. When correction is made, it may improve to as high as 90% or higher, leading.

As we have seen in Chapter 1, the basic types of ballasts are designed to operate either one lamp or two lamps simultaneously. Ballasts for multilamp operation are available, but they are not popular because of economics. From a pf standpoint, however, all fluorescent ballasts can be divided into four types, namely,

1. lag type,
2. lead type,
3. lead-lag type,
4. series-lead type.

The lag-type ballasts are created to operate with a single lamp and at low wattage (usually between 6 and 30 watts, nominal). With no special correctional means, these offer a low, lagging pf and are usually the least expensive of all fluorescent ballasts. Although a lagging, low pf contributes to a lower overall pf, because these ballasts are used only for lower wattage and special lamps, their effect in overall building pf is usually insignificant.

A lead-type ballast also operates on a single lamp but offers leading, high pf. The improvement in pf results when a capacitor is placed across the primary or across a winding to step up the voltage of the primary. Some have the capacitor placed in series with the lamp, causing a leading current in the lamp. This not only improves the pf by offsetting the necessary magnetizing current; it also provides a better-regulated lamp current should the voltage vary. The lead-type ballasts are created mostly for the use of 40-watt and higher wattage lamps. A high pf means a pf greater than 90%.

A lead-lag-type ballast is used for a pair of lamps. These are generally the preheat or switch-start types, consisting of a reactor or autotransformer that provides low-pf operation of the first lamp, with a lagging current, and then provides the second lamp through a capacitor, with a leading current. The leading current tends to cancel out the lagging current and thus offers a near-0° phase difference between the voltage and current. The result is a high pf. Being so connected, if one lamp fails, the other remains unaffected, with no damage to the ballast. However, depending upon which lamp failed, the overall pf may be either leading, offering higher pf, or lagging, offering lower pf.

The series-lead (or series-sequence) type of ballast is also designed to operate a pair of lamps—however, with a significant difference. In this type of circuit, because the lamps are connected in series, with the failure of one lamp, the other one also goes out. However, these ballasts offer a high, leading-type pf, since they make use of capacitors. The series-lead ballasts are the most popular types since they save in space and cost, and provide optimum light output. The basic principle starts one lamp at a time and then puts them in series with a capacitor in the line to improve the pf and to regulate the current.

Different combinations of ballasts and lamps produce different types of electrical characteristics. Each ballast manufacturer shows the input wattage, voltage, line current, etc. as they apply to specific lamp-ballast combinations. While the type of ballast circuit determines whether the pf is leading or lagging, the numerical value of the pf is obtained by calculation. This is illustrated as follows: Suppose we are to calculate the pf of a series-lead type of ballast connected to two F40 T12/RS lamps. From the ballast manufacturer's catalog, we find the following:

Input circuit voltage = 120 volts
Line current = 0.80 amperes
Input watts = 95 watts
Ballast circuit type = series-lead

From the basic formula for power factor (pf), we now find P.F. = Watt/Volt-amp. = 95/120 × 0.8 = 0.98. So, the pf of the system is 98%. It is a high pf, since it is higher than 90%. Also, the pf is leading, since the type of ballast circuit selected is a series-lead type.

Sample Problem for Power Factor Determination

We have seen the effect of pf on different types of lighting systems. A high pf is always desirable because of the advantages mentioned earlier. If a building's overall pf is low, the contribution from one luminaire with a very high pf will have insignificant effect on the resultant pf. However, if all the luminaires have high pf, collectively they may bring significant improvement to the overall pf. We will take an example to illustrate how the pf of a typical application can be calculated, and how the luminaires affect the overall pf.

Suppose that the building has the following:

1. An induction motor, 20 hp, @0.8 pf (lagging)
2. Resistance heater, 5000 watts, 1.0 pf
3. Fractional hp motors, 30 hp total, 0.4 pf (lagging)
4. Fluorescent lighting, 25,000 watts total, 0.92 pf (leading)
5. HID lighting, 3000 watts total, 0.9 pf (lagging)
6. Incandescent lighting, 1000 watts total, 1.0 pf

The first step is to form a table as shown in Table 6-2, converting horsepower ratings to wattage (W), determining the volt-amperes (VA) and the reactive volt-ampere (VAR).

Since 1 hp = 746 watts,
20 hp = 746 × 20 = 14,920 watts
and 30 hp = 746 × 30 = 22,380 watts

Table 6-2. Power Factor Calculating Table.

Load	hp	Watt	pf = $\cos\theta$	$VA = W/\cos\theta$	$VAR = \sqrt{VA^2 - W^2}$
Induc. motor	20	14,920	0.8 (lag)	18,650	11,190 (lag)
Resist. heat		5,000	1.0	5,000	0
Frac. hp motors	30	22,380	0.4 (lag)	55,950	51,279 (lag)
Total		42,300			62,469 (lag)
Fluorescents		25,000	0.92 (lead)	27,174	10,650 (lead)
HID lighting		3,000	0.90 (lag)	3,333	1,452 (lag)
Incand. lighting		1,000	1.0	1,000	0
Grand total		71,300			53,271 (lag)

Note that the total obtained for the VAR column is the algebraic sum of the different leading and lagging types of load. The lead-type loads are added together and subtracted from the addition of the lag-type loads. The net result is lag since the total of the lag loads is larger than the total of the lead loads.

To determine the pf of the building before the lighting loads are added, we draw the power triangle, shown in Figure 6-7a. Note that the VAR is drawn downwards since the load is lagging. So, from the triangle,

$$\begin{aligned} \text{VA} &= \sqrt{\text{W}^2 + \text{VAR}^2} \\ &= \sqrt{42{,}300^2 + 62{,}469^2} \\ &= 75{,}443. \end{aligned}$$

and

$$\cos\theta_1 = \text{W/VA} = 42{,}300/75{,}443 = 0.56 \text{ (lagging)}.$$

Thus, the overall pf of the building, before the lighting loads are added, is 0.56, or 56%, and lagging type. This is certainly a low pf.

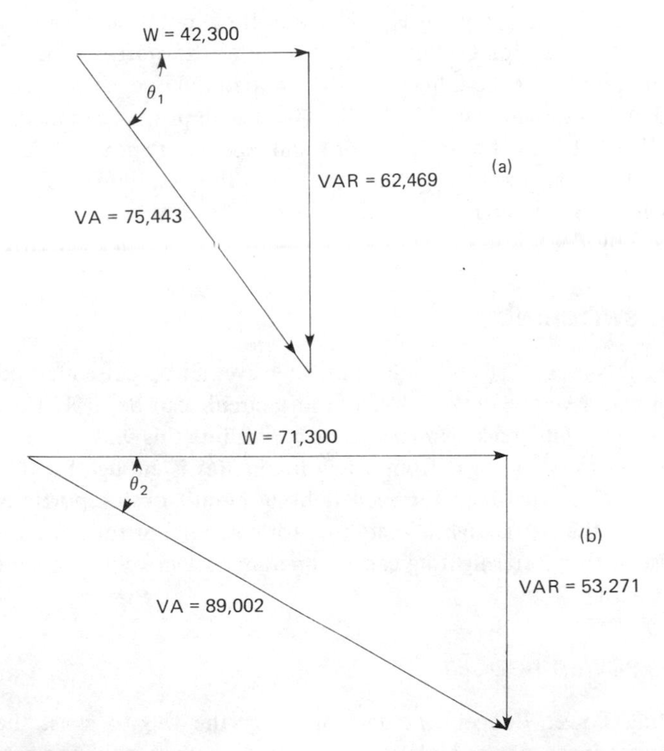

Fig. 6-7. (a) Power triangle of the depicted example before the lighting load is added. (b) After the lighting load is added.

Let us now see the effect of adding the lighting loads. From Table 6-2, the grand total of W = 71,300 and that of VAR = 53,271 (lagging). Drawing the power triangle again (this is shown in Figure 6-7b),

$$\begin{aligned} VA &= \sqrt{W^2 + VAR^2} \\ &= \sqrt{71{,}300^2 + 53{,}271^2} \\ &= 89{,}002. \end{aligned}$$

so

$$\cos\theta_2 = W/VA = 71{,}300/89{,}002 = 0.80 \text{ (lagging)}.$$

The new pf is thus 0.80, or 80% (lagging), which is a significant improvement over the original 56%.

The power factor can also be determined by graphical method. This method adopts the technique of algebraically adding all the power triangles of individual loads, as shown in Figure 6-8. Starting from the first load, all other loads are consecutively added from the connecting point of the VAR and the VA of the previous load. For lagging loads this connecting point is always below the wattage line, whereas for leading loads it is above the wattage line. For unity pf loads this point is at the other end of the horizontal line, since the magnitude of VA and W is the same and VAR = 0. The overall pf now can be obtained by drawing a line between the starting point and connecting point of the VAR and the VA of the final load. This line represents the resultant VA of the total system. The pf of the complete system now can be determined by dividing the total wattage by the VA.

LIGHTING SWITCHING

This is probably the most convenient, inexpensive, and precise method of lighting control and saving energy. All lights in a circuit can be easily turned on by operating the circuit breaker in the panel board, but this will waste energy for those applications where light from a few luminaires is enough for the purpose. Switching must be provided for each lighting circuit or for portions of each circuit, so that the partial lighting required for custodial or for effective complementary use with natural lighting can be operated selectively. This is discussed further in Chapter 7.

LIGHTING POWER BUDGET

The Lighting Power Budget was developed by the IES to assist the lighting designer in setting a power budget for the lighting portion of the total design.

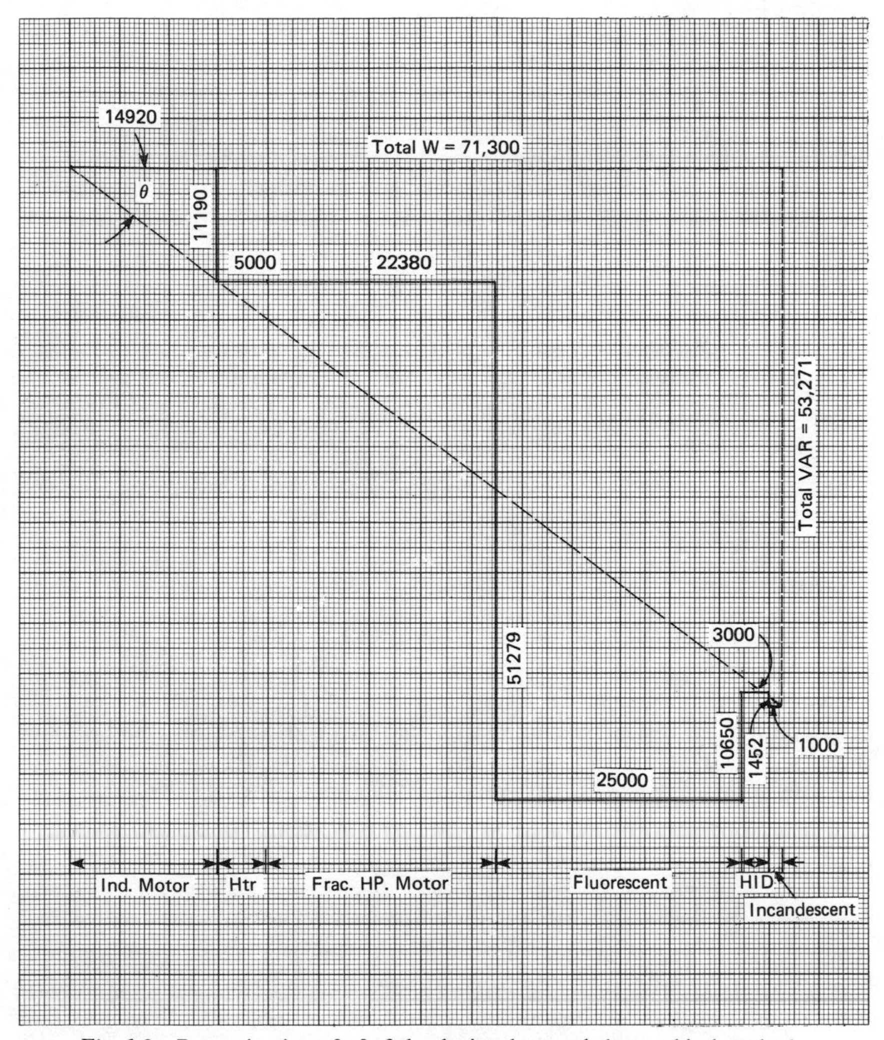

Fig. 6-8. Determination of pf of the depicted example by graphical method.

The idea was to get rid of the conventional, mostly wasteful practice of providing rooms full of lighting. The budget is developed, based on the necessary task lighting and ambient lighting around the task areas, with the use of efficient luminaires and efficient room-surface reflectances. Once the budget is set within the specific guidelines, the designer has the freedom of developing the final design any way that he or she finds most efficient and effective, staying within the limits.

The procedure has been officially accepted by several states and is treated as one of the mandatory standards to obtain a building permit. The Lighting Power

Budget, in general, is applicable for new buildings, where an abundance of light is usually used and where lights are left on continuously for long periods of time, e.g., offices, schools, banks, factories, and warehouses. However, it does not apply to certain areas of the buildings, such as the hallways, stairways, lobbies, or utility rooms. Other buildings, such as theaters, auditoriums, residences, and churches, are excluded from the Lighting Power Budget. Applicable areas of the Lighting Power Budget varies from state to state, so it is advisable for the designer to be familiar with the up-to-date rules on the matter, before making any calculations.

Two things are to be remembered here. First, the procedure actually deals with the power connected, rather than with energy consumption. In other words, the calculation procedure dictates only the maximum power that will be allowed at any instant for a specific lighting application; it does not limit the duration of usage. Second, the calculation procedure allows the designer a ceiling of total power usage; it does not interfere with the actual design of the lighting system.

Calculation Procedure

The calculation procedure is based on some preset parameters.

Determination of Illuminance Levels and Areas. The rooms or spaces under consideration are basically assumed to have three types of areas: the task area, the general area, and the noncritical area. As was mentioned earlier, the concept of Lighting Power Budget is based on providing adequate light in task areas and relatively lower levels of light in nontask areas, thus avoiding the conventional approach of providing a room full of task lighting. The task area is the area where the main work is done in the room. It may be on a desk located several feet above the floor or on the floor. There may be several task locations in a room, each needing a different lighting level. However, there is a restraint on how much of the total space may be considered to require task lighting. A maximum of 50% of the total area in any given space may be considered task area; the remaining area is considered nontask area, consisting of general area and noncritical area. The level of illuminance for each task area may be found from the IES handbook. When the actual task area is not known, each task area is assumed to occupy 50 square feet, with a maximum of one-half the room area. If the calculated value of the total task area is found to be more than 50% of the room area, then individual task areas are reassumed to be proportionally smaller than 50 square feet to satisfy the 50% area criterion. This is explained in the following example.

1. Suppose a room's dimensions are 40 × 30 feet and there are 10 task locations. The total area of the room is 40 × 30 = 1200 square feet. Assuming

50 square feet per task area, the total task area is 50 X 10 = 500 square feet. Since 500 is less than 50% of 1200, the assumption of 50 square feet for the task area will be valid.

2. Let us now assume that the total number of task areas is 12. So the total new task area is 50 X 12 = 600 square feet. Since 600 is exactly equal to 50% of 1200, the assumption of 50 square feet per task area is still valid.
3. Let us now assume that the total number of task areas is 14. The total new task area is 50 X 14 = 700 square feet. Since 700 is larger than 50% of 1200, the assumption of 50 square feet per task is not valid anymore. It has to be readjusted, to be smaller than 50 square feet. So the new assumed area per task location is 600/14 = 42.85 square feet, so that 42.85 square feet X 14 task locations = 600 square feet.

The general area is the area surrounding the task area, which provides general ambient lighting and probably even supplement the lighting on the task areas. For budgetary purposes, if the non-task area is larger than task area, then general area will be equal to task area. The maximum general area of a room is limited to 50% of the total room area, and its illuminance level can be no larger than one-third that of the task area and no less than 20 fc. If there are several task areas of different illuminance level requirements, the general area lighting level will be one-third of the weighted average of the specific task levels.

If the total area is smaller than 50% of the room area, and an equal amount is the general area, then the remaining will be considered to be noncritical area. Noncritical areas are the circulating or sitting areas where no specific visual tasks are done. The lighting level in these areas is to be one-third that of the general area (or one ninth that of the task area); under no circumstances is less than 10 fc. If it is found that a room has 50% of its area as task area, the remaining 50% is automatically assumed to be general area. In such circumstances there will be no noncritical areas.

Using the previous examples, we will determine the general and noncritical areas in each case. In the first example, the total room area is 1200 square feet, and total task area is 500 square feet. Since 500 is smaller than 50% of 1200, the general area is equal to 500 square feet (equal to the task area), and the remaining 1200 - 500 - 500 = 200 square feet is noncritical area. In the second example, the total area is again 1200 square feet, but the total task area is 600 square feet. So the remaining area, which is equal to 50% of the total room area, will be the general area. In this case, there is no noncritical area. In the last example, the adjusted total task area is 600 square feet; as in the second example, all of the remaining area is general area and there is no noncritical area.

Determination of Lighting Systems Data. For the calculation of power budget purposes, the lamps selected are to have the following initial lumen output per watt, including ballast losses.

Application	Lumens per watt
Where moderate color rendition is appropriate	55
Where good color rendition is appropriate	40
Where high color rendition is appropriate, spaces are less than 50 sq. ft., or where low-wattage HID lamps under 250 watts or fluorescent lamps under 40 W are appropriate	25

The luminaire CU can be obtained from the IES handbook or RP-15. If the luminaire selected is not listed, manufacturer's data can be used. However, in all cases, no luminaire shall have a CU for RCR = 1 of less than that given in the following table.

Space Use	Min. CU @RCR = 1
For space with tasks subject to veiling reflections	0.55
For spaces without tasks, or with tasks not subjected to veiling reflections but where visual comfort is important	0.63
For spaces without tasks and where visual comfort is not a criterion	0.70

For the purpose of a selection, the reflectances of ceiling, wall, and floor cavities are 80/50/20%, respectively. The LLF in all cases is 0.70.

In determining the Power Budget, the following formula can be used:

$$P = \frac{A \times E}{0.7 \times F \times \mathrm{CU}}$$

where

P = Power consumed by all lights in the room, including ballast losses
A = Area in square feet
E = Illuminance in footcandles
CU = Coefficient of utilization
F = Efficacy or lumens per watt

Note that this formula is derived from the basic formula of lumen method, as follows:

$$N = \frac{A \times E}{L \times \mathrm{LLF} \times \mathrm{CU}}$$

where

N = Total number of luminaires
A = Area in square feet
E = Illuminance in footcandles
L = Total initial lumens
LLF = Light loss factor
CU = Coefficient of utilization

Multiplying each side of this equation by the power consumed by each luminaire (including losses in ballasts), and calling it W,

$$N \times W = \frac{A \times E \times W}{L \times \text{LLF} \times \text{CU}}$$

or

$$\text{Total power} = \frac{A \times E}{\text{LLF} \times \text{CU}} \times \frac{W}{L}$$

The factor W/L is actually the reciprocal of lumens per watt, or the efficacy of the lamps used for budget purposes. If F is the value used for lamp efficacy, then $W/L = 1/F$. Also, since the LLF = 0.7 (constant), the formula now can be written, as above,

$$P = \frac{A \times E}{0.7 \times F \times \text{CU}}$$

Let us take an example to illustrate the determination of the Lighting Power Budget. Figure 6-9 shows a typical floor plan of an office building. The dimensions of various rooms are as shown in the figure. The ceiling is 8 feet above the

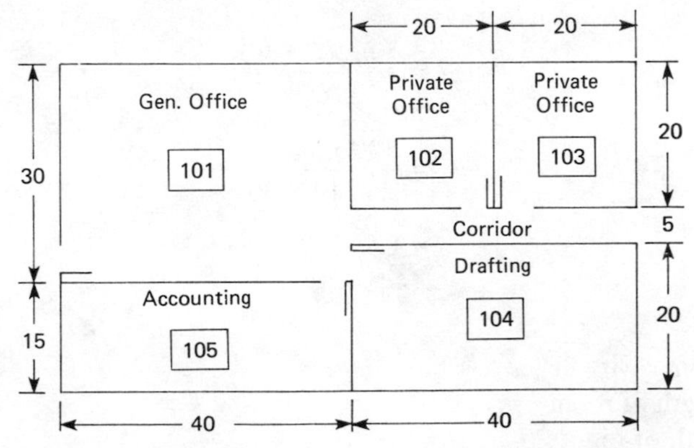

Fig. 6-9. Floor plan of depicted example for calculation of the lighting power budget.

Table 6-3. Types of Task and Illuminance Level.

Room No.	Name	Task (T)	No. of Task	Illuminance (E) (in fc.)
101	General office	T_1: Medium pencil/ink	12	E_1: 70
		T_2: Active filing	2	E_2: 100
102	Private office	T_1: Medium pencil/ink	2	E_1: 70
103	Private office	T_1: Medium pencil/ink	2	E_1: 70
104	Drafting	T_1: Detailed drafting, designing	8	E_1: 200
105	Accounting	T_1: Auditing, bookkeeping	6	E_1: 150

floor. The most important and possibly the most difficult information to be obtained at this point concerns the various types and number of tasks involved in each room. This can usually be obtained from the architect, the owner, or the tenant of the space. If information is not available, a realistic assumption has to be made. Let us suppose the information is as in Table 6-3.

Room 101, General Office: $L = 40$ ft.; $W = 30$ ft.; task ht. = 2.5 ft.; room ht. = 8 ft.; gross area, $A_g = 1200$ sq. ft.; room cavity height, $h_{rc} = 8 - 2.5 = 5.5$ ft.

With these parameters, the next step is to determine the appropriate CU. For this, two pieces of information are needed; the type of luminaires to be used and the RCR. Let us assume the luminaires to be 2 X 4-foot troffers with acrylic prismatic lens, and four rapid-start fluorescent lamps, as shown in Figure 2-1 in Chapter 2. This luminaire qualifies for the Lighting Power Budget calculations since its CU value, at RCR = 1 and room surface reflectance = 80/50/20, is larger than 0.55. The RCR can be determined by using the following formula:

$$\text{RCR} = \frac{5\, h_{rc}(L + W)}{L \times W}$$

$$= \frac{5 \times 5.5\,(40 + 30)}{40 \times 30}$$

$$= 1.6$$

where

h_{rc} = Room cavity height
L = Length of room
W = Width of room

From the CU table of Figure 2-1, CU = 0.66 (interpolated). Since exact task-area dimensions are not known, each task area is assumed to be 50 square feet. The total task area is $(12 + 2) \times 50 = 700$ sq. ft. This is larger than 50% of the gross area, A_g (1200 sq. ft.), so the area per task has to be readjusted so that total task area will not exceed 600 sq. ft. The maximum allowable area per task is $600/14 = 42.8$ sq. ft., rounded to 40 sq. ft.

We can readjust the task areas as follows:

Task 1: 12 work locations, $E_1 = 70$ fc.
task area, A_1 is $12 \times 40 = 480$ sq. ft.

Task 2: 2 work locations, $E_2 = 100$ fc.
task area$_1$, A_2 is $2 \times 40 = 80$ sq. ft.

The power allowed per task type is, then,

$$\text{Task 1:} \quad P_1 = \frac{A \times E}{0.7 \times F \times \text{CU}}$$

$$= \frac{480 \times 70}{0.7 \times 55 \times 0.66} = 1322 \text{ watts;}$$

$$\text{Task 2:} \quad P_2 = \frac{80 \times 100}{0.7 \times 55 \times 0.66} = 315 \text{ watts.}$$

The total task area is $480 + 80 = 560$ sq. ft., and total nontask area is $1200 - 560 = 640$ sq. ft. Since the nontask area is larger than the task area, the general area is equal to the task area (560 sq. ft.), and the remaining is the noncritical area ($1200 - 560 - 560 = 80$ sq. ft.).

The illuminance for the general area (E_g) is one-third of the weighted average of the total task area and in no case less than 20 fc. So

$$E_g = \frac{(E_1 \times A_1) + (E_2 \times A_2)}{3\,(A_1 + A_2)}$$

$$= \frac{(70 \times 480) + (100 \times 80)}{3\,(480 + 80)}$$

$$= \frac{33600 + 8000}{1680} = 24.7 \text{ fc.}$$

The power allowed for the general area is, then,

$$P_g = \frac{560 \times 24.7}{0.7 \times 55 \times 0.66} = 544 \text{ watts.}$$

The illuminance for the noncritical area is either 10 fc. or one-third of that in the general area, whichever is larger. The calculated value is $\frac{1}{3} \times 24.7 = 8.23$ fc. So noncritical illuminance = 10 fc. The noncritical area power is

$$P_{nc} = \frac{80 \times 10}{0.7 \times 55 \times 0.66} = 32 \text{ watts.}$$

So the total power in Room 101 is (1322 + 315) + 544 + 32 = 2213 watts.

Room 102, Private Office: $L = 20$ ft.; $W = 20$ ft.; $A_g = 20 \times 20 = 400$ sq. ft.; $h_{rc} = 5.5$ ft.

$$\text{RCR} = \frac{5 \times 5.5\,(20 + 20)}{20 \times 20} = 2.75,$$

so CU = 0.58 (interpolated). The total task area is $2 \times 50 = 100$ sq. ft. No adjustment in the task area is required, since it is less than one-half the gross area.

$$\text{Task area power, } P_1 = \frac{100 \times 70}{0.7 \times 55 \times 0.58} = 313 \text{ watts.}$$

The nontask area is 400 - 100 = 300 sq. ft. Since the nontask area is larger than the task area, the general area is equal to the task area. The noncritical area is the remaining area of the room, given by (400 - 100 - 100) = 200 sq. ft.

The illuminance of the general area is $\frac{1}{3} \times 70 = 23.3$ fc., and that for the noncritical area = 10 fc., since $\frac{1}{3} \times 23.3 = 7.7 < 10$ fc.

$$\text{General area power, } P_g = \frac{100 \times 23.3}{0.7 \times 55 \times 0.58} = 104 \text{ watts.}$$

$$\text{Critical area power, } P_{cr} = \frac{200 \times 10}{0.7 \times 55 \times 0.58} = 89 \text{ watts.}$$

So total power is 313 + 104 + 89 = 506 watts in Room 102.

Room 103, Private Office: The dimensions and task requirements in this room are identical to those of Room 102. The total power in this room is, then, 506 watts.

Room 104, Drafting: $L = 40$ ft.; $W = 20$ ft.; task ht. = 3.5 ft. above floor; $A_g = 40 \times 20 = 80$ sq. ft.; $h_{rc} = 8 - 3.5 = 4.5$ ft.

$$\text{RCR} = \frac{5 \times 4.5\,(40 + 20)}{40 \times 20} = 1.6.$$

CU is then 0.66 (interpolated). The total task area is 8 X 50 = 400 sq. ft. No adjustment in the task area is required, since it is equal to one-half the gross area (800 sq. ft.).

$$\text{Task area power, } P_1 = \frac{400 \times 200}{0.7 \times 55 \times 0.66} = 3148 \text{ watts.}$$

The nontask area is 800 - 400 = 400 sq. ft. Note that since the nontask area is equal to the task area and is one-half the gross area, this space is all general area. No critical area exists in this situation.

Illuminance for the general area is $\frac{1}{3} \times 200 = 66.7$ fc.

$$\text{General area power, } P_g = \frac{400 \times 66.7}{0.7 \times 55 \times 0.66} = 1050 \text{ watts.}$$

So the total power is 3148 + 1050 = 4198 watts in Room 104.

Room 105, Accounting: $L = 40$ ft.; $W = 15$ ft.; $A_g = 40 \times 15 = 600$ sq. ft.; $h_{rc} = 5.5$ ft.

$$\text{RCR} = \frac{5 \times 5.5\,(40 + 15)}{40 \times 15} = 2.5.$$

The CU = 0.59 (interpolated). The total task area is 6 X 50 = 300 sq. ft. No adjustment in the task area is required, since it is equal to one-half the gross area, A_g.

$$\text{Task area power, } P_1 = \frac{300 \times 150}{0.7 \times 55 \times 0.59} = 1981 \text{ watts.}$$

The nontask area is 600 - 300 = 300 sq. ft. This whole area is thus general area, and there is no noncritical area. E_g for the general area is $\frac{1}{3} \times 150 = 50$ fc. So

$$P_g = \frac{300 \times 50}{0.7 \times 55 \times 0.59} = 660 \text{ watts.}$$

So total power is 1981 + 660 = 2641 watts in Room 105.

Room 106, Corridor: $L = 40$ ft.; $W = 5$ ft.; task ht. = 0 ft. (floor); $h_{rc} =$ 8.0 ft.

$$\text{RCR} = \frac{5 \times 8\,(40 + 5)}{40 \times 5} = 9.$$

So, CU = 0.29. With illuminance 20 fc, and the whole floor as the task area,

$$\text{power} = \frac{200 \times 20}{0.7 \times 55 \times 0.29} = 358 \text{ watts}$$

The total Lighting Power Budget = 2213 + 506 + 506 + 4198 + 2641 + 358
= 10,422 watts.

LIGHTING POWER LIMITS

At the time of publishing this book, IES introduced a modified and simpler technique of determining the Lighting Power Budget, under the name Lighting Power Limits (UPD Procedure). The formula used in determining the Lighting Power Limits, is:

$$P_r = A_r \times P_b \times \text{RF} \times \text{SUF}$$

where

P_r = power budget of the room or space in watts.
A_r = area of the room in square feet (square meter)
P_b = base unit power density (UPD) in watts per square feet (square meter)
RF = room factor
SUF = space utilization factor

P_b is a predetermined value given in watts per square foot or square meter, for each specific type of task (see Table 6-4). The criteria used in determining the base UPD are the same as those used in the Lighting Power Budget discussed earlier. It provides sufficient power to satisfy the lighting requirements of the listed visual tasks for the space, assuming the power is utilized effectively in a large unobstructed space. If the room has multiple work stations, P_b shall be the weighted average of the individual task UPDs. The technique used in determining the weighted P_b is shown in the example given at the end of this chapter.

RF is a multiplication factor (between 1.00 and 2.00) which adjusts the P_b for spaces of various dimensions to account for the effect of room configuration on lighting efficiency. In selecting RF, major room and ceiling dimensions have to

Table 6-4. Partial List of Base UPD (P_b). (Reprinted with permission from EMS-6, July 1980, Illuminating Engineering Society of North America.)

Task or Area	Base UPD (W/SF)	Note
Common Areas		
Boiler room	0.7	d
Conference room	1.3	a
Corridor	0.6	
Garage, parking	0.2	d
Lobby, reception, waiting	1.0	
Toilet and washroom	0.7	
Office		
Accounting	3.2	f
Drafting	4.7	f
Filing (active)	2.0	
Filing (inactive)	0.8	
Graphic arts	3.0	f
Typing and reading	2.2	f
Commercial and Institutional		
Art galleries	1.6	a
Banks		
Lobby, general	2.3	
Posting, and keypunch	4.7	
Tellers' stations	4.7	
Bar (Lounge)	1.1	b
Courtrooms	0.9	
Hospitals		
Autopsy	3.2	
Nursing areas	0.6	
Surgical and lab area	0.8	
Critical care areas	3.8	
Emergency outpatient	3.8	
Library		
Audio listening areas, general	0.7	
Audiovisual areas	1.7	
Book stacks (active)	0.9	
Book stacks (inactive)	0.4	
Card files	3.2	
Cataloging	2.2	
Microfilm areas	2.2	
Reading areas	2.2	
Post Offices		
Lobby	0.7	
Sorting, mailing, etc.	2.8	
Schools		
Art	3.0	
Classrooms	2.2	
Drafting	3.2	
Laboratories	2.8	
Sewing	4.1	
Industrial		
Brewing		
General production	1.7	
Filling (bottles, cans, kegs)	1.3	
Foundries		
Core making and inspection	2.5	
Cupola area	0.6	
Fine inspection	5.0	
General production	1.0	
Molding and grinding	2.5	
Garage-service		
Repair area	2.5	
Traffic area	0.6	
Inspection		
Difficult	2.8	
Fine	5.5	
Ordinary	1.4	
Metal fabrication (bulk)	0.6	
Paint manufacturing	0.7	
Paper manufacturing		
General production	1.0	
Inspection	2.5	
Rewinder	3.7	
Printing		
Composing room	2.8	
Electrotyping	1.9	
Photoengraving	1.9	
Printing plants	5.1	
Type foundries	1.9	
Sheet metal works		
General production	1.3	
Inspection and scribing	5.5	
Soap manufacturing	1.0	

Table 6-4. *(Continued)*

Task or Area	Base UPD (W/SF)	Note	Task or Area	Base UPD (W/SF)	Note
Textile mills			Welding		
Drying and finishing	3.8		General illumination	1.3	
General production	1.0		Woodworking (general)	1.3	
Warping, weaving, grading	3.8				

[a] Includes 0.5 W/SF for special tasks.
[b] Allow additional lighting for clean-up.
[f] Determine task area within the space by the size and number of work locations.

be used; it has no relation to the mounting methodology of the luminaires (see Table 6-5).

SUF is a multiplication factor (between 0.4 and 1.0) which adjusts the power budget to account for the task areas. If the total task area is 50% of the room or larger, SUF is equal to unity (1.0). For other values they are smaller, but no less than 0.4 (see Table 6-6). Since the majority of well-planned spaces utilize more than half the room for work stations, the SUF is equal to 1.00, and as such, this factor may be eliminated. For areas with isolated work stations, however, SUF provides the necessary power budget adjustments.

Let us use the preceding example to determine the Lighting Power Limit.

Room 101, General Office:

Reading/typing . . . 12 locations
Accounting . . . 2 locations

A_r: $A_r = 40 \times 30 = 1200$ square feet

P_b: There are two types of tasks involved. P_b shall be the weighted average of the two individual tasks UPDs. Referring to Table 6-4,

$$P_b = \frac{(2.2 \times 12) + (2.0 \times 2)}{12 + 2} = 2.17 \text{ watts per square foot}$$

RF: From Table 6-5, corresponding to room dimensions of 40 ft. (L), 30 ft. (W), and 8 ft. (H), RF = 1.05

SUF: To determine SUF, total task area must be divided by the area of the room. Since actual task area is not known, they are assumed to be 50 square feet each.

So, $$\frac{\text{total task area}}{\text{room area}} = \frac{50 \times 14}{1200} = 0.58$$

From Table 6-6, SUF = 1.00

So, $$P_r = A_r \times P_b \times \text{RF} \times \text{SUF}$$
$$= 1200 \times 2.17 \times 1.05 \times 1.0$$
$$= 2734 \text{ watts.}$$

Room 102, Office:
Reading/typing . . . 2 locations

$A_r = 20 \times 20 = 400$ square feet
$P_b = 2.2$ (from Table 6-4)
RF = 1.20 (from Table 6-5)
SUF = 0.55 (from Table 6-6). Note that this value is obtained corresponding to the ratio of total task area to the room area, which is less than 0.3.

$$\left(\frac{50 \times 2}{400} = 0.25\right)$$

So, $P_r = 400 \times 2.2 \times 1.20 \times 0.55 = 580$ watts.

Room 103, Identical to room 102. So, $P_r = 580$ watts.

Room 104, Drafting: 8 work locations

$A_r = 40 \times 20 = 800$ square feet
$P_b = 4.7$ (from Table 6-4)
RF = 1.1 (from Table 6-5)
SUF = 1.0 (from Table 6-6), since total task area is larger than 50% of the room area.

So, $P_r = 800 \times 4.7 \times 1.1 \times 1.0 = 4136$ watts.

Room 105, Accounting: 6 work locations

$A_r = 15 \times 40 = 600$ square feet
$P_b = 3.2$ (from Table 6-4)
RF = 1.15 (from Table 6-5)
SUF = 1.0 (from Table 6-6), since total task area is larger than 50% of the room area.

Table 6-5. Room Factor (RF). (Reprinted with permission from EMS-6, July 1980, Illuminating Engineering Society of North America.)

DIMEN.*		CEILING HEIGHT (FT.)*									
W	L	8	8.5	9	10	11	12	14	16	18	20+
6	6	2.00	2.00	2.00	2.00	2.00	2.00	2.00	2.00	2.00	2.00
6	9	2.00	2.00	2.00	2.00	2.00	2.00	2.00	2.00	2.00	2.00
6	12	1.85	2.00	2.00	2.00	2.00	2.00	2.00	2.00	2.00	2.00
6	15	1.75	1.90	2.00	2.00	2.00	2.00	2.00	2.00	2.00	2.00
6	18	1.70	1.80	1.95	2.00	2.00	2.00	2.00	2.00	2.00	2.00
6	24	1.65	1.75	1.85	2.00	2.00	2.00	2.00	2.00	2.00	2.00
6	30	1.60	1.70	1.80	2.00	2.00	2.00	2.00	2.00	2.00	2.00
6	36	1.60	1.65	1.75	1.95	2.00	2.00	2.00	2.00	2.00	2.00
6	60	1.55	1.60	1.70	1.85	2.00	2.00	2.00	2.00	2.00	2.00
6	60+	1.45	1.50	1.60	1.75	1.90	2.00	2.00	2.00	2.00	2.00
8	8	1.85	2.00	2.00	2.00	2.00	2.00	2.00	2.00	2.00	2.00
8	12	1.65	1.75	1.85	2.00	2.00	2.00	2.00	2.00	2.00	2.00
8	16	1.55	1.65	1.70	1.90	2.00	2.00	2.00	2.00	2.00	2.00
8	20	1.50	1.55	1.65	1.80	1.95	2.00	2.00	2.00	2.00	2.00
8	24	1.45	1.50	1.60	1.75	1.90	2.00	2.00	2.00	2.00	2.00
8	32	1.40	1.45	1.55	1.65	1.80	1.95	2.00	2.00	2.00	2.00
8	40	1.40	1.45	1.50	1.65	1.75	1.90	2.00	2.00	2.00	2.00
8	48	1.35	1.45	1.50	1.60	1.75	1.85	2.00	2.00	2.00	2.00
8	80	1.35	1.40	1.45	1.55	1.65	1.80	2.00	2.00	2.00	2.00
8	80+	1.30	1.35	1.40	1.50	1.60	1.70	1.90	2.00	2.00	2.00
10	10	1.60	1.70	1.80	2.00	2.00	2.00	2.00	2.00	2.00	2.00
10	15	1.45	1.50	1.60	1.75	1.90	2.00	2.00	2.00	2.00	2.00
10	20	1.40	1.45	1.50	1.65	1.75	1.90	2.00	2.00	2.00	2.00
10	25	1.35	1.40	1.45	1.55	1.70	1.80	2.00	2.00	2.00	2.00
10	30	1.30	1.35	1.40	1.50	1.65	1.75	2.00	2.00	2.00	2.00
10	40	1.30	1.35	1.40	1.45	1.60	1.70	1.90	2.00	2.00	2.00
10	50	1.25	1.30	1.35	1.45	1.55	1.65	1.85	2.00	2.00	2.00
10	60	1.25	1.30	1.35	1.45	1.50	1.60	1.80	2.00	2.00	2.00
10	100	1.25	1.25	1.30	1.40	1.45	1.55	1.75	1.95	2.00	2.00
10	100+	1.20	1.25	1.25	1.35	1.40	1.50	1.65	1.85	2.00	2.00
12	12	1.45	1.50	1.60	1.75	1.90	2.00	2.00	2.00	2.00	2.00
12	18	1.35	1.40	1.45	1.55	1.70	1.80	2.00	2.00	2.00	2.00
12	24	1.30	1.35	1.40	1.45	1.60	1.70	1.90	2.00	2.00	2.00
12	30	1.25	1.30	1.35	1.45	1.50	1.60	1.80	2.00	2.00	2.00
12	36	1.25	1.30	1.30	1.40	1.50	1.55	1.75	2.00	2.00	2.00
12	48	1.20	1.25	1.30	1.35	1.45	1.50	1.70	1.90	2.00	2.00
12	60	1.20	1.25	1.25	1.35	1.40	1.50	1.65	1.85	2.00	2.00
12	72	1.20	1.20	1.25	1.30	1.40	1.45	1.60	1.80	2.00	2.00
12	120	1.15	1.20	1.25	1.30	1.35	1.40	1.55	1.75	1.90	2.00
12	120+	1.15	1.15	1.20	1.25	1.30	1.35	1.50	1.65	1.80	1.95
16	16	1.30	1.35	1.40	1.45	1.60	1.70	1.90	2.00	2.00	2.00
16	24	1.20	1.25	1.30	1.35	1.45	1.50	1.70	1.90	2.00	2.00
16	32	1.20	1.20	1.25	1.30	1.35	1.45	1.60	1.75	1.95	2.00
16	40	1.15	1.20	1.20	1.25	1.35	1.40	1.55	1.70	1.85	2.00
16	48	1.15	1.15	1.20	1.25	1.30	1.35	1.50	1.65	1.80	1.95
16	64	1.15	1.15	1.20	1.25	1.30	1.35	1.45	1.55	1.70	1.85
16	80	1.10	1.15	1.15	1.20	1.25	1.30	1.40	1.55	1.65	1.80
16	96	1.10	1.15	1.15	1.20	1.25	1.30	1.40	1.50	1.65	1.75
16	160	1.10	1.10	1.15	1.20	1.20	1.25	1.35	1.45	1.60	1.70
16	160+	1.10	1.10	1.10	1.15	1.20	1.25	1.30	1.40	1.50	1.60

2.00

TO CONVERT ROOM DIMENSION FROM (FT.) TO (METER) MULTIPLY GIVEN VALUE BY 0.305.

DIMEN.*		CEILING HEIGHT (FT.)*									
W	L	8	8.5	9	10	11	12	14	16	18	20+
20	20	1.20	1.25	1.25	1.35	1.40	1.50	1.65	1.85	2.00	2.00
20	30	1.15	1.15	1.20	1.25	1.30	1.35	1.50	1.65	1.80	1.95
20	40	1.10	1.15	1.15	1.20	1.25	1.30	1.40	1.55	1.65	1.80
20	50	1.10	1.10	1.15	1.20	1.25	1.30	1.40	1.50	1.60	1.70
20	60	1.10	1.10	1.15	1.15	1.20	1.25	1.35	1.45	1.55	1.65
20	80	1.10	1.10	1.10	1.15	1.20	1.25	1.30	1.40	1.50	1.60
20	100	1.05	1.10	1.10	1.15	1.20	1.20	1.30	1.40	1.45	1.55
20	120	1.05	1.10	1.10	1.15	1.15	1.20	1.30	1.35	1.45	1.55
20	200	1.05	1.05	1.10	1.10	1.15	1.20	1.25	1.35	1.40	1.50
20	200+	1.05	1.05	1.05	1.10	1.15	1.15	1.20	1.30	1.35	1.45
24	24	1.15	1.15	1.20	1.25	1.30	1.35	1.50	1.65	1.80	1.95
24	36	1.10	1.10	1.15	1.20	1.25	1.25	1.35	1.45	1.60	1.70
24	48	1.10	1.10	1.10	1.15	1.20	1.25	1.30	1.40	1.50	1.60
24	60	1.05	1.10	1.10	1.15	1.15	1.20	1.30	1.35	1.45	1.55
24	72	1.05	1.05	1.10	1.10	1.15	1.20	1.25	1.35	1.40	1.50
24	96	1.05	1.05	1.10	1.10	1.15	1.15	1.25	1.30	1.40	1.45
24	120	1.05	1.05	1.05	1.10	1.15	1.15	1.20	1.30	1.35	1.45
24	144	1.05	1.05	1.05	1.10	1.10	1.15	1.20	1.25	1.35	1.40
24	240	1.05	1.05	1.05	1.10	1.10	1.15	1.20	1.25	1.30	1.35
24	240+	1.00	1.05	1.05	1.05	1.10	1.10	1.15	1.20	1.25	1.30
30	30	1.10	1.10	1.15	1.15	1.20	1.25	1.35	1.45	1.55	1.65
30	45	1.05	1.05	1.10	1.10	1.15	1.20	1.25	1.35	1.40	1.50
30	60	1.05	1.05	1.05	1.10	1.15	1.15	1.20	1.30	1.35	1.45
30	75	1.05	1.05	1.05	1.10	1.10	1.15	1.20	1.25	1.30	1.40
30	90	1.00	1.05	1.05	1.05	1.10	1.10	1.20	1.25	1.30	1.35
30	120	1.00	1.05	1.05	1.05	1.10	1.10	1.15	1.20	1.25	1.30
30	150	1.00	1.00	1.05	1.05	1.10	1.10	1.15	1.20	1.25	1.30
30	180	1.00	1.00	1.05	1.05	1.05	1.10	1.15	1.20	1.25	1.30
30	300	1.00	1.00	1.00	1.05	1.05	1.10	1.10	1.15	1.20	1.25
30	300+	1.00	1.00	1.00	1.05	1.05	1.05	1.10	1.15	1.20	1.20
40	40	1.05	1.05	1.05	1.10	1.15	1.15	1.20	1.30	1.35	1.45
40	60	1.00	1.05	1.05	1.05	1.10	1.10	1.15	1.20	1.25	1.30
40	80	1.00	1.00	1.00	1.05	1.05	1.10	1.15	1.20	1.20	1.25
40	100	1.00	1.00	1.00	1.05	1.05	1.05	1.10	1.15	1.20	1.25
40	120	1.00	1.00	1.00	1.05	1.05	1.05	1.10	1.15	1.20	1.20
40	160	1.00	1.00	1.00	1.00	1.05	1.05	1.10	1.10	1.15	1.20
40	200	1.00	1.00	1.00	1.00	1.05	1.05	1.10	1.10	1.15	1.20
40	240	1.00	1.00	1.00	1.00	1.05	1.05	1.05	1.10	1.15	1.20
40	400	1.00	1.00	1.00	1.00	1.00	1.05	1.05	1.10	1.15	1.15
40	400+	1.00	1.00	1.00	1.00	1.00	1.00	1.05	1.10	1.10	1.15
60	60	1.00	1.00	1.00	1.05	1.05	1.05	1.10	1.15	1.20	1.20
60	90	1.00	1.00	1.00	1.00	1.00	1.05	1.05	1.10	1.15	1.15
60	120	1.00	1.00	1.00	1.00	1.00	1.00	1.05	1.10	1.10	1.15
60	150	1.00	1.00	1.00	1.00	1.00	1.00	1.05	1.05	1.10	1.10
60	180	1.00	1.00	1.00	1.00	1.00	1.00	1.05	1.05	1.10	1.10
60	240	1.00	1.00	1.00	1.00	1.00	1.00	1.00	1.05	1.05	1.10
60	300	1.00	1.00	1.00	1.00	1.00	1.00	1.00	1.05	1.05	1.10
60	360	1.00	1.00	1.00	1.00	1.00	1.00	1.00	1.05	1.05	1.10
60	600	1.00	1.00	1.00	1.00	1.00	1.00	1.00	1.05	1.05	1.05
60+	600+	1.00	1.00	1.00	1.00	1.00	1.00	1.00	1.00	1.05	1.05

1.00

Table 6-6. Space Utilization Factor (SUF).

$\dfrac{\text{Task Area } (A_t)}{\text{Room Area } (A_r)}$	SUF
$\geqslant 0.5$	1.00
<0.5	0.85
<0.4	0.70
<0.3	0.55
<0.2	0.40

So, $P_r = 600 \times 3.2 \times 1.15 \times 1.0 = 2208$ watts

Room 106, Corridor (no specific task locations)

$A_r = 50 \times 6 = 300$ square feet
$P_r = 0.6$ (from Table 6-4)
RF = 1.6 (from Table 6-5)
SUF = 1.0, since the whole area is the task (walking) area.

So, $P_b = 300 \times 0.6 \times 1.6 \times 1.0 = 288$ watts

So, the Lighting Power Limit of the whole office area is:

$$= 2734 + 580 + 4136 + 2208 + 288$$
$$= 9946 \text{ watts.}$$

Note that although both methods are used to determine the upper limit of lighting power, their values may be different. The newer method is much simpler to calculate, and its values are more conservative. The reader is advised to refer to the IES Lighting Handbook, 1981, Application Volume, or the IES publication number EMS-6 for a complete understanding of this system.

Once the budget is determined, the designer may use any type of luminaire and design procedure that he finds most suitable without exceeding the calculated maximum value. In states where the Lighting Power Budget (or Limit) has been accepted as a mandatory standard, the complete calculation is to be presented in tabular form, along with the proposed design plans to the responsible agency for approval. For most states, it is a partial requirement toward obtaining the building permit. The reader should be aware of any changes adopted by local authority in determining the Lighting Power Budget or the Limit.

Chapter Seven
Energy Saving By Controls

A considerable amount of electrical energy can be saved by various means of lighting control. The main idea is to provide only that portion of the total light that satisfies a specific working condition, for a required duration, and is then turned off when not needed. Use of the proper type of control will provide this flexibility. The operation can be manual or automatic, depending on the type of control selected.

The different types of lighting-control devices can be loosely divided into two groups: on-off and level-control types. The on-off controls are those that can turn lights on or off directly. The level controls are those that can provide intermediate levels in addition to the on-off function. The operation can be manual or automatic, depending on the type selected. Devices that may be termed *on-off controls* are as follows:

Circuit breakers
Wall switches
Interval timers
Low-voltage switches
Time switches
Photocells
Ultrasonic detectors

Level controls include

Dimmers
Multilevel ballasts
Microprocessor controlled systems

ON-OFF CONTROLS

Circuit Breakers

The main purpose of a circuit breaker is to protect the conductors leading to the electrical load. Usually the wire size selected is either equal to or somewhat larger than the circuit breaker used. If there is a fault at the load end of the system, the circuit breaker recognizes the overload in terms of the drawing of a larger current than it is designed to carry and trips immediately to protect the conductors.

Circuit breakers are common for turning lights on and off directly in certain applications. This practice is mostly popular in warehouses, industrial areas, big office buildings, and the like, where a large number of luminaires are to be turned on or off for convenience. A disadvantage of this method, with respect to energy savings, is that more lights are either left on or turned on than are needed for after-hours use; e.g., for cleaning and maintenance.

Circuit breakers used for lighting controls can be either 120- or 277-volt, single-phase type. They are either 15- or 20-ampere and are allowed to carry a maximum of 80% of this amperage, as specified in the National Electrical Code. The procedure for determining the maximum wattage per circuit breaker is as follows:

	120-Volt System	277-Volt System
for 20-amp. circuit breaker:	120 volts X 20 amperes X 80% = 2400 X 0.8 = 1920 watts.	277 volts X 20 amperes X 80% = 5540 X 0.80 = 4432 watts.
for 15-amp. circuit breaker:	120 volts X 15 amperes X 80% = 1800 X 0.8 = 1440 watts.	277 volts X 15 amperes X 80% = 4155 X 0.80 = 3324 watts.

With these figures, a maximum allowed number of luminaires may also be determined. A 2 X 4-foot luminaire with four F40T12 rapid-start fluorescent lamps or an 8-foot strip luminaire with two F96T12 slim-line fluorescent lamps are usually considered as being 200 watts each. The maximum number of luminaires connected per circuit breaker can be determined as follows:

	120-Volt System		277-Volt System	
20-amp. circuit breaker:	1920/200	9 luminaires	4432/200	22 luminaires
15-amp. circuit breaker:	1440/200	7 luminaires	3324/200	16 luminaires

Figure 7-1 shows how a number of luminaires are connected to various circuit breakers, which are attached to the bus bars of the lighting panel board.

From the standpoint of saving energy it is advisable that the designer preselects the crucial luminaires that may have to be left on for security purposes and also isolates one or more separate circuits that operate a minimum number of luminaires to provide sufficient light for cleaning purposes. While the security lighting circuit has to be directly connected to the building's emergency system,

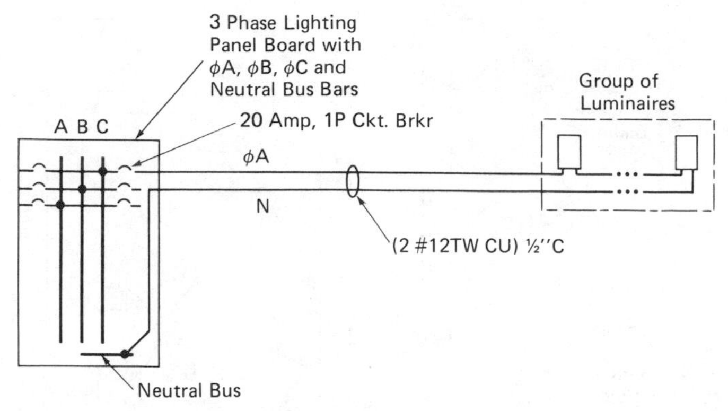

Fig. 7-1. A typical connection diagram of luminaires to panel board, through circuit breakers.

the cleaning-purpose lighting circuit may be connected directly to an isolated circuit breaker in the lighting panel board or through a wall switch, conveniently located at the entrance. If operation is directly from a panel board, it is highly recommended that only circuit breakers rated for "switching usage" be used, so as to retain the breakers' longevity of reliable operation.

Circuit breakers are usually rated for a minimum of 10,000 amperes interrupting capacity. This means that as long as the available fault circuit at the panel board location does not exceed 10,000 amperes, the fault will be cleared without damage to circuit breaker or its surroundings. When higher values of fault current are possible, circuit breakers having higher interrupting capacities, such as 22,000 amperes, must be installed.

Wall Switches

Wall switches are the most inexpensive, convenient, and popular means of operating lights in rooms. The biggest advantage is that with these devices the total number of luminaires can be subdivided among many switches, although all of them may originate from one circuit breaker. With several switches, lighting can be controlled by zones and sometimes by different levels.

Wall switches are available in different types. Besides the conventional single pole, they are also available in 3-way or 4-way models, which are used to control the same luminaires from two or more locations. Figure 7-2 shows a typical lighting circuit connected from one 20-ampere, 1-pole circuit breaker, but equipped with several wall switches for zonal and level lighting control. A total of 20 luminaires, each having four lamps, are connected to the circuit breaker operating at 277 volts. In addition to the individual room control, the lighting of Room 100 has been divided into two levels, each switch operating one-half the light emitted by each luminaire. This is accomplished by connecting one of

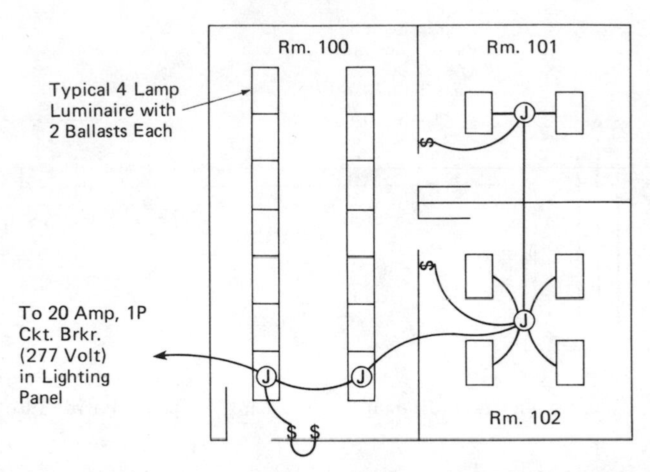

Fig. 7-2. A typical lighting circuit connected from one 20-ampere, one-pole circuit breaker, having several wall switches for zonal and level lighting control.

the two ballasts of each luminaire to the first switch and the remaining to the second. In three-lamp luminaires, however, this division of half-and-half lighting level is not possible. A three-lamp luminaire has two ballasts; the first controls two lamps, and the other, one lamp. With this set-up, two separate switches will provide full, two-thirds, or one-third of the total lighting.

A further energy saving can be accomplished if two adjacent three-lamp luminaires are connected with three twin-lamp ballasts, instead of using the conventional set-up with two ballasts each. This type of arrangement sometimes is known as the *master/slave* or *master/satellite* system, where the master luminaire accommodates two ballasts controlling the outer two lamps of each luminaire, and the slave has the remaining ballast controlling the middle lamp of each luminaire. This saves energy since two single-lamp ballasts consume more energy than one twin-lamp ballast. This is shown as follows:

Power consumed by two luminaires with conventional system
= power consumed by two twin-lamp ballasts
+ two single-lamp ballasts
$= (2 \times 96) + (2 \times 53)$
$= 192 + 106$
= 298 watts.

Power consumed by the same luminaires in a master/slave system
= power consumed by three twin-lamp ballasts
$= 96 \times 3$
= 288 watts.

One disadvantage with this system is that since the master controls the slave, a room partition cannot be erected between the two. If a partition is placed between the two, one of the luminaires must be removed, or another ballast must be installed in the slave unit and wiring rearranged. The other disadvantage is that if the luminaire manufacturer wires the two luminaires together without a plug-in arrangement, it may become quite difficult to install the luminaires in the grid system. Both luminaires must go through the same opening in the grid system and then be arranged at their respective locations.

The purpose of using 3-way switches is to provide the convenience of operating a group of luminaires from two different locations, and that of the 4-way switches from three or more locations. Figure 7-3 shows a typical wiring diagram using 3-way and 4-way switches. If the luminaires are to be controlled from numerous locations, a special circuitry using low-voltage switches and relays is often more economical to wire.

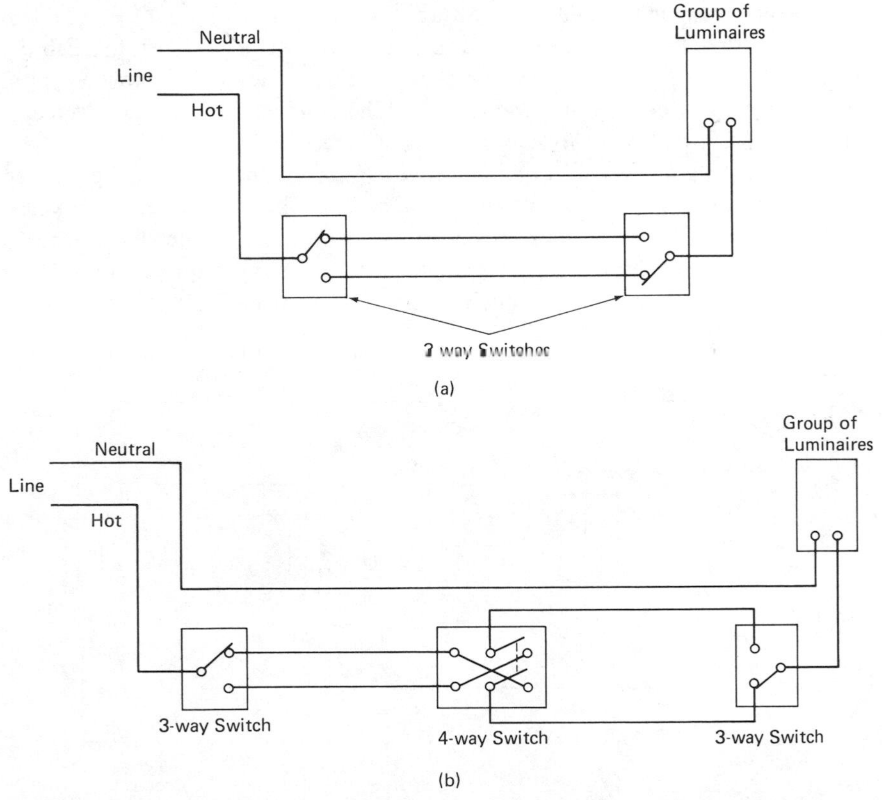

Fig. 7-3. (a) Circuit diagram with three-way switches. (b) Circuit diagram with four-way switches.

Interval Timer

Interval timers are equipped with a mechanical winding system that can provide automatic turn-off, once the timing and turning on of the device has been manually set. These devices are suitable to replace wall switches for applications where lights or appliances have to be turned off in a predictable manner. Some models are equipped with time cycles of 0-5 minutes, 0-15 minutes, 0-30 minutes, 0-60 minutes, 0-6 hours, and 0-12 hours. Others can be timed for 0-3 minutes, 0-60 minutes, and 0-12 hours, and they also feature a "hold" position for keeping the circuitry closed continuously. The main advantage of the use of these devices is to compensate for the human factor of forgetting to turn off lights when they are not required.

Low-Voltage Switching

The general principle behind low-voltage switching lies in controlling a line voltage lighting circuit by means of a low-voltage switching circuit, through the use of low-voltage switches, transformers, and relays. Figure 7-4 shows a typical circuit diagram, incorporating two groups of lighting loads that are connected to two circuits through a low-voltage switching arrangement.

There are several advantages with low-voltage switching: (1) Large groups of luminaires can be controlled by fewer switches; (2) there is reduced shock hazard; (3) the use of conventional 3-way or 4-way switches can be eliminated; (4) the wiring does not have to be enclosed in conduits in most applications. None of these factors, however, is conducive to the turning off of lights when they are

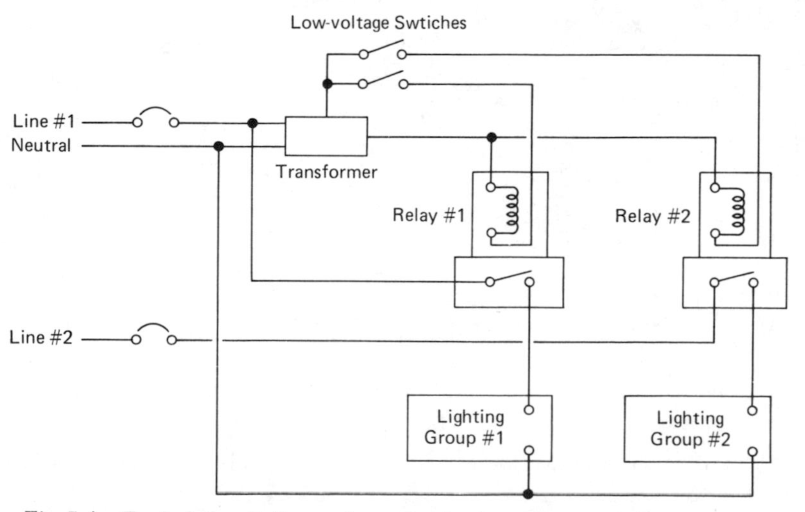

Fig. 7-4. Typical circuit diagram incorporating low-voltage switching system.

not in use. For most applications, a high level of light is still required for maintaining productivity, but buildings continue to "burn the midnight oil" and keep unnecessary lights on long after workers have left. Typically, lighting remains on anywhere between 12 to 18 hours a day at many buildings–sometimes all night. Statistically, if lights are turned off totally for unnecessary periods, and levels are reduced to only what is required, many such facilities may easily trim the equivalent time of operation to anywhere from 8 to 6 hours a day! The problem can be minimized with the use of programmable control systems or an automatic control system; these are discussed at the end of this chapter.

Time Switches

Time switches are mostly used for outdoor lighting control, although their use is becoming increasingly popular for interiors as well. These are sometimes commonly used for corridor lighting control in apartment or condominium buildings. The elementary time switch has a single on-off operation for each day of the week. For exterior applications, it is necessary that the timing of operation adjust with the season. During winter, dusk arrives much faster than it does in summer. With an elementary type of time switch, it is necessary that changes in the time of operation be done manually since the units do not have automatic adjusting mechanisms.

Time switches are available to switch directly one, two, three, or four circuits simultaneously. In applications where there is a large number of luminaires to be controlled on the same program, the use of multipole-type time switches is more economical than the use of a single-pole type of time switch for each circuit. Figure 7-5 shows a typical interior and connecting diagrams of single-pole and multipole time-switches.

Time-switches that are designed to adjust automatically for the seasonal changes are known as astronomic time switches. The basic idea behind these is that manual operation is not required for seasonal changes. Depending on the geographical area of the application, these time switches are equipped with the suitable driving gears that compensate for the seasonal changes. The unit turns on automatically at dusk and goes off at dawn, consistently, and also makes appropriate changes for the seasons. In addition, these time switches also can be adjusted to turn off the lights before the dawn at a predetermined time.

Photocells

The use of photocells is chiefly popular for outdoor lighting, particularly for parking areas and street lighting. Some photocells can be adjusted for a change in level of operation. This is simply done by sliding a movable plate across the face of the light-sensitive cell. The wider the face is open, the more light enters

Typ. Time Clock Interior

(a)

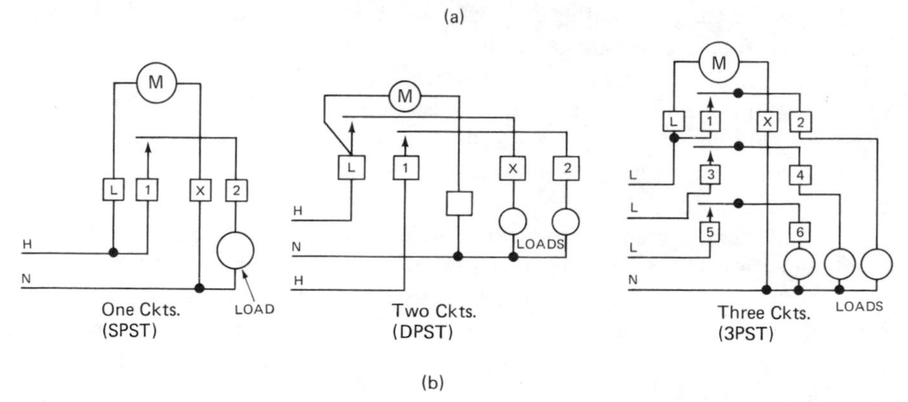

(b)

Fig. 7-5. (a) A typical time-switch interior (Courtesy of Tork.) (b) Connection diagrams of one, two, and three circuit time-switches operating a group of luminaires.

the cell, causing it to turn on later (darker) and off earlier. If the plate is slid to make a smaller opening, the unit operates faster, simulating earlier darkness. The biggest advantage of photocells is that they are virtually maintenance-free, relatively inexpensive, and self-adjusting to seasonal changes. Use of a photocell indoors for automatic-lighting energy control is an excellent choice and an effective means of saving energy. When crucially located on the ceiling, a photocell will read the level of lighting, incorporating daylight influence, and automatically signal the system to adjust the artificial lighting level. This system is described in detail at the end of this chapter.

Photocells can be directly installed in a luminaire for individual control, or mounted separately to control a group of luminaires (see Figure 7-6).

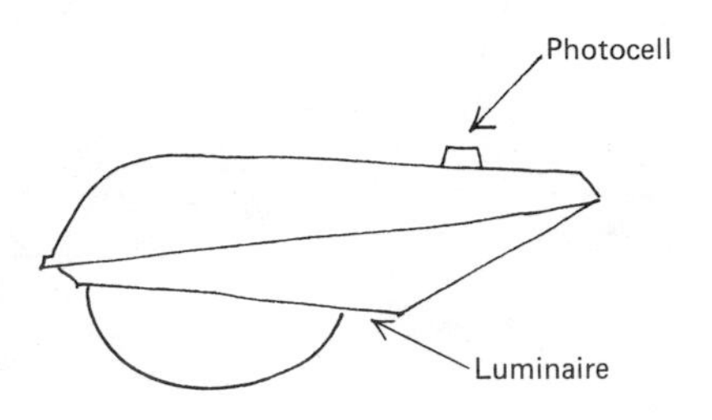

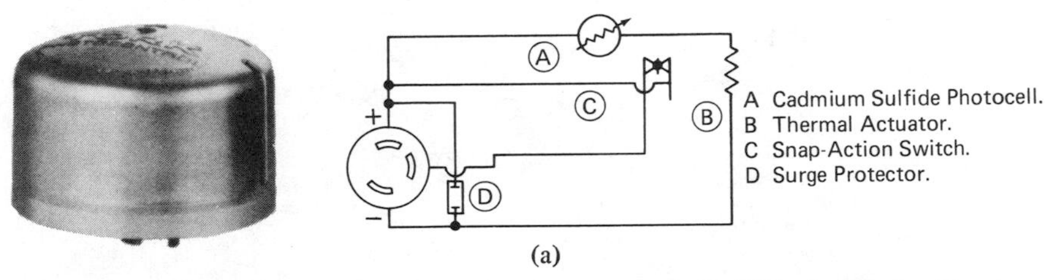

(a)

Fig. 7-6. (a) Photocells directly mounted on luminaires for individual control. (Courtesy of Tork.) (b) Photocell mounted separately to control a group of luminaires. (Courtesy of Tork.)

Ultrasonic Detectors

The basic idea behind the ultrasonic detector is to have lights on only when room is occupied. The lights will turn on automatically as soon as people enter the room, and go off with their departure. Overriding manual switching will probably be required if lights are to be turned off temporarily when people are within the room. A multipurpose auditorium where slide projections or movies are shown will fit into this category. Although a high degree of energy saving is possible, the use of ultrasonic detectors has been restricted to situations where its initial high expense can be justified. The use of ultrasonic detectors is popular in security alarm systems, where some selected luminaires turn on immediately upon the entrance of intruders.

LEVEL CONTROLS

Dimmers

Dimmers are the most popular and provide maximum flexibility for intermediate-level control applications. The three types of dimmer include (a) incandescents, (b) fluorescents, and (c) HID dimmers.

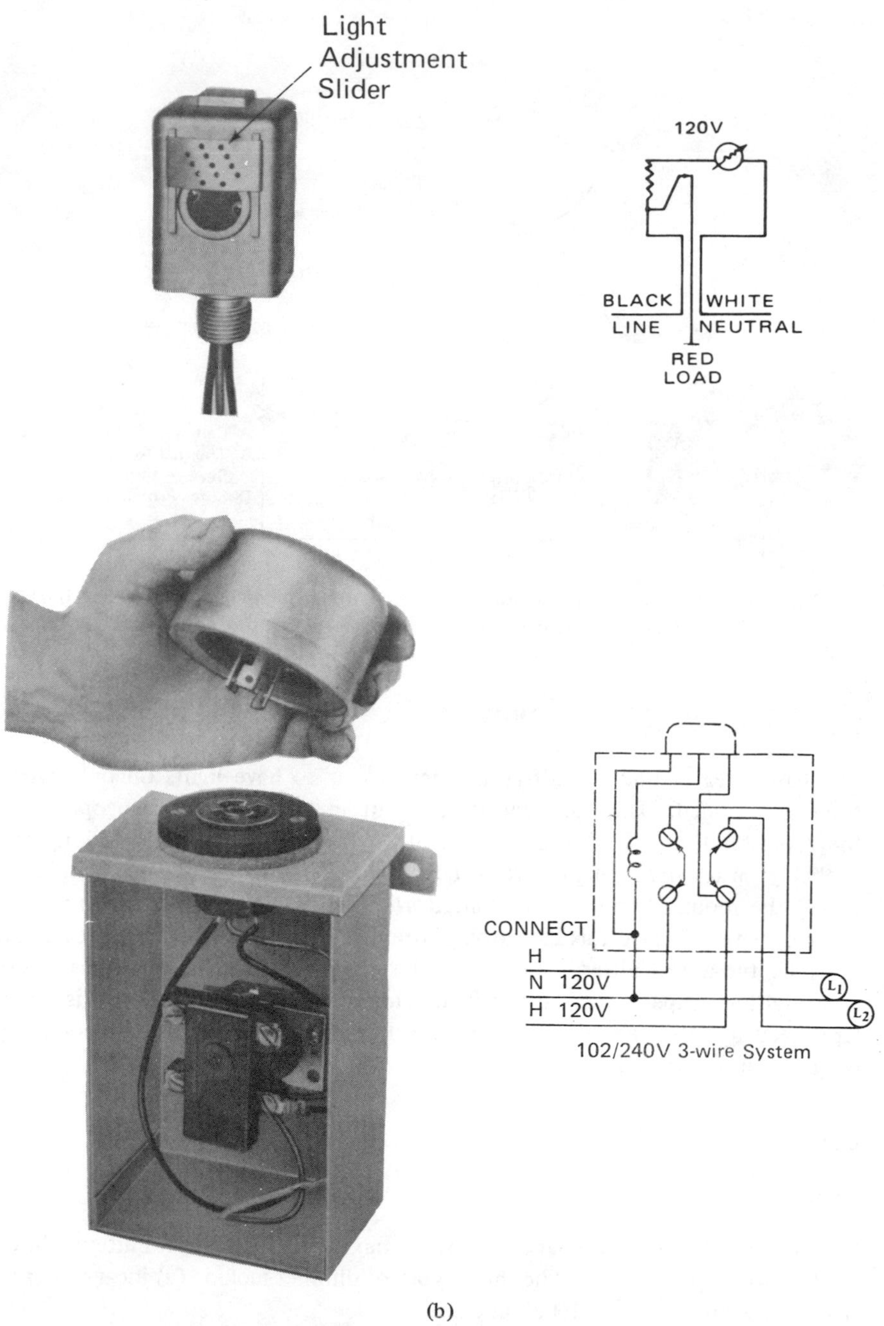

(b)
Fig. 7.6. (*Continued*)

Incandescent Dimmers. Incandescent dimmers have been in use for a number of decades, although it was not until 10 years ago that an energy-conserving device was developed. Originally, dimmers were of the rheostat, autotransformer, or thyratron types; these had several disadvantages, including a high degree of power loss, large space requirements, and high initial cost. Figure 7-7 shows the general circuit diagram of an autotransformer dimmer connected to a lighting load. Based on the characteristics of autotransformers, there is a large loss of power as heat in the transformer coil. The heat dissipation is the product of the square of the current times the resistance in the coil. Because of large space requirements, these dimmers were basically used for theatrical purposes and were hardly popular for residential or commercial applications.

With the development of solid-state dimmer devices, the problems of energy waste and space requirements have been solved. The credit for energy efficiency (the dimmer consumes insignificant energy, almost nil) goes to the use of the solid-state semiconductor, known as the thyristor, which is a member of the transistor group. It performs two important functions: (1) It prevents current flow with high voltage applied, and (2) upon receipt of a trigger pulse, it will conduct current copiously and effectively.

Figure 7-8 shows a sine wave representing the voltage input to the lighting circuit. From A to B, as the voltage rises, the thyristor prevents current flow. The net power during this period is:

$$\text{Power}_{AB} = \text{volts} \times \text{amperes}$$
$$= \text{volts} \times 0 = 0 \text{ watts.}$$

From B to C the thyristor conducts current copiously, and the shaded area represents the power that is used by the luminaires. Travel between points C to B′ also draws nil power, and the travel between B′ to C′ is a repeat of B to C,

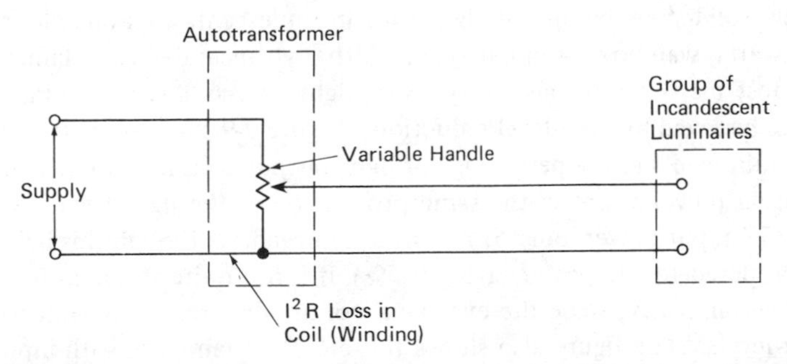

Fig. 7-7. Autotransformer dimming system for incandescent lights. A good portion of the energy is wasted, appearing as heat.

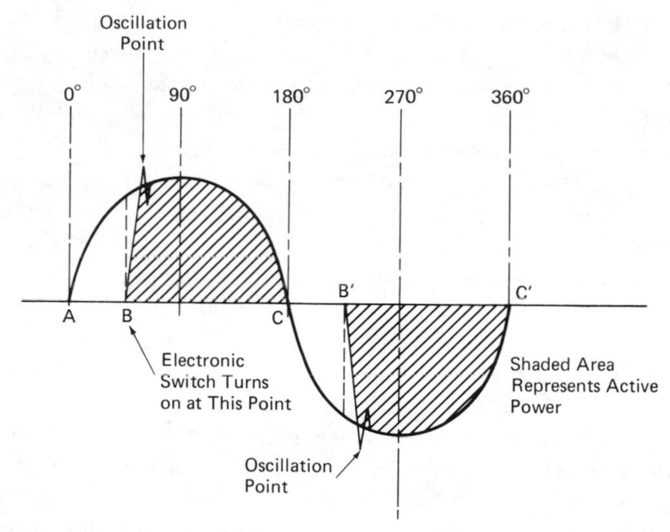

Fig. 7-8. Sine wave of a solid-state dimming device using thyristor. The shaded area represents power at a dimmed condition.

drawing the working power. The net power requirement for the dimming purposes, representing the total shaded area, is much less than would be required for an undimmed situation.

The slope shown at the rising point of current at points B and B′ represents the short time interval required to start from 0 to the first peak value in amperes. In reality, the current tends to overshoot the curve and somewhat oscillate before reaching a steady state. The phenomenon repeats twice each cycle. This results in a radio frequency (RF) noise and lamp filament ringing. Many manufacturers provide RF suppression and have chokes available that will minimize filament ringing.

The new solid-state dimmers designed for incandescents are available from 600 to 2000 watts, wall-box mounted type. Although incandescent filament lamps are the most inefficient of all the types of light source, a significant amount of power can be saved by light-level reduction. Figure 7-9 shows the effect of input power requirements as the percentage of light output is varied. In incandescents, the saving in power is not in the same proportion as the light level. As can be seen, at 75% input power, only 52% light is achieved. Although this light output is low compared to the power saving (25%), it is not quite as bad as it sounds in terms of visual acuity, since the eye can accommodate relatively wide variations in lumen levels. The figure also shows the effect of lamp life with input power variation. At 75% input power, along with 25% energy savings, the life of the bulb shows an increment of 11.3 times the normal rated life.

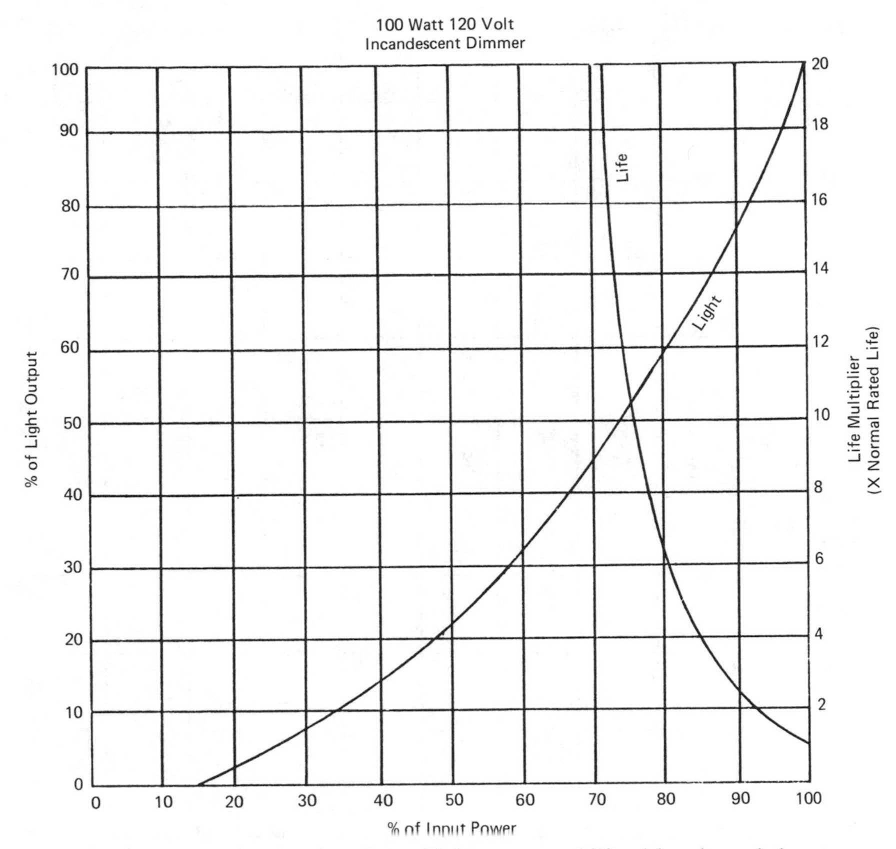

Fig. 7-9. Curves representing the effect of light output and life with a change in input power for incandescent lamps.

Fluorescent Dimmers. The basic concepts and requirements for fluorescent dimming are quite different from those for incandescents. These require a special dimming ballast, different types of electronic dimmers (not interchangeable with the incandescent dimmers discussed above), different circuitry, and special lamp holders. A conventional wall-box type of fluorescent dimmer can operate up to 30 lamps of 40 watts each, whereas, a dimming module can operate up to 80 lamps. Fluorescent lamps, which are the most widely used commercially, can produce a significant energy savings when they are dimmed. Figure 7-10 shows the effect on lighting output with input power variation. At 60% of input power, the light output is close to 62%. The minimum light output produced by a fluorescent system is about 1/500 of the maximum, and it is never equal to zero, when the dimmer is turned to its minimum position. For all

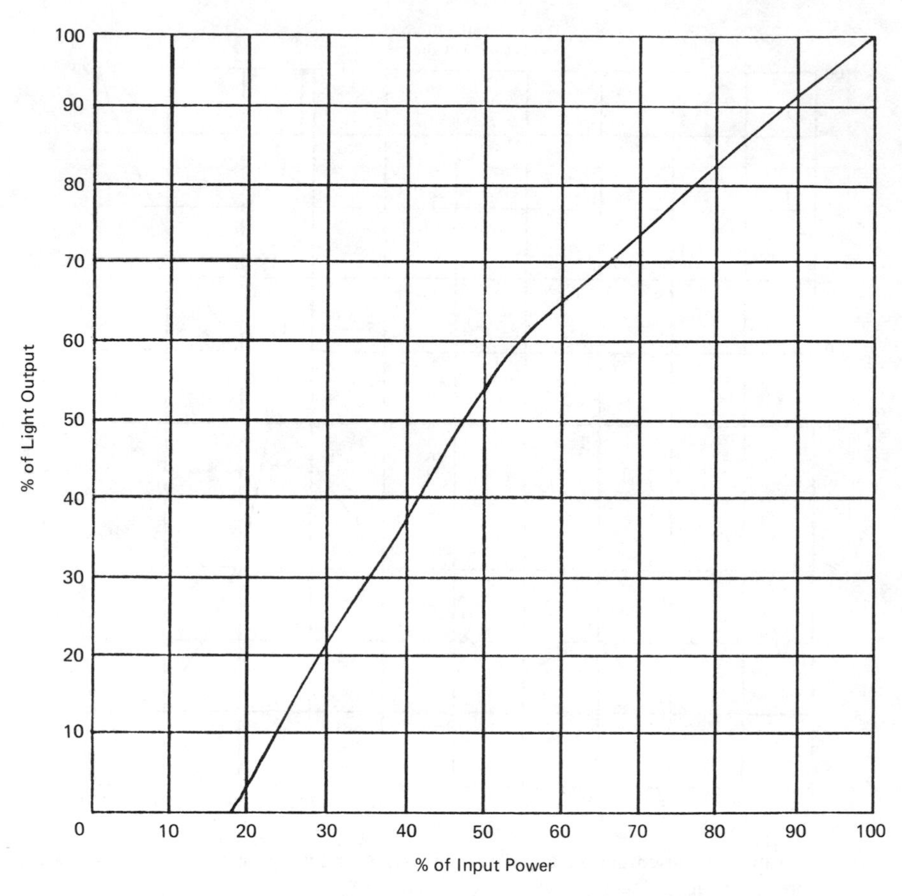

Fig. 7-10. Lamp intensity versus input power of a 40-watt, 120-volt rapid-start fluorescent dimmer.

practical purposes, however, a minimum level of 1/500 of maximum light output is the same as being completely "off."

HID Dimmers. Use of HID luminaires for exterior and interior lighting is becoming increasingly popular because of their energy-saving characteristics. Most exterior lighting, e.g., roadway or parking lighting, has made successful use of HID for a number of years now. In interiors, however, their use is limited to applications where color rendering is not of importance. From this respect, use of HID dimming also has very limited applications. Use of HID dimming may be considered at areas such as gymnasiums or rentable exhibition halls where multilevel lighting is required for various activities. For exteriors, their use may be considered for the wee hours of morning or evening when lighting level does not have to be 100%.

Mercury vapor dimming systems have been available for about a decade, although significant improvement has been recently achieved in metal halide and HPS. There are two types of mercury vapor dimming systems available. The first system works directly with the conventional ballast associated with the luminaire, and the other system eliminates the ballast and works much quieter. This system requires a dimmer module with remote intensity control and may be used with a preset control package. Figure 7-11 shows the effect of light output with a variation in input power for a mercury vapor dimmer. At 80% power, the measured light is 82%.

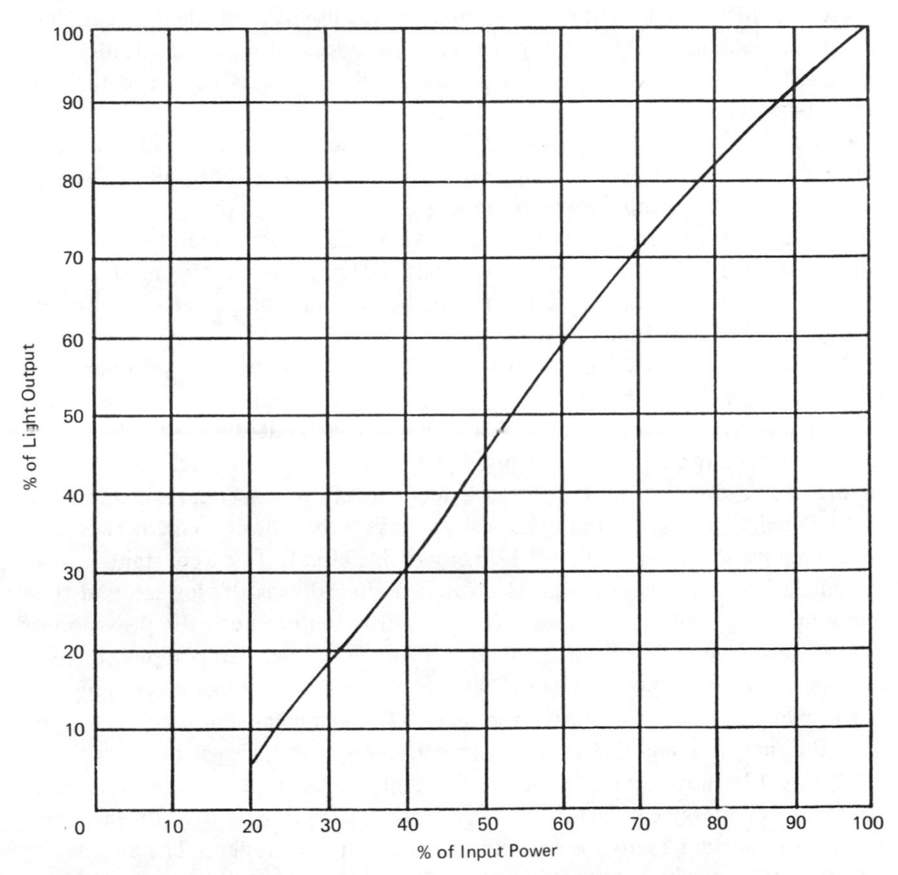

Fig. 7-11. Light intensity versus input power of a 100-watt, 120-volt, mercury vapor dimmer.

There has been significant improvement in metal halide and HPS dimming systems. One main defect in metal halide lamps is the lack of a consistent color rendering of all lamps at the same time. In all HID sources, a minimum light output of 1/50 of the maximum must be preserved to maintain the arc. Operation under this level will extinguish the arc and the cycle has to start over after awaiting the restrike time. The color of HPS tends to shift to monochromatic yellow as the intensity is dimmed.

Automatic Energy Control with Dimming Systems. Systems for dimming luminaires can be effectively used for substantial energy savings, especially when they are equipped with an automatic energy-control system. Two of the major advantages with this system are (1) the compensation for the wasted power involved in high initial lighting levels, and (2) the use of available daylight. The automatic reading of the lighting level in the room is done by a photoelectric cell known as the sensor, which sends signals for necessary action to the controlling device.

As seen in Chapter 3, any lighting design practice includes an LLF that considers the depreciation of lamp lumens with time and the accumulation of dirt on luminaires and lamps, resulting in a gradual reduction of light output. A suitable LLF is used in calculations to arrive at the required maintained illuminance level, just before the lamp replacement and luminaire cleaning requirements (see Figure 7-12a). This results in a higher amount of power consumption throughout the lamp life.

The sensor is crucially located in the ceiling; it senses the change in light output at the task level and around it, and then sends appropriate signals to the control unit to raise or lower the light level to a predetermined value. The dimmer automatically adjusts to provide the required light level in the room. In the course of time, as lamp lumens decrease and dirt accumulates on lamps and luminaires, the system automatically allows more power to maintain a constant illuminance level. Figure 7-12b shows this effect. For a constantly maintained illuminance level, the initial power consumption is the lowest, and then consumption gradually increases. The process continues until the provision of further power does not maintain the lighting level; then lamp replacement is necessary. At this point, after group relamping and luminaire cleaning, the power consumption cycle starts over again. The shaded area in the figure represents the energy savings due to the automatic energy-control system.

Figure 7-12c shows the effect when daylight is present in the room. Substantial savings of energy can be achieved if daylighting is properly used on the task.

Several important factors are involved in this type of design. The first is the selection of rooms by location. Rooms with southern or southeastern exposures seem to have the greatest daylighting advantage, while a northern exposure has the least. Room size and the number of windows, the direction and brightness

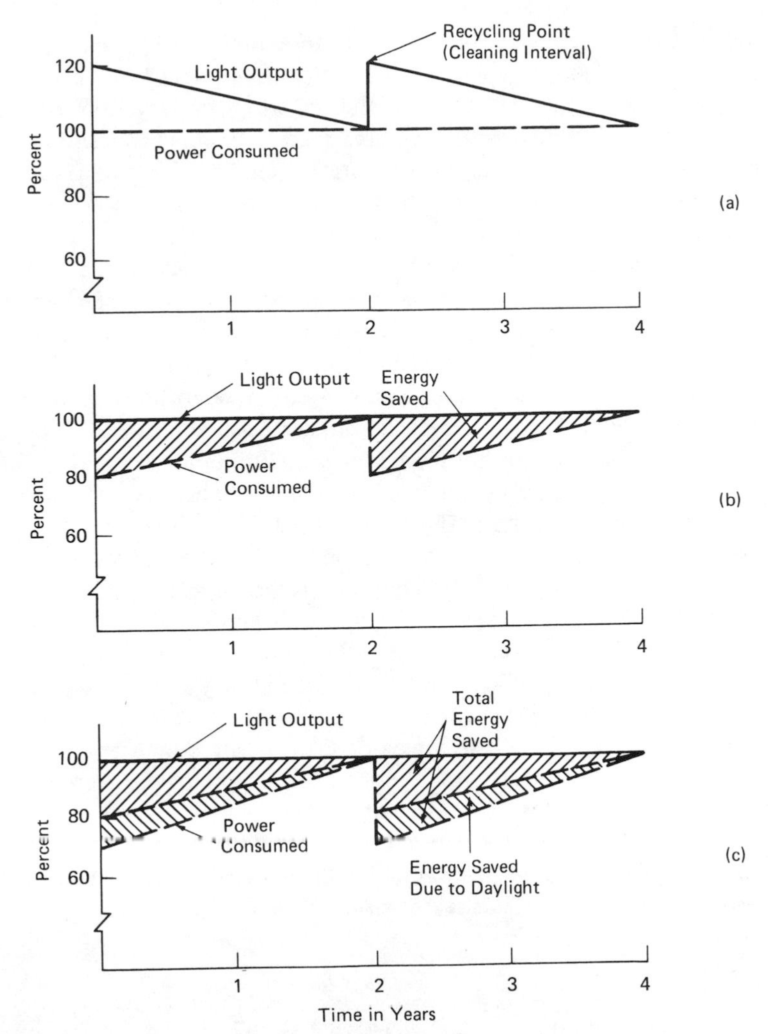

Fig. 7-12. (a) Typical situation without automatic energy-control system. (b) The same system with automatic energy-control system. The shaded area shows the savings in power. (c) The use of daylight will produce additional savings.

of the sunlight, etc. also have a significant effect on the level adjustment. Depending upon the size of the window opening, a photocell daylight sensor crucially located on the ceiling senses the natural light contribution and sends signals to the control unit for artificial lighting adjustment. If direct sunlight falls on the task, a translucent curtain or blind on the window will diffuse the daylight and avoid the concentrated brightness. Distracting sunlight glare from windows is best prevented through use of close-mesh fabric draperies in front of the windows to provide a comfortable lighting environment.

There are three types of automatic fluorescent lighting energy controlling devices available. The first system requires special dimming ballasts in the fluorescent luminaires and is developed to control a number of luminaires at a time. The second system does not require dimming ballasts and is developed to control a number of luminaires at a time. The third system also does not require dimming ballasts, but is developed to control luminaires on an individual basis.

Figure 7-13 is a typical circuit diagram representing the first system and shows the different components involved. The sensor is located on the ceiling. If used to incorporate daylight, it is located at approximately twice the height of the window from the edge of the ceiling above the window; for the measurement of the average interior illuminance level, the sensor may be located in the middle of the room. The power control module is usually mounted next to a distribution panel. The luminaires are equipped with dimming ballasts and are installed by the luminaire manufacturer. The on-off switch is used to operate the power-control module, which is fed from the power source. Each of these modules is capable of controlling 10 to 80 4-foot rapid-start (F40T12) lamps. With a capacity of 80 lamps, the system can handle 20 luminaires (each having four lamps), which can be distributed with two 20-ampere, 120-volt, or one 20-ampere, 277-volt circuit. With this system, the lighting level can be uniformly varied from 10 to 100% of light output.

The major limitation and disadvantage of this system is that it works successfully only with conventional, 4-foot, rapid-start lamps that are equipped with dimming ballasts. Low-loss energy-saving lamps, low pf and other non-rapid-start lamps are not recommended to be used with this system. The initial cost is substantial. The requirement of dimming ballasts and the special circuitry for the power-control module and sensor usually make it more suitable for new

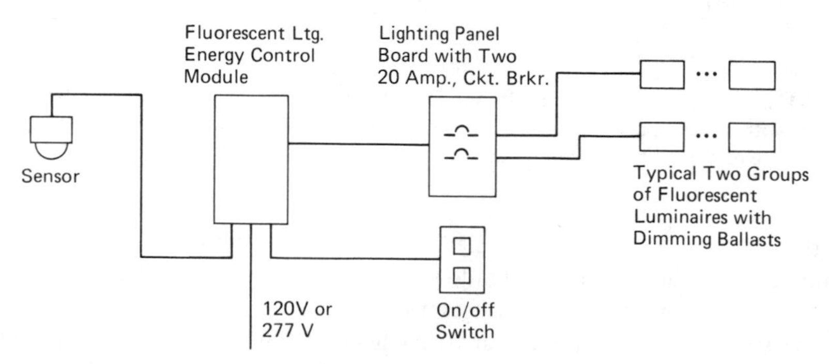

Fig. 7-13. Typical diagram of an automatic fluorescent-lighting energy-control system using a control module, sensor, and luminaires with dimming ballasts.

constructions rather than for retrofits. If installed in a suitable application, the system will typically pay for itself in about 3 to 5 years.

The second fluorescent dimming system brings a major breakthrough by operating with luminaires that have conventional ballasts. This relatively new system introduces a new type of solid-state, electronic power-control module that consumes insignificant power and is capable of reducing energy consumption up to 50%. The basic principle behind the new module is that it lowers the light level by reducing the power supplied to the existing ballasts and lamps. Figure 7-14 shows the typical connection diagram of this system. The unit is connected to the power source of the fluorescent luminaires and has a mechanical scale that can be adjusted manually to change the lighting level from 100% to a minimum of 40%. This adjustment can be made easily with a screwdriver. When a change is made, the unit will automatically lower the light level to any point from full to 40% of the lighting system's rated output. When first turned on, lights will be at their maximum intensity for proper striking. After approximately 30 seconds, the illuminance level will gradually fade to the required level. The system can also be operated automatically with the help of a sensor located on the ceiling. The energy saved is directly proportional to the degree of light reduction. This is shown in Figure 7-15. The module can be easily installed in any convenient location, e.g., in the plenum, or the electrical distribution room. Each module is capable of controlling only one 20-ampere circuit of lighting, either on 277- or 120-volt. On 277-volt a maximum of 90 rapid-start (F40T12) lamps and on 120-volt a maximum of 36 similar lamps can be used with each module. The area coverage can vary anywhere from 1200 to 2500 square feet.

The main disadvantage with this system is that the illuminance level cannot be adjusted below 40% of the maximum. Another major disadvantage is its relatively high cost, which may limit its application to installations where saving energy is of prime importance and payback period is reasonable. However, relative ease in installation makes these suitable for most retrofits.

The third fluorescent dimming system, working on the same basic principles, is available for individual luminaire control. The power-control modules of this

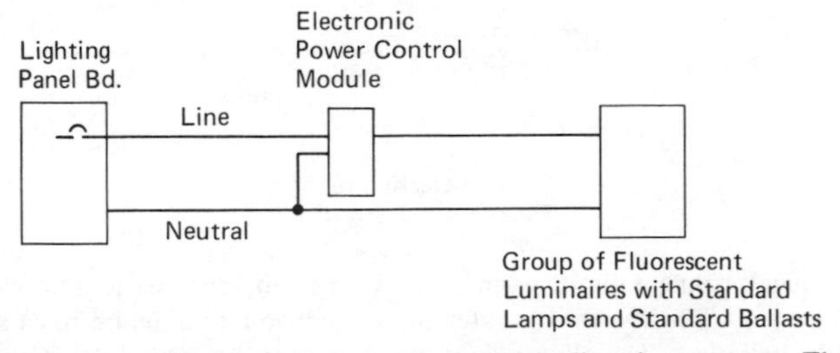

Fig. 7-14. Connection diagram of a newer fluorescent dimming system. The system does not require dimming ballasts; it can adjust light output from full to 40% of rated value.

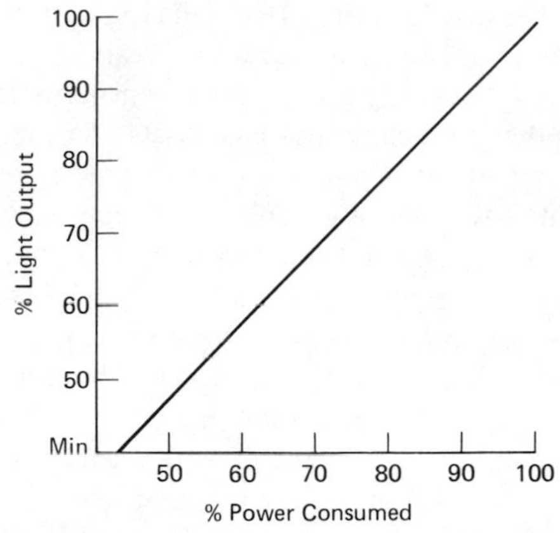

Fig. 7-15. Light output versus power consumption with an electronic power-control module. **The illuminance level cannot be adjusted below 40% of the rated output.**

system are much smaller (about 6″ × 4″ × 2″) and are installed on top of new or existing luminaires to project in the plenum space. These too, are to be used only with four-foot-long rapid-start (F40T12) lamps and do not require dimming ballasts. Figure 7-16 shows the typical connection diagram of this system. The

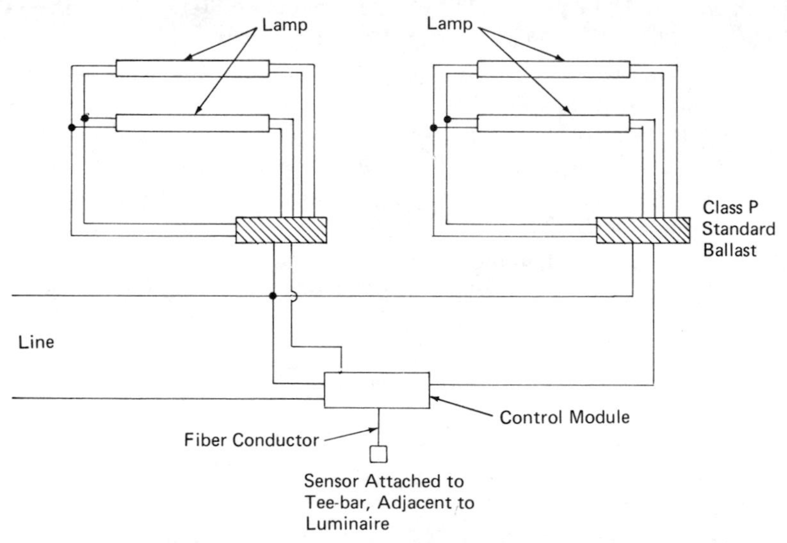

Fig. 7-16. A typical connection diagram of a four-lamp, automatic fluorescent-lighting energy-control device. This system operates on an individual-luminaire basis and does not require dimming ballasts. **The illuminance level cannot be adjusted below 40% of the** rated output.

sensor is installed on the ceiling (usually attached to the tee-bar) adjacent to the luminaire. It contains a small diffusing lens that averages the light levels that exist across the entire lighted area underneath the luminaire and sends signals to the control module through a fiberoptic conductor. It has the mechanism for manually presetting the required illuminance level, which is done with the help of a footcandle meter. Any change in light level because of daylighting or luminaire losses is received by the light-sensitive surface of the sensor, which then generates an electric pulse (voltage) that is compared with a fixed reference voltage within the control module. The results of the comparison automatically adjust the output of the lighting level. The circuitry provides special precautions against lamp flickering or arc extinguishing, which will happen otherwise if attempts are made to operate the system at voltages that are too low. For this reason, as in the previous module, the minimum light output that can be obtained with this system is about 40%. When the voltage called for by the sensor is lower than the minimum operating voltage, the results of the feedback signal comparison are ignored. The circuit also provides maximum voltage during start-up to insure proper establishment of the arc.

The main advantage of this system is that it can control light and output on a luminaire-to-luminaire basis, individually controlled, with daylighting variations. Individual optical control allows its use for smaller and larger rooms alike.

Multilevel Ballasts

A simple wiring change in these ballasts will result in different levels of illuminance from the luminaires. These are in general available in two- or three-level models. Connected to a pair of F40T12 rapid start lamps, the two-level ballast will produce 100% light with a consumption of approximately 98-watt input, and 55% light at 55 watts. The three-level ballast, on the other hand, will produce 100%, 55%, and 38% light output at about 98 watts, 55 watts, and 38 watts, respectively.

Microcomputer-Controlled Systems

The latest breakthrough in lighting energy control is the microcomputer-equipped system, which combines remote on-off control for all types of lighting and dimming control for incandescents. The system consists of a master controller that transmits commands in the form of a low-voltage, high-frequency signal (10 volts, 121 kHz) superimposed on the standard 60 Hz AC sinusoidal power curve, and receivers that are decoders with relay switches (see Figure 7-17). The luminaires to be controlled are connected to the receiving decoders, which may be wall-receptacle modules, wall-switch modules, or box-mounted switching

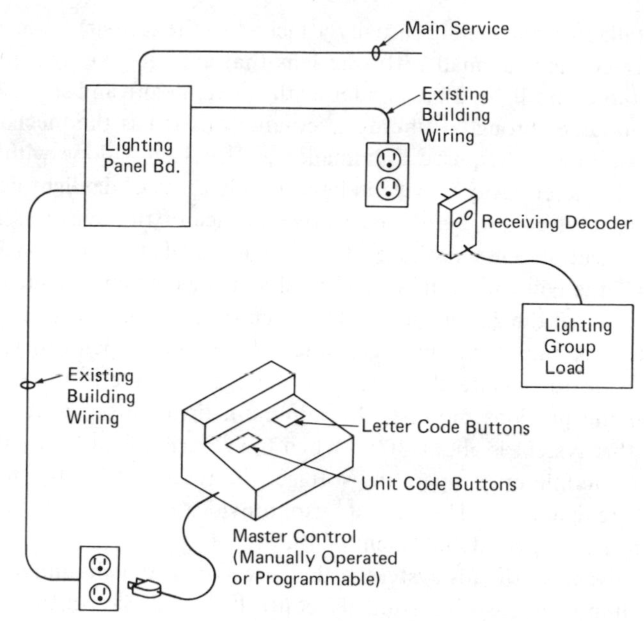

Fig. 7-17. Microcomputer-controlled systems. These do not require separate wiring, and they use the existing building wiring to transmit a command signal.

modules, each of which has two dials for coded address. The first dial usually has letter codes and the other specifies unit codes; a combination of these dials offers a wide range of addresses. Once the dials have been set for a receiver, it will respond only to those commands intended specifically for it. In reality, more than one receiving decoder may be assigned the same address. If each of these stations is connected to a group of luminaires, all luminaires at all stations will operate together upon the command of the master controller at that coded address. The signals are transmitted within 200 microseconds from the zero magnitude of the sinusoidal AC curve where susceptibility to electrical interference is almost eliminated.

The master controller may be manually operated or programmable. The manually operating types have an alarming system to remind the operator of load conditions. The programmable controllers use microprocessors to give instantaneous selection of any combination of all basic control points. Once programmed, this unit will carry out scheduled operations automatically without the help of the operator.

The greatest advantage of this system is that it does not require any additional wiring other than the existing wiring system of the building. If the unit cost of the master controller and receiving decoder is low, each luminaire in a projcct can be equipped with the system; this will save energy and operating costs, and offer maximum flexibility in lighting. The limitations of this system are as follows:

1. Dimming is possible for incandescents only (500-watt maximum)
2. The master control and receiving decoder must be in phase.
3. The master control is designed to operate at 120-volt AC only. Phase-coupling capacitors may be required for operating on 208- or 277-volt modules.

Chapter Eight
Lighting and Air-Conditioning

Heat is a necessary byproduct of any lighting system. Each watt consumed produces 3.413 Btu of heat per hour that can be used for spacing heating. For space cooling, however, this heat must be removed or counterbalanced by providing additional cold air. As can be realized, although the heat of light offers a great advantage in energy saving in one season, it can be a major energy waster in another. An ideal situation will be one where a lighting system produces maximum heat during winter and possibly nil during summer. Unfortunately, this will be impossible, since the luminaires will require equal amounts of power in both seasons. The use of energy-efficient luminaires equipped with energy-saving lamps and ballasts will obviously help in cooling seasons, but it will not contribute as much to space heating during other seasons. A lighting designer thus must be careful to select a lighting system by evaluating its effect on the mechanical system as well.

Exactly how much influence does heat of light have in energy savings, with its impact on heating or cooling systems? To evaluate this, let us consider the energy condition of this nation. As of the writing of this book, according to the U.S. Department of Energy, 93% of the resources used in this country are fossil fuels, i.e., coal, oil, and natural gas. Out of these, oil and natural gas are considered to be critical fuels since their total world reserve is quickly diminishing and they are nonreplenishable. Of the total resources consumed, approximately 38% is used for generating electricity. Individually, it consists of 64% of total coal, 20% of the total oil, 18% of total natural gas, and 100% of hydro and nuclear fuels. From a critical-fuel standpoint, only 19% of the total oil and gas is used for generating electricity.

Of the total electrical energy, approximately 20% is used to produce light. This means $0.38 \times 0.20 = 0.076$, or 7.6% of the total energy, and $0.19 \times 0.20 =$ 3.8% of the total critical fuel is being used to produce light today. By the same token, heating and cooling requires 24% and 3% of the total energy, respectively.

It is obvious that lighting is not responsible for the energy crisis. In fact, if all lighting were eliminated, energy would not be saved from its critical condition.

Space heating on the other hand, requires almost one-fourth of the total energy; involving more than 95% of it are oil and natural gas, the critical fuels. Cooling systems mainly use electrical energy, affecting 3% of the total. Curtailment of all interior lighting will actually increase the space heating by two points, while saving only a half point in cooling energy. So if all lighting were eliminated, the net saving in energy would be only (7.6% - 2.0% + 0.5%) = 6.1%.

Why is it important to save energy for lighting? There are two main reasons. First, for the current existing conditions, we must look to all sections of energy use for a contribution to energy conservation. The production of electricity still depends upon 18% and 20% of gas and oil, respectively. From the standpoint of the limited world reserve, the use of oil and gas for electricity generation should be phased out, and more use of coal should be emphasized. This has been one major consideration of the Department of Energy towards critical energy saving, which has been successful in dropping the combined percentage by 8% since 1973. The other main reason for saving lighting energy is to minimize the sky-rocketing operating costs. Currently, a survey shows, in terms of final end user, lighting alone is responsible for 30–50% of the total energy cost of a building. If all electricity is available from coal, hydro, or nuclear fuel, use of heat from lighting systems will make an important contribution to reducing oil and gas consumption during the winter. By the same token, taking lamps out or replacing existing lighting by energy-efficient luminaires may actually increase oil and gas consumption in many instances; this important point, unfortunately, still has not been recognized by many.

HEAT OF LIGHT

The general term *heat of light* relates to the contribution of the lamps, the luminaires, and the lighting systems as a whole, as heat sources.

Lamps as Heat Sources

All electric lamps, starting from the filament type to gaseous discharge, are efficient converters of electric power to heat energy. Although the amount of light produced by different types varies for the same wattage, each watt consumed is equal to 3.413 Btu of heat per hour, just as for any electric heating device. The energy generated by a lamp may be divided into two parts: (1) conduction-convection energy and (2) radiant energy, including infrared and ultraviolet light. Light is a part of the total energy generated by the lamp; although it does not contribute as much as the convection sources, light produces heat and increases room temperature as it falls on a surface and is absorbed. Different types of lamps produce different proportions of each type of energy. Tables 8-1, 8-2, and 8-3 compare the amount of each type of energy emanated from four types

Table 8-1. Energy Output for Some Fluorescent Lamps of Cool White Color (Lamps Operated at Rated Watts on High Power Factor, 120 Volt, 2-Lamp Ballasts, Ambient Temperature 77° F, Still Air).

Type of Energy	40WT12	96 Inch T12 (800 mA)	PG17† (1500 mA)	T12 (1500 mA)
Light	19.0%	19.4%	17.5%	17.5%
Infrared (est.)*	30.7	30.2	41.9	29.5
Ultraviolet	0.4	0.5	0.5	0.5
Conduction-convection (est.)	36.1	36.1	27.9	40.3
Ballast	13.8	13.8	12.2	12.2
Approximate average bulb wall temperature	106°F	113°F	140°F	

*Principally far infrared (wavelengths beyond 5000 nanometers).
†Grooves sideways.

Table 8-2. Energy Output for Some Merc. Vap., Met.-Halide, HPS and LPS Lamps.

Type of Energy	400-Watt Mercury	400-Watt Metal Halide	400-Watt High Pressure Sodium	135-Watt Low Pressure Sodium
Light	14.6%	20.6%	25.5%	35.5%
Infrared	46.4	31.9	37.2	4.5
Ultraviolet	1.9	2.7	0.2	0
Conduction-convection	27.0	31.1	22.2	42
Ballast	10.1	13.7	14.9	18

Table 8-3. Energy Output for Some Incandescent Lamps.

Type of Energy	100-Watt* (750-hour life)	300-Watt (1000-hour life)	500-Watt (1000-hour life)	400-Watt‡ (2000-hour life)
Light	10.0%	11.1%	12.0%	13.7%
Infrared†	72.0	68.7	70.3	67.2
Conduction-convection	18.0	20.2	17.7	19.1

*Coiled-coil filament.
†Principally near infrared (wavelengths from 700 to 5000 nanometers).
‡Tungsten-halogen lamp.

of sources (incandescent, fluorescent, LPS, and HID). The values are for lamps and ballasts operated under specific conditions. The energy distribution from a single luminaire housing the lamp or from a group of luminaires representing the total lighting system is different from that of the lamps alone.

Luminaires as Heat Sources

Luminaires containing lamps emit a wide range of energy, from near ultraviolet through far infrared wavelengths. Although the lighting-design aspects generally lie in the visible portion of the spectrum, it is of equal importance to understand the invisible part of the spectrum, since this is a major portion of the energy emitted. As seen in Tables 8-1 through 8-3, a large percentage of energy converted by electric lamps is radiation lying predominently in the near or far infrared regions. The amount of infrared a material can absorb determines how efficiently the material can act as a heat source. Different materials used in luminaires have varying characteristics of infrared absorption, lighting reflection, and lighting transmission. Table 8-4 compares properties of materials commonly used for luminaires, in terms of percentage of reflectance (R) and transmittance (T) at selected wavelengths. A study of these data will be helpful in understanding to what extent the luminaire material will absorb the temperature radiation of various lamps and become radiation sources of heat.

Lighting Systems as Heat Sources

A lighting system is a combination of the lamps, luminaires, and surroundings, and visual and thermal effects. The visual comfort is largely dependent upon the quantity of light and the quality of the lighting arrangement, a detailed discussion of this subject can be found in Chapter 3. Thermal comfort, on the other hand, is primarily dependent upon the proper balance between humidity, temperature, and air motion in the space. A full evaluation of these factors, thus, will relate to the total energy distribution, their relationship to room surfaces, and the type of air conditioning. The lighting and heating characteristics are influenced by luminaire performance, ambient temperature, surrounding materials and surface reflectances.

Calculations involved in evaluating the accurate amount of sensible heat produced by a lighting system is not straight forward. The rate of heat gain to the air caused by light can be quite different from the total power supplied. Some of the major factors that would determine the sensible heat are the amount of heat absorbed by surrounding materials, the length of time the lights are on, and the energy distribution pattern of the luminaires. Radiating energy of lighting affects the air after it has been absorbed by walls, floor and furniture, and has raised their temperature higher than the room temperature. This stored

Table 8-4. Properties of Lighting Materials (Per Cent Reflectance (R) and Transmittance (T) at Selected Wavelengths).

	Visible Wavelengths						Near Infrared Wavelengths						Far Infrared Wavelengths							
	400 nm		500 nm		600 nm		1000 nm		2000 nm		4000 nm		7000 nm		10,000 nm		12,000 nm		15,000 nm	
Material	R	T	R	T	R	T	R	T	R	T	R	T	R	T	R	T	R	T	R	T
Specular aluminum	87	0	82	0	86	0	97	0	94	0	88	0	84	0	27	T	16	0	14	0
Diffuse aluminum	79	0	75	0	84	0	86	0	95	0	88	0	81	0	68	0	49	0	44	0
White synthetic enamel	48	0	85	0	84	0	90	0	45	0	8	0	4	0	4	0	2	0	9	0
White porcelain enamel	56	0	84	0	83	0	76	0	38	0	4	0	2	0	22	0	8	0	9	0
Clear glass (.125 inch)	8	91	8	92	7	92	5	92	23	90	2	0	0	0	24	0	6	0	5	0
Opal glass (.155 inch)	28	36	26	39	24	42	12	59	16	71	2	0	0	0	24	0	6	0	5	0
Clear acrylic (.120 inch)	7	92	7	92	7	92	4	90	8	53	3	0	2	0	2	0	3	0	3	0
Clear polystyrene (.120 inch)	9	87	9	89	8	90	6	90	11	61	4	0	4	0	4	0	4	0	5	0
White acrylic (.125 inch)	18	15	34	32	30	34	13	59	6	40	2	0	3	0	3	0	3	0	3	0
White polystyrene (.120 inch)	26	18	32	29	30	30	22	48	9	35	3	0	3	0	3	0	3	0	4	0
White vinyl (.030 inch)	8	72	8	78	8	76	6	85	17	75	3	0	2	0	3	0	3	0	3	0

Note: (a) Measurements in visible range made with General Electric Recording Spectrophotometer. Reflectance with black velvet backing for samples. (b) Measurements at 1000 nm and 2000 nm made with Beckman DK2-R Spectrophotometer. (c) Measurements at wavelengths greater than 2000 nm made with Perkin-Elmer Spectrophotometer. (d) Reflectances in infrared relative to evaporative aluminum on glass.

Source: Reprinted with permission from the IES Lighting Handbook, 5th Edition.

energy contributes to the space cooling load after a time lag and is present after the lights are switched off. The other determining factor is the "use factor," that is, the percentage of the luminaires in use. In evaluating the amount of sensible heat produced by a lighting system, the following formula from the 1977 Fundamentals Handbook, American Society of Heating, Refrigerating and Air-Conditioning (ASHRAE) can be used.

$$q = 3.413 \times \text{NOL} \times W \times \text{UF} \times \text{CLF}$$

where

q = sensible lighting heat load in Btuh/hr
NOL = number of luminaires
W = total watts per luminaire
UF = use factor
CLF = cooling load factor

Watts per luminaire, (W) include the total power used by lamps and ballasts. For incandescents, this value is directly equal to the rated value of the lamp; for fluorescents, it is the total lamp power multiplied by a ballast loss factor ranging from 1.08 to 1.30; and for HID, it is the total lamp power multiplied by a ballast loss factor ranging from 1.04 to 1.37. The variation in ballast loss factor is mainly due to different design techniques used by different manufacturers.

The use factor, UF, represents the ratio of wattage in use under which the load estimate is being made to the total installed wattage. For stores or shopping centers where lights are left on continuously during business hours, this factor usually is unity.

The cooling load factor, CLF, takes into account the type of luminaires and their energy distribution pattern, the length of time the lights are left on, furnishings, type of room construction, etc.

Total energy distribution involves all three basic mechanism of heat transfer: convection, conduction, and radiation. The type of luminaire, mounting method, and the ceiling material have an important role in the distribution of thermal energy. Figure 8-1 shows the effect of the ceiling-to-luminaire relationship upon the lighting system heat transfer. Three different popular luminaire-mounting arrangements have been taken into consideration: suspended from the ceiling, surface-mounted on the ceiling, and recessed in the ceiling. In the suspended-luminaire, convection and radiation heat transfer take place. If the ceiling material is a good heat insulator, practically all of the input will remain within the occupied space thereby contributing to a higher CLF value. Figure 8-1b shows the luminaire surface-mounted on the ceiling. All three modes of heat transfer are present. In addition to radiation and convection, conduction takes place through the ceiling. If the ceiling material is a good thermal insula-

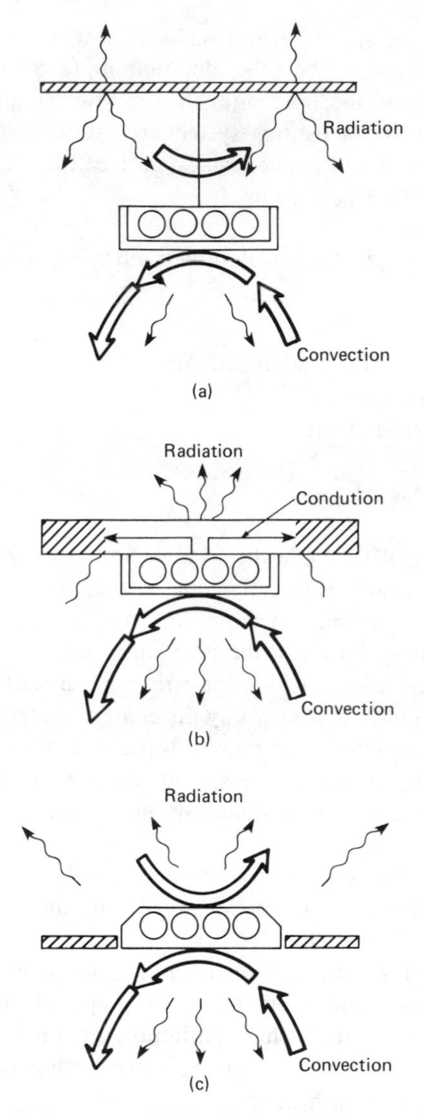

Fig. 8-1. (a) Heat transfer in a suspended luminaire. (b) Heat transfer in a surface-mounted luminaire. (c) Heat transfer in a recessed luminaire.

tor, negligible conduction of heat will occur and it may be assumed that all input energy will remain in the occupied space. The temperature within the luminaire will be higher compared to a situation where the ceiling material is a good conductor. In this condition the CLF will tend to be higher. When luminaires are recessed as is shown in Figure 8-1c, the distribution of heat is significantly different from the previous examples. Since a major portion of the luminaire is above the suspended ceiling, much of the input power remains there. The distribution of energy is a function of luminaire design and plenum and ambient conditions. For most recessed static luminaires, this ratio is nearly 50% above and 50% below the ceiling, and as such it contributes to a relatively lower CLF value.

Task/ambient lighting systems will have a different lighting energy distribution since it may employ more than one of the systems above. For most cases, where ambient lighting is done with recessed luminaires and the task lighting is either an integral part of the furniture or mounted separately near the task, the heat load must be figured separately as only the task lighting load is entirely instantaneous space heat. The CLF, under these circumstances, will be lower than if the ambient lighting was done with surface mounted or suspended luminaires.

Lighting Heat Control

Heat from lighting may be controlled by the use of three types of materials for the heat-transfer mechanism, namely, gaseous, liquid, and solid. The most popular is air. When it is desired to keep infrared radiation to a minimum, a filtering material applied to the lens allows the light to transmit, but reflects infrared back inside the luminaire. This may cause excessive heat build-up inside the luminaire, resulting in poor performance of the lamps and ballasts. Movement of air in and around the luminaires must be provided to carry this heat out.

Figures 8-2a and 8-2b show suspended and surface-mounted luminaires with air flow over the lamps and the heated surfaces. In either case, installing an air-return at the ceiling, will prevent stratification of air, lower ceiling temperature, and carry out much of the lighting heat. Figure 8-2c shows a recessed luminaire, subject to similar conditions. The amount of energy removed through internal flows can be substantial if the heated air is carried to the plenum and thence to the building's air-handling system. Much of these benefits can be accomplished with the help of air-handling luminaires, frequently known as the *heat-removal* types and sometimes referred to as the *air-return* types. The main purpose of these luminaires is to allow the room air to enter the lamp cavity and carry out much of the lighting heat to the plenum. For a 2 X 4-foot fluorescent troffer, the two smaller ends usually have the air-intake openings around the door. During air-return operation, room air enters through air-inlet openings in the door frame, passes through the lamp cavity, and exits through the heat removal slots on the top (see Figure 8-3). To control the amount of air return (in terms of cubic feet per minute, or CFM), some luminaires are available with adjustable dampers in the heat-removal slots. Other heat-removal types of luminaire are also suitable for air supply as well; these are frequently known as the *supply-return* or the *combination* type. In addition to providing all the services

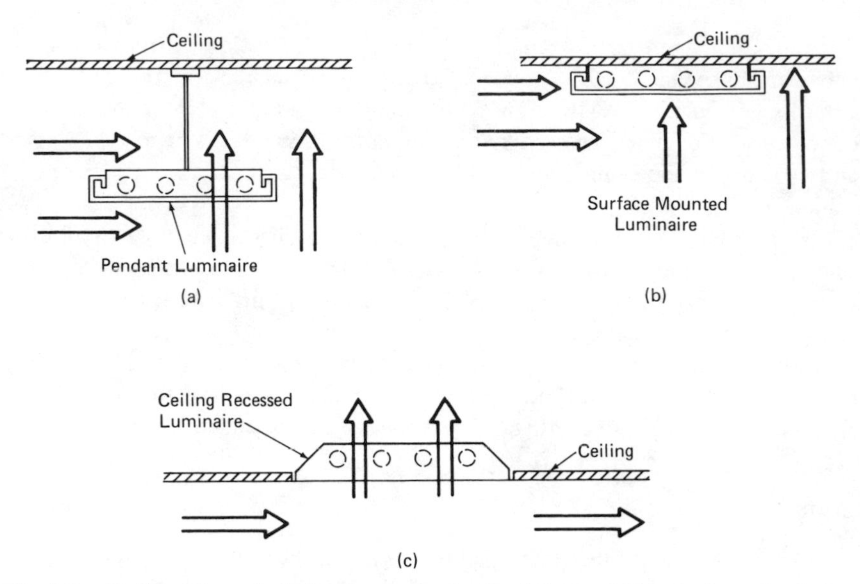

Fig. 8.2. Air flows through the lamp cavity and around the luminaires for heat circulation.

of the heat-removal type, these provide air supply with the help of an air-supply boot installed at the top of the luminaire, as shown in Figure 8-4. A flexible duct carries the supply air into the boot, which in turn discharges through the supply slots in the door frame. For a 2 X 4-foot troffer, the larger sides of the door have the air-supply slots. Unlike the heat-removing function, the air supply does not enter the lamp cavity. In essence, all types of air-handling luminaires

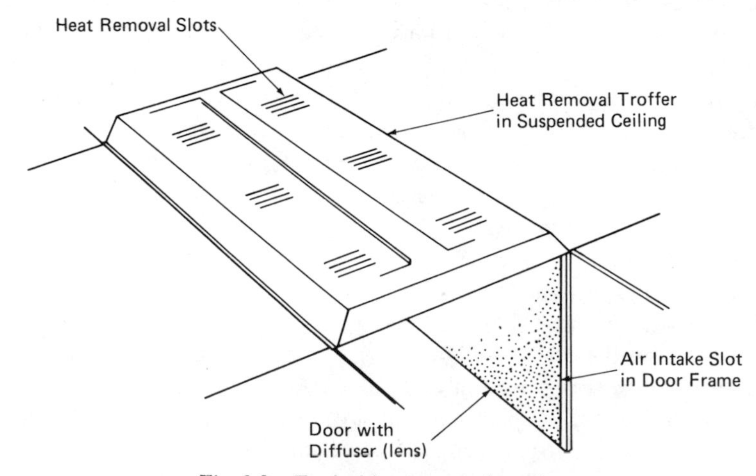

Fig. 8-3. Typical heat-removal troffer.

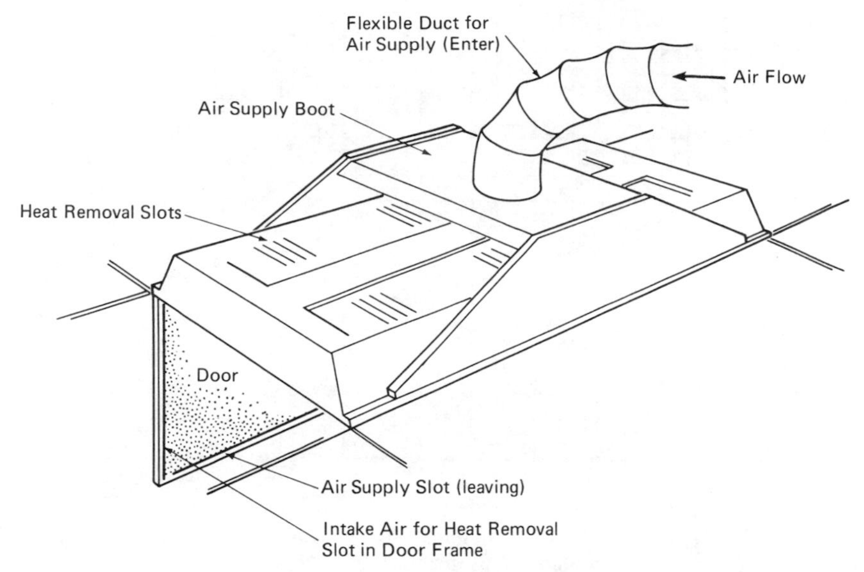

Fig. 8-4. Typical supply-return (combination) troffer.

provide the same basic function. The heat produced by lighting is extracted by mechanical equipment through a ceiling cavity and returned to the mechanical room. The heat can be recycled to reduce heating energy needs in cold weather or discarded to reduce the cooling energy needs during warmer weather.

In addition to heat removal, a great advantage of these luminaires is the increase in lighting efficiency, especially when fluorescent lamps are used. In fact, a fluorescent lamp is the only major light source that is sensitive to ambient temperature, in terms of light output and input power. Figure 4-8 in Chapter 4 shows these characteristics of a typical indoor-type fluorescent lamp. The amount of light produced by a fluorescent lamp is directly proportional to the ultraviolet energy in the lamp, which excites the phosphor. When the temperature at some spot or spots on the bulb is reduced, the mercury tends to condense at these areas, resulting in a low vapor pressure in the tube. This causes less ultraviolet energy and hence low light output. Most fluorescent lamps are designed to peak in light output at about 100°F bulb-wall temperature. If the temperature exceeds this value, too much mercury vapor is present in the lamp and ultraviolet energy is absorbed, with a reduction in both lighting output and power consumption.

Depending upon the number of lamps and the type of luminaire used, operating bulb-wall temperature may vary anywhere from 100°F to 145°F. Ambient room temperature, air-draft condition, luminaire mounting details, ceiling insulation, etc. are also responsible for the temperature variance. Figure 8-5 shows

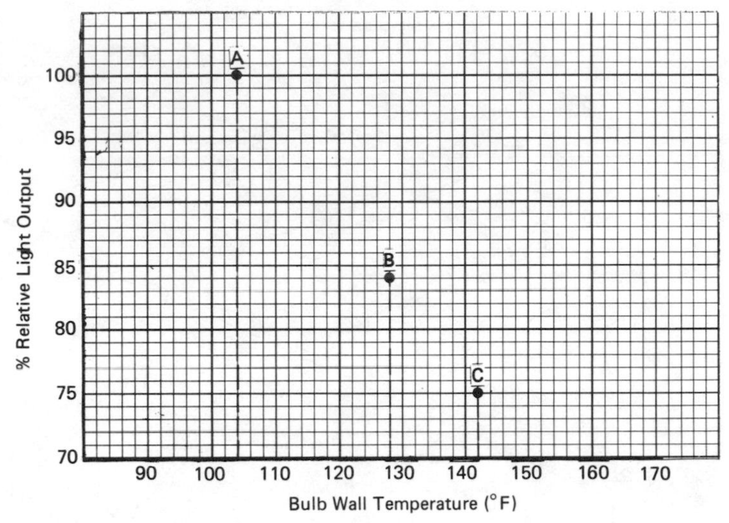

Fig. 8-5. Bulb operating temperature of three types of luminaire equipped with standard lamps and ballasts at a room ambient temperature of 77°F. (A) Two lamps, 4-foot-long fluorescent strip, suspended. (B) Four lamps, 2 X 4-foot troffer, solid acrylic lens, plenum temperature at 95°F. (C) Two lamps, surface-mounted wraparound, 4-feet-long, against low-density ceiling.

typical bulb-wall temperatures of different types of commonly used static (non–air-handling) luminaires equipped with standard lamps and ballasts, all operated at a still-air, room ambient temperature of 77°F. As can be seen, an air-suspended open-strip luminaire operating directly at a room ambient temperature of 77°F offers the maximum light output at a bulb-wall temperature of 104°F, and the surface-mounted wraparound against a low-density ceiling produces the least light at a bulb wall temperature of 142°F. Apparently, a maximum light output is achieved when the lamp is operating in an ambient temperature of 77°F (see Figure 8-6).

When return air flows through the lamp cavity of a heat-removal type of luminaire to the plenum, the room ambient temperature lowers the bulb-operating temperature. The amount of the air motion is measured in cubic feet per minute. If the CFM through a luminaire is increased, heat extraction lowers the bulb-wall temperature, offering a peak light output for a certain CFM value at 100°F, and then decreasing light output with lower temperatures. Peak light-producing CFM will vary from luminaire to luminaire, depending upon the number of lamps used and the configuration of the lamp cavity. A 2 X 4-foot luminaire with four lamps operating at 128°F bulb-wall temperature will require a higher CFM value for peak light output than a 2 X 2-foot luminaire equipped with two lamps operating at 110°F bulb-wall temperature (see Figure 8-7). This increase in light output because of return air flow has the effect of increasing

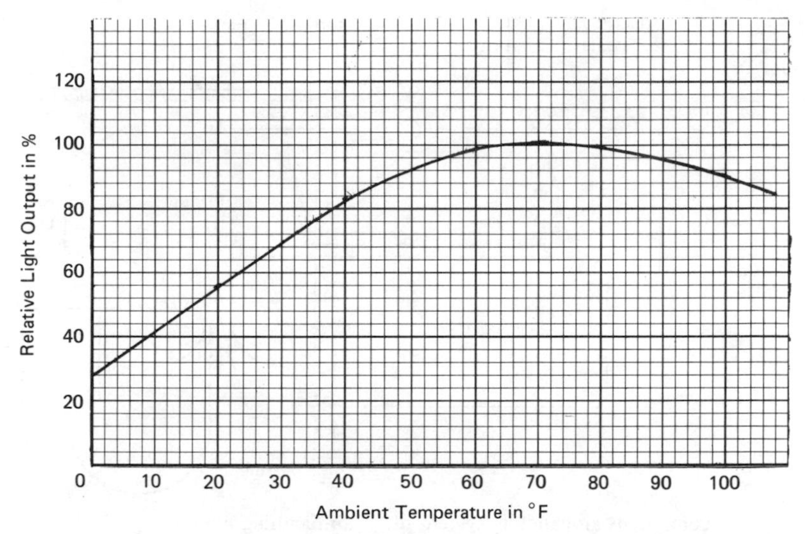

Fig. 8-6. Maximum light output is achieved when a conventional fluorescent lamp is operated at an ambient temperature of 77° F.

the luminaire's CU, which is normally measured under static conditions. For example, if the CU value in a static condition is, say, 0.60, then at a flow of, say, 50 CFM, the relative light-output changes will be reflected by a new CU value of $0.60 \times 1.20 = 0.72$.

Light-output improvement because of heat removed by air flow is usually overrated by luminaire manufacturers. Several different factors are to be considered here. First, even if the improvement factors are realistic, they are only true while the air-return system is in operation. In many systems, air return flow occurs only when a space thermostat requires heating or cooling. The change in

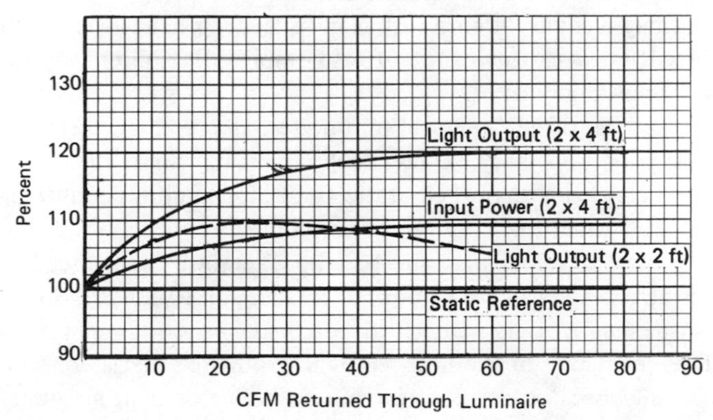

Fig. 8-7. Effect in light output and input power for heat-removal troffers.

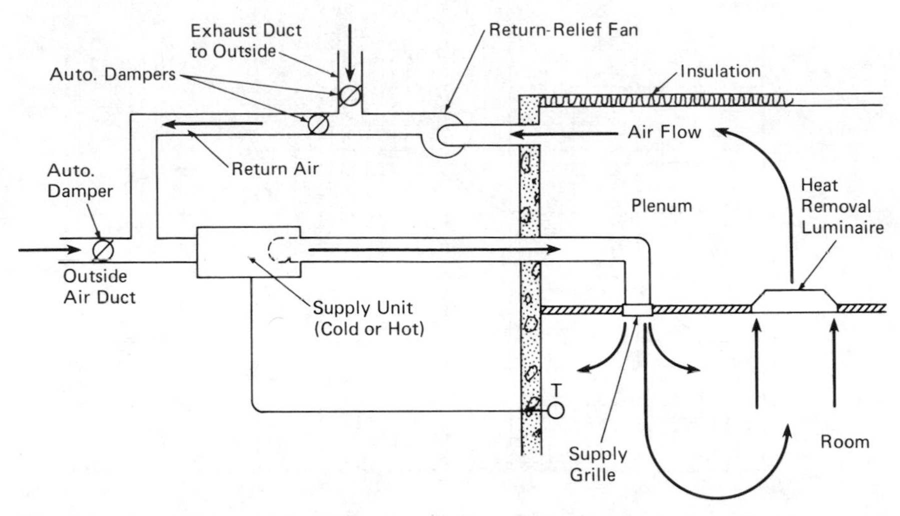

Fig. 8-8. A continuous air-handling system using air-handling luminaires. Heat is removed from the lamp cavity and circulated by the mechanical system. During the continuous process of air handling, a portion of it is exhausted out with an equal amount of fresh air introduced by automatic temperature control. In warmer seasons, most of this heated air is discarded, while in winter it is recycled.

bulb-operating temperature is a slow process; there may be an insignificant or no change in light output during the short periods of air flow. In some air-handling systems, however, air return is a continuous process, as a part of the air circulation, with or without heating or cooling (see Figure 8-8). During the continuous circulation of the building's ambient air, a portion of it is exhausted and an equal amount of fresh air is introduced by automatic temperature control. In a warm season most of this heated air is discarded; in winter it is recycled, resulting in an improvement in lighting efficiency as well as energy savings. The use of the ceiling plenum for air return is more desirable than the use of a ducted system, since it is less expensive and offers complete flexibility as to the location of the heat-removal luminaires.

Photometric data, of heat-removal luminaires, as presented by manufacturers, are for static conditions only. Under such circumstances, these luminaires offer slightly lower efficiency than their static (non air-handling) counterparts. This can be observed when their CU values are compared. Figure 8-9 shows curves representing the CU values of static and heat removal luminaires for different RCR values at ceiling, wall, and floor reflectances of 80/50/20. Both are 2 X 4-foot grid-suspended troffers with acrylic prismatic lens and equipped with four lamps. The deficiency of the heat-removal luminaire's static condition performance is mainly because of its difference in reflector contour design. Static luminaires have flat reflecting surfaces except for the ballast compartment,

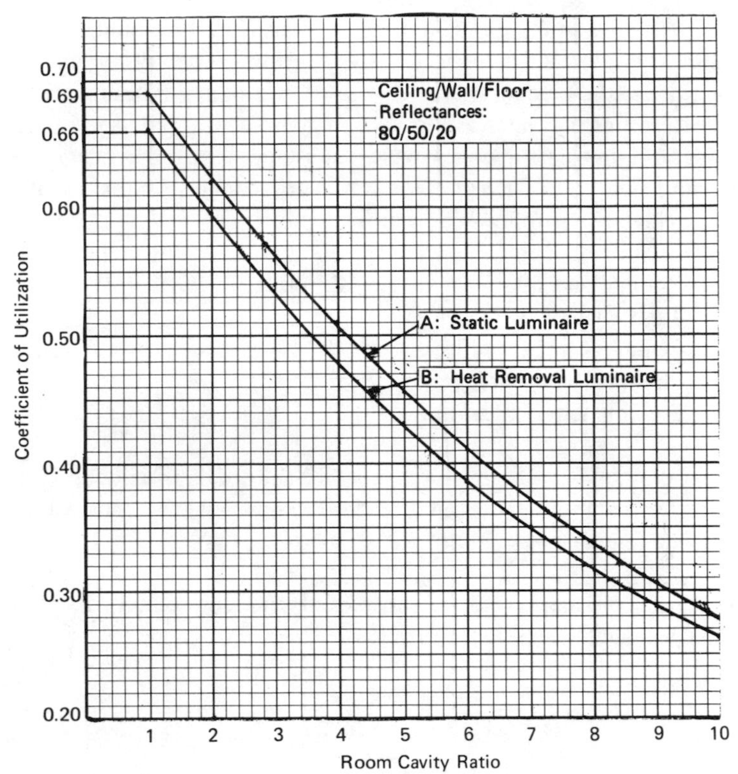

Fig. 8-9. CU comparison of a static luminaire and a heat-removal luminaire at static operation.

which usually bulges down towards the diffuser. In heat-removal luminaires there are slots or openings that do not reflect light and actually leak light to the plenum. This, along with some light trapping along the air inlet on the door, causes approximately 5% lower CU values. A possibility for lower LLF is also an important consideration here. Although continuous air flow will tend to keep the luminaire clean and possibly carry out any loose dirt entering through the heat-removal slots, a dirtier environment will drag in dirt and leave a film or coating of dirt on the lamps, lens, and the reflecting surfaces, especially if they are sticky. Use of these luminaires in office rooms located adjacent to manufacturing facilities may experience such disadvantages.

Although improvement in lighting efficiency may result in fewer luminaires, any claim of energy saving should be thoroughly studied. The most important thing to remember is that, along with the increase in lighting output with heat removal, the input power increases. A luminaire when first turned on, draws the near bench-test power as claimed by lamp/ballast manufacturers. As the luminaire remains on for awhile, the temperature rises inside and the input power decreases, as shown in Figure 4-8 in Chapter 4. As heat is continuously removed from luminaires, for different values of CFM, the input power increases and possibly reaches the bench-test values at a certain point. Although the percentage of light output may increase to its maximum at this point, the reduction in the number of luminaires may not make a substantial savings in energy consumption as compared to the use of static luminaires for the same amount of light. This is explained in the following example.

Given: Room dimension: $L = 60$ ft.; $W = 60$ ft.; $H = 9$ ft.
Task height: 3 ft. above the floor.
Ceiling/wall/floor reflectances: 80/50/20.
Illuminance level required: 70 fc.
Environment: Clean; LLF = 0.70.

Determine the number of luminaires required and total power consumed by 2 X 4-foot troffers equipped with four standard lamps and two standard ballasts in the following cases.

Case I—Static Luminaires. Using zonal cavity lumen method, the formula for determining the number of luminaires is:

$$\text{NOL} = \frac{A \times E}{L \times \text{LLF} \times \text{CU}}$$

$$= \frac{60 \times 60 \times 70}{(4 \times 3150) \times 0.7 \times 0.69}$$

$$= 41 \text{ luminaires.}$$

where

A = Area in square feet.
E = Illuminance in footcandles
L = Total initial lumens
LLF = Light loss factor
CU = Coefficient of utilization

Note that the value of CU = 0.69 is obtained from Figure 8-9 for RCR = 1, at the given surface reflectances. The value of RCR can be found directly from the cavity ratio table, Table 2-1 in Chapter 2. With a continuous operation of the luminaire, the bulb wall temperature settles at approximately 128°F and offers a lower amount of light with the heat build-up. The loss of light is automatically adjusted by the given CU value, representing the system efficiency of the luminaires at that condition. The input power also reduces with heat build up and settles at approximately 176 watts, which can be read from a wattmeter in the circuit. The initial reading before heat build-up is 192 watts, the same as the values at bench-test condition. The total power consumption in steady state is $41 \times 176 = 7216$ watts.

Case II—Heat-Removal Luminaires in Static Condition.

$$\text{NOL} = \frac{60 \times 60 \times 70}{(4 \times 3150) \times 0.7 \times 0.66} = 43 \text{ luminaires.}$$

The value of $CU = 0.66$ is obtained from Figure 8-9, curve B. Total power consumed in steady state is $43 \times 176 = 7568$ watts.

Case III—Heat-Removal Luminaires Operating at 50 CFM.

$$\text{NOL} = \frac{60 \times 60 \times 70}{(4 \times 3150) \times 0.7 \times 0.79} = 36 \text{ luminaires.}$$

Note that the modified CU value of 0.79 is obtained by a light improvement factor of 1.2, obtained from Figure 8-7 for CFM = 50, multiplied by the CU = 0.66. The input power at this condition can be read directly from a wattmeter in the circuit or found on the curve shown in Figure 8-7. The increase factor of input power at this condition is 1.09. So the total input power in steady state is $36 \times 176 \times 1.09 = 36 \times 192 = 6912$ watts.

Conclusion. Case III, requiring the fewest of luminaires, saves 12% and 14% of luminaires when compared with Cases I and II, respectively. Savings in power consumption, however, is only 4.2% and 8.6% in Cases I and II, respectively. The reduction in the quantity is significant, but the energy savings is marginal. Using the heat-removal types of luminaire for return-air purposes is much more advantageous than using them for static purposes only.

A true picture of the overall energy savings cannot be realized until the savings in the mechanical system has been analyzed. As explained earlier, a continuous removal or recovering of lighting heat through return air may show substantial energy savings in the building's air-handling system. For this, the heat-vent-air-conditioning (HVAC) system must be analyzed, with the influence of lighting heat evaluated by consulting with the mechanical engineer. Air-handling luminaires are more expensive than their static counterparts. An initial high cost may be justified if there is a significant energy savings and if the pay-back period is reasonable.

Another advantage of these luminaires is that lower operating temperature of the ballasts will extend the life of ballasts considerably beyond their normal rated span. Although there are no official claims, the old rule of thumb assumes that for each 18°F (10°C) reduction in ballast temperature, the ballast life is doubled. Ballast heat normally dissipates upwards, warming the luminaire's top. However, a proper dissipating effect may be attained if the contact between ballasts and the luminaire is done by metallic fasteners. A full urethane foam gasket around the door of the heat-removal luminaire would prevent light leakage and also reduce vibrations.

A number of different types of energy-saving lamps and ballasts are availablc today. Different combinations of these products produce varying results. While some may minimize input watts and produce lower light output, others reduce input watts and maintain the same or produce higher light output. From the standpoint of true energy savings, the combination producing the most light for least power should be selected. The inherent low power consumption of many of these products offers relatively higher light output because of lower bulb-wall temperature. For instance, when the luminaire is equipped with four lite-white lamps and two wave modified low-loss ballasts, there is an approximate 104% light output with 84% watts of input power, at a bulb-wall temperature of 120°F. Under such circumstances, during heat-removal operation, this system will not offer as much improvement in light as with standard lamps and ballasts. Increment in light output as presented by luminaire manufacturers relates to standard systems only; the same factors cannot be used for luminaires with energy saving components, as the peak light output is attained at relatively lower CFM values.

In a very crude sense, heat extraction from luminaires, by passing 77°F room air around the lamps approximates the bench-test operating conditions; therefore, the maximum light output and input power stay close to their rated values. In fact, even if lamps do operate at 100°F bulb-wall temperature, the net light output will still be restricted by the luminaire efficiency, whereas the power requirement may reach the rated maximum. Under such circumstances, the efficacy of the system may rise but will not exceed that found with bare lamps and ballasts operating under bench-test conditions.

Precautions advised by the manufacturers of energy-saving lamps should be carefully followed especially when air-handling system is incorporated. The most important caution that applies is the restriction of operation below 60°F room temperature. Reduced-wattage lamps containing krypton gas lower the bulb-wall temperature sufficiently for them to be recommended for use at 60°F ambient and above, compared to the 50°F recommendation for standard lamps. Besides, the peak light output occurring at somewhat higher bulb-wall temperatures, the thermal effects are also somewhat different.

Light-heat recovery from luminaires either for discarding or recycling is an excellent way of saving energy. The system will be most effective when the luminaires operated in a stable ambient air temperature. The Emergency Building Temperature Restriction Regulations of 1979, which placed restrictions on heating and cooling in nonresidential buildings, affect that assumption. According to these rules, the room temperature can be no lower than 78°F for cooling and no higher than 65°F for heating. In addition, these regulations require temperature set-backs during unoccupied periods of 8 hours or more. The designer must take these into consideration and design to otpimize savings throughout the year. A well-balanced design would enable luminaires and lamps to operate more efficiently, reduce fan horse-power requirements of the air circulation system, and permit uncluttered ceiling design with fewer visible grilles.

REFERENCES

American Society of Heating, Refrigerating and Air-Conditioning. 1977 Fundamental Handbook, Chapter 25, New York.

Bonvallet, G. G. "Method of Determining Energy Distribution Characteristics of Fluorescent Luminaires." *Illuminating Engineering*, LVIII, February, 1963, p. 69.

National Electrical Code. National Fire Protection Association, Boston, 1981.

Sorcar, Prafulla C. "A Study in Energy Savings with Heat-Removal Luminaires." *Electrical Construction and Maintenance*, November, 1981, pp. 66–69.

Henderson, Therman A. "Heat Transfer From Ceiling Plenums." Air Conditioning, *Heating and Ventilating*, November, 1966, pp. 72–77.

Chapter Nine
Exterior Lighting

The primary reasons for exterior lighting are safety and security. All exterior luminaires can be loosely divided into two categories: the static and the adjustable. Luminaires permanently located on the top of a pole to produce light in a predetermined manner are of the first kind. Luminaires mounted with the help of adjustable brackets or knuckles are of the second. Static-type luminaires are mainly used for roadway lighting. Movable or adjustable type luminaires have been mostly used for flood-lighting and area lighting, although use of static-type luminaires for area lighting is a commonly accepted method. In general, all exterior lighting for average roadway and areas makes use of pole heights that can be serviced from ground. For multiple highway intersection areas or larger parking areas, there are high-mast lighting systems that are generally equipped with a lowering system. There is no set rule that determines a height to qualify as a high-mast lighting system. In general, lighting systems with 50-foot or higher poles usually require a high-mast system. Although lowering systems are associated only with high-mast systems, every high-mast system is not necessarily equipped with a lowering system.

The basic design technique for all area- and roadway-lighting systems is the same, whether they are high mast or regular pole mounted. Design criteria involve terminology that is unique to exterior lighting; these terms should be understood before we get into design. They are based on the concept of having the luminaire installed on a pole with an arm, located on the curb, to produce light on the roadway beneath. Figure 9-1 shows this relationship. The various definitions are as follows:

Origin. The point on the road, directly underneath the luminaire.

Longitudinal roadway line (LRL). Imaginary lines running parallel to the curb, sometimes expressed in terms of multiples of luminaire mounting height (MH).

Transverse roadway line (TRL). Imaginary lines running perpendicular to the curb or the LRL, expressed in terms of multiples of luminaire MH.

Reference lines (0 MH LRL and 0 MH TRL). Reference lines passing through the origin. 0 MH indicates luminaire mounting height at MH = 0.

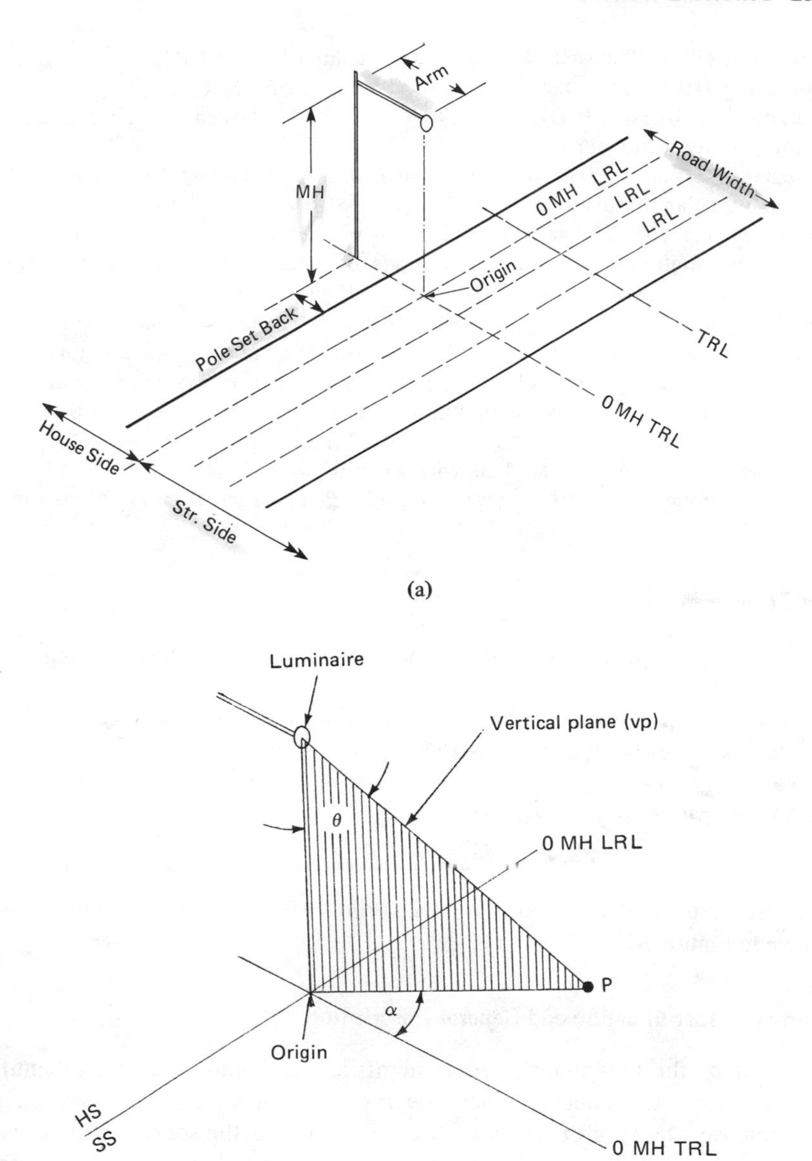

Fig. 9-1. (a) Roadway-lighting terminology. (b) If point P represents maximum candlepower location, then shaded plane is Maximum Vertical Plane and angle θ represents Maximum Cone.

Street side (SS). The space located at street side of 0 MH LRL.
House side (HS). The space located at house side of 0 MH LRL.
Longitudinal distance (LD). A distance measured between two TRL, in a direction parallel to the curb.
Transverse distance (TD). A distance measured between two LRL, in a direction perpendicular to the curb.
Elevation angle, θ. The vertical angle measured from nadir. (See Figure 9-1b).
Horizontal angle, α. The horizontal angle between vertical planes measured in the horizontal plane of the roadway. (See Figure 9-1b).
Vertical plane (VP). Any plane perpendicular to the horizontal plane of the roadway, passing through the origin, at a horizontal angle α. (See Figure 9-1b).
Maximum vertical plane. A vertical plane in which the maximum candlepower occurs. If point P represents maximum candlepower location, then the shaded plane is the maximum vertical plane. (See Figure 9-1b).
0° vertical plane. A vertical plane where $\alpha = 0°$.
Maximum cone. This is the elevation angle, θ, at which the maximum candlepower occurs.

PHOTOMETRICS

A photometric report of an exterior type luminaire includes the following:

Luminaire identification and general description
Classification of luminaire (ANSI/IES type)
Utilization curve
Isofootcandle curves or diagram
Luminous flux distribution or light flux values

For the purpose of explaining, we will refer to the typical photometric data shown in Figure 9-2.

Luminaire Identification and General Description

This part of the photometric report identifies the luminaire with its manufacturer's name and product number, the type of lamp or lamps suitable for the photometric, the type of reflector, the refractor, and the socket position used, and shows cross-sectional details of the luminaire, with dimensions. The general description may also include other useful information, such as test distance, maximum candela, maximum cone, maximum vertical plane, maximum candela at 80 and 90°, nadir candela, and nadir footcandle. If the photometric data are suitable for more than one type of lamp, the values are expressed in terms of 1000 lamp lumens. In this case, all lumen, candela, and footcandle values must be multiplied by the ratio of actual lamp lumens to 1000 lumens.

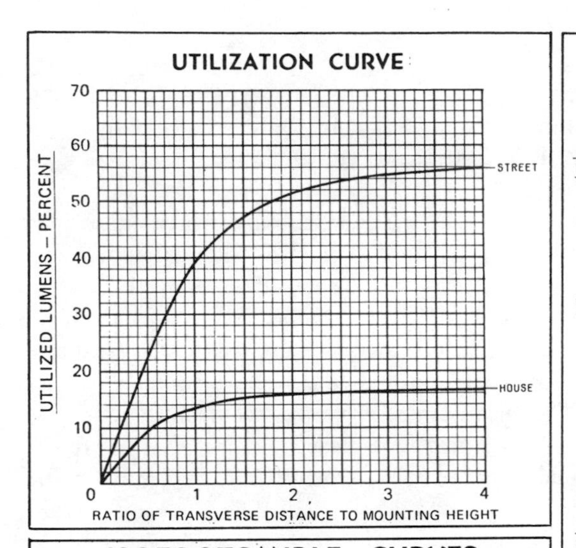

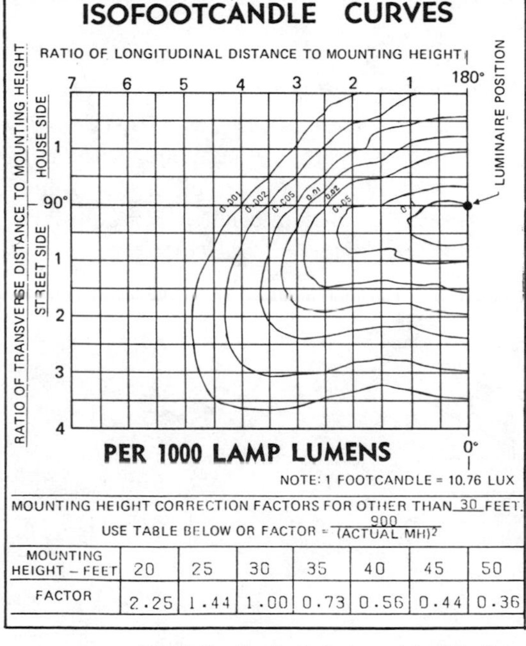

MOUNTING HEIGHT CORRECTION FACTORS FOR OTHER THAN 30 FEET.
USE TABLE BELOW OR FACTOR = $\frac{900}{(\text{ACTUAL MH})^2}$

MOUNTING HEIGHT – FEET	20	25	30	35	40	45	50
FACTOR	2.25	1.44	1.00	0.73	0.56	0.44	0.36

PHOTOMETRIC DATA

PER 1000 LAMP LUMENS

LUMINAIRE

XYZ MANUFACTURING CO.
CAT. NO. ABC123

LAMP

200, 250 OR 400 WATT HPS
GE NO. LU200, LU250 OR LU400
ANSI NO S66,S50 OR S51

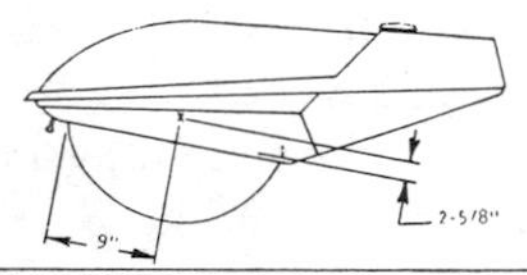

GENERAL INFORMATION

TEST DISTANCE	25 FEET
MAX CANDELA	851
MAX CONE	67.5°
MAX VERTICAL PLANE	72.5°/287.5°
MAX CANDELA AT 90°	24
MAX CANDELA AT 80°	177
NADIR FOOTCANDLES	0.09590
NADIR CANDELA	86

MULTIPLY ALL LUMEN, CANDELA, AND FOOTCANDLE VALUES BY THIS RATIO

$$\text{RATIO} = \frac{\text{ACTUAL LAMP LUMENS}}{1000}$$

PHOTOMETRIC TEST IN ACCORDANCE WITH IES GUIDE

ANSI/IES TYPE

MEDIUM/SEMI CUT-OFF/TYPE III

CIE TYPE

NON CUT-OFF

LIGHT FLUX VALUES

	LUMENS	PERCENT OF LAMP
DOWNWARD STREET SIDE	568	56.8
UPWARD STREET SIDE	15	1.5
DOWNWARD HOUSE SIDE	165	16.5
UPWARD HOUSE SIDE	7	0.7
TOTAL	755	75.5

Fig. 9-2. Typical photometric data for an exterior luminaire.

Classification of Luminaires

All roadway and area luminaires are classified into three types of light distribution, namely, (1) vertical light distribution, or simply, the spread; (2) lateral light distribution; and (3) control of light distribution at high angles.

Vertical Light Distribution or Spread. This is expressed in terms of either short (S), medium (M), or long (L); identity depends upon where the maximum candlepower occurs. Figure 9-3 shows the three types of spread.

The first zone, bounded by TRL 1 MH and 2.25 MH is the S zone. If the maximum candlepower occurs in this zone, the luminaire will be classified as S, or short spread. The second zone, representing the M zone, is bounded by TRL at 2.25 MH and 3.75 MH. If the maximum candlepower occurs in this zone, the luminaire will be classified as M, or a medium spread. The last zone, representing the L zone, is the area bounded by the 3.75-6 MH TRL. If maximum candlepower occurs in this zone, the luminaire will be classified as L, or long spread. The luminaire shown in the figure is of medium spread, as depicted in this example.

Lateral Light Distribution. The lateral distribution of a luminaire is classified into five types, usually indicated by the roman numerals, I, II, III, IV, and V. These represent roughly the shape of the beam pattern occupying a space on the horizontal surface of the roadway. The exact type of distribution is determined by knowing the exact location of the half-maximum candlepower contour on the ground. In determining the contour, all points of the lighted area receiving

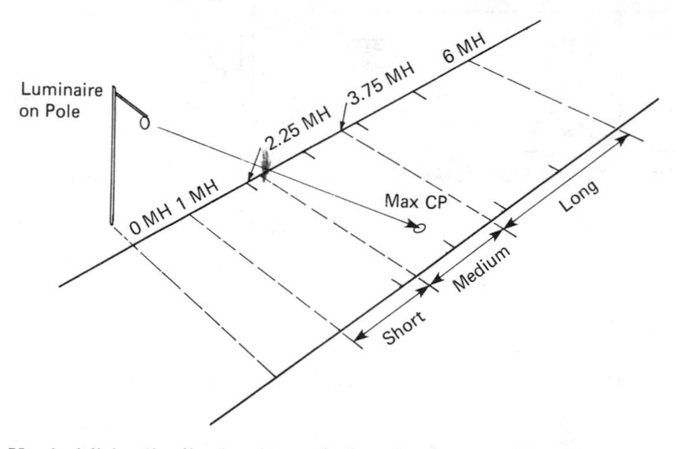

Fig. 9-3. Vertical light distribution (spread) given by short, medium, or long classifications. The luminaire shown has a medium spread, since the maximum CP falls between 2.25 and 3.75 MH.

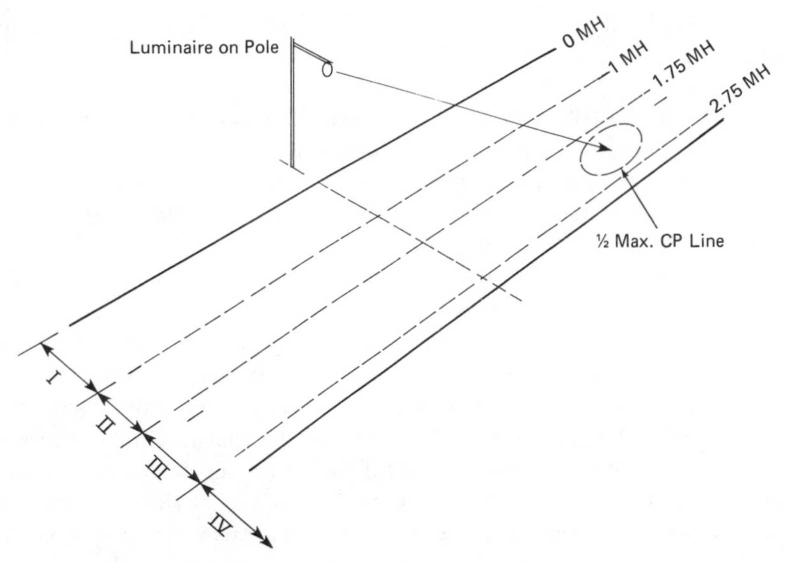

Fig. 9-4. Lateral light distribution. The luminaire is classified to be a type III distribution since the $\frac{1}{2}$ maximum CP contour occurs within 1.75-2.75 MH LRL.

one-half the maximum candlepower reading are joined together. (See Figure 9-4).

Distribution type I: The half-maximum candlepower contour occurs within 0-1 MH LRL

Distribution type II: The half-maximum candlepower contour occurs within 1-1.75 MH LRL

Distribution type III: The half-maximum candlepower contour occurs within 1.75-2.75 MH LRL

Distribution type IV: The half-maximum candlepower contour occurs beyond 2.75 MH LRL

Distribution type V: Circular symmetrical candlepower distribution at all lateral angles around the luminaire

A change from type-I to type-IV distribution characteristics is usually done by making a change in the reflector design or simply by changing the position of the lamp socket. While some manufacturers produce a family of luminaires with various reflector designs to accomplish the distributions, others use the same luminaire and field reset the reflector or socket position. The same flexibility can sometimes be accomplished by altering the angle of the luminaire position. However, this is only true for luminaires with an adjustable mounting bracket or knuckle.

In this example, the distribution is type III, since the half-maximum candlepower contour is within 1.75-2.75 MH LRL.

Control of Light Distribution at High Angles. Control of light distribution at high angles, in general, relates to the amount of candelas above maximum cone. They are classified into three types: cut-off, semi-cut-off, and non-cut-off. Each situation is defined by the amount of candelas emitted at 80-90° zone, compared to the rated lumens of the lamp. This is illustrated in Figure 9-5. The control is classified as a cut-off if a maximum of 25 candelas per 1000 lamp lumens (2.5% of rated lumens) occurs at or above 90° and a maximum of 100 candelas per 1000 lamp lumens (10% of rated lumens) occurs at or above 80°. The control is classified as a semi-cut-off if a maximum of 50 candelas per 1000 lamp lumens (5% of rated lumens) occurs at or above 90°, and a maximum of 200 candelas per 1000 lamp lumens (20% of rated lumens) occurs at or above 80°. The control is classified as a non-cut-off if the maximum candlepower does not fit into any of the limits above and primarily has no candlepower limitation. For this example, the control is a semi-cut-off, since its candlepower at 90° is 24 per 1000 lumens and that at 80° is 177 per 1000 lumens, which are less than 50 and 200 candelas, respectively.

These classifications directly relating to the candlepower distribution associated with high angles (70-90°) indicate the amount of glare. The greater the amount of candelas, the more brightness that occurs.

Utilization Curve

The utilization curve is a graphical representation of a collection of CU data. It represents the amount of light falling on the ground as compared to the bare-lamp lumens. Usually there are two utilization curves associated with a roadway luminaire: one representing the street side, the other house side or curb side.

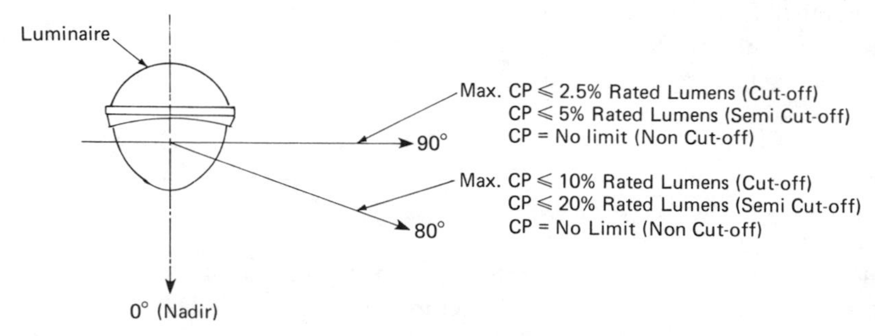

Fig. 9-5. Distinction between cut-off, semi-cut-off and non-cut-off is solely based upon how much candlepower is emitted at or about 80° and 90° angles, expressed as a percentage of rated lumens.

Most luminaires usually have a higher utilization curve for the street side than for the curb side. A luminaire having a distribution type V, however, has only one utilization curve representing both, since the amount of light falling on each side is the same. If two curves are shown, they are exactly the same and are mirror opposites.

In order to determine CU value, it is necessary to determine the ratio of TD to MH for street and curb (house) side and then look up the corresponding CU values from the utilization curves. The net CU value is the addition of the two. This is explained in the following example.

Suppose the luminaire of the depicted example is mounted at 25 feet above ground and at the end of a 2-foot arm, as shown in Figure 9-6. Let us assume that the roadway is 30 feet wide. Note that since the luminaire is mounted 2 feet beyond the pole, part of the light on street is due to the light on the HS of the luminaire. The steps involved in determining the net CU that determines the amount of light falling on the street is as follows:

Step 1: Determine TD/MH ratio:
House-side TD/MH: 2/25 = 0.08
Street-side TD/MH: 28/25 = 1.12

Step 2: Determination of CU (from utilization curves in Figure 9-2):
CU for house side: 0.12
CU for street side: 0.42
So the net CU: 0.12 + 0.42 = 0.54

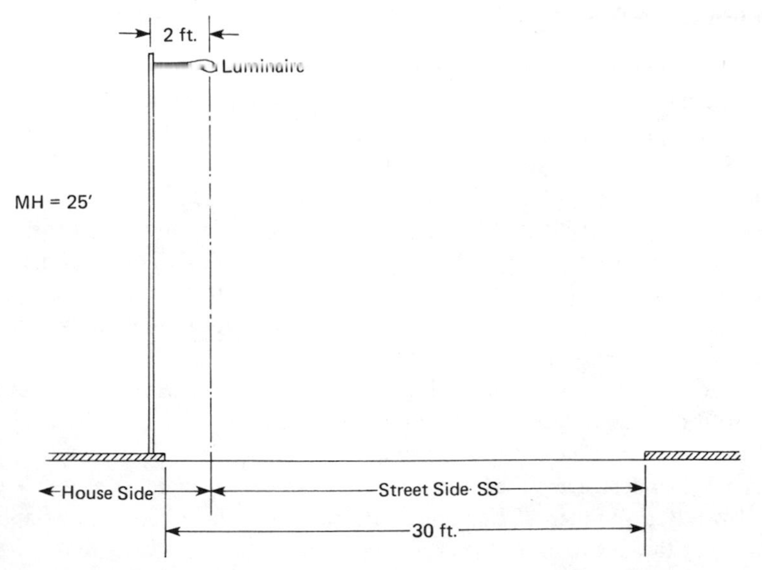

Fig. 9-6. Street-lighting arrangement of the depicted example.

In selecting a luminaire, a thorough study of its utilization curve will be extremely helpful, although this should not be the only criterion of selection. A CU value determining the actual amount of useful light falling on the ground represents the energy effectiveness. A luminaire with high efficiency may have much of its light lying in nonuseful areas, e.g., as with ballard or globe types, which produce light in all 360°; the CU values for these luminaires will obviously be low. A straight utilization curve would indicate uniform distribution of lamp lumens across the road; however, this sort of curve seldom exists in reality. Most luminaires produce a large uniform amount of light to a certain transverse distance and then gradually drop in value. The portion of the utilization curve that swings the upward curve to a gradual flattening off represents this phenomenon and is frequently known as the "knee" of the curve. The sharper this bend and the flatter the curve, the more drastic and sudden the drop in light in the transverse direction. A study of the knee of the curve will be helpful in determining the maximum width of the roadway that can be uniformly illuminated. By the same token, the knee is equally useful in determining the MH of the luminaire. For instance, if the knee of the curve of streetside occurs at a TD/MH = 2, as in our example, this will indicate that the mounting height of the luminiare should be a minimum of MH = TD/2, or one-half the transverse distance. Note that if the luminaire is located at the edge of the roadway meeting the curb, the transverse distance will be equal to the width of the roadway, and no calculation for the house side will be necessary.

Isofootcandle Curves or Diagram

The isofootcandle diagram represents the actual illuminance levels on the ground. The diagram consists of a number of curves, each representing the contour of a specific footcandle level. In determining these curves, all illuminances of the same level are joined together. Note that, as shown in the example, the isofootcandle diagram of a photometric report is based on a specific MH. If it is required to determine these values for a different MH, the footcandle values shown in the diagram must be multiplied by an appropriate factor, as shown in the MH correction table. Also note that in determining the actual illuminance at a point on the ground, a further correction is necessary by multiplying the value by the ratio of total lumens per lamp to 1000.

Isofootcandle diagrams usually show only one side of the illuminance distribution, as in our example. The other side is assumed to be exactly identical but in a mirror-image configuration. A complete isofootcandle diagram can be of great advantage to a designer if these are drawn to a scale that matches the scale of the site plan. If a number of luminaires are involved, these can be overlapped as needed and the net illuminance at a point or area can be directly determined by adding the individual footcandle values. Several luminaire manufacturers provide such "design templates." These are developed on transparent paper and are available in different pole height, lamps, and scale. These are also helpful in quickly determining the farthest spacing possible for an even illuminance level.

Luminous Flux Distribution

Luminous flux distribution shows the amount of flux (lumens) available in different zones of the luminaire. Figure 9-7 shows the four quadrants, each representing a zone. The two zones below 90° are the downward houseside (DNHS) and downward streetside (DNSS), and those above 90° are the upward houseside (UPHS) and upward streetside (UPSS). The zonal lumens may be expressed as a percentage of total lumens or per 1000 lamp lumens. For this example, these are shown per 1000 lamp lumens. The efficiency of the luminaire can be directly determined by adding the percentage of lumens of each zone or by dividing the total zonal lumens by bare lamp lumens. If the zonal lumens are expressed in terms of per 1000 lamp lumens, then the total should be divided by 1000 instead of by the actual lamp lumens. In this example, the efficiency of the luminaire can be found in two ways. It can either be found directly as 75.5%, as shown in the total of percentage of the lamp, or by dividing the total of lumens (755) by 1000 = 0.755, or 75.5%.

To minimize energy waste, it is necessary to have the least or no light output above 90°. The amount of light produced between 70° and 90° will have a serious impact on glare.

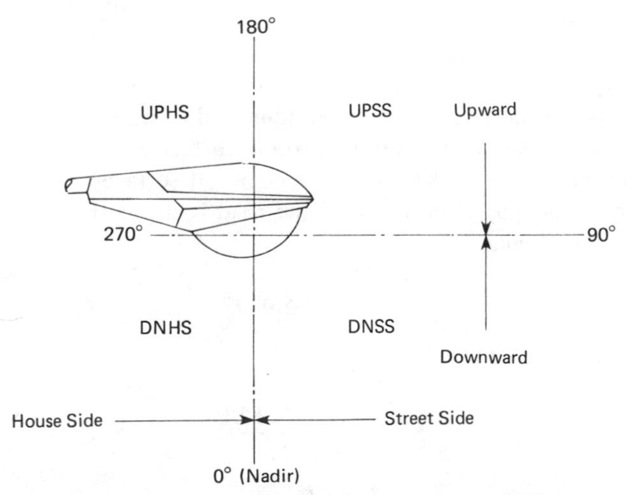

Fig. 9-7. Four quadrants, consisting of upward houseside (UPHS), upward streetside (UPSS), downward houseside (DNHS), and downward streetside (DNSS).

OTHER DESIGN CONSIDERATIONS

Lamp Factor

The rated number of lumens per lamp varies with the manufacturer. The lumens of coated lamps are different from those of clear lamps. If the lamp used is different from the one used in developing photometrics, a lamp factor (LF) must be used to compensate for the change in light output. This is given by

$$\text{LF} = \frac{\text{Actual lamp lumens}}{\text{Test lumens}}$$

Spacing and Uniformity

The spacing required between luminaires can be determined with the help of the lumen formula:

$$E = \frac{L \times \text{CU} \times \text{LLF}}{A}$$

where

E = Illuminance in footcandles
L = Total initial lumens
A = Area in square feet
LLF = Light Loss Factor
CU = Coefficient of utilization

Note that the formula is the same as used for indoor applications. The major difference between the two is in the technique of determining the CU factor. In indoor systems, the CU values vary with room-surface reflectances. In outdoor systems, however, these values are independent of surface reflectances. For a roadway-lighting design, if the width of the road is W and the spacing is S, then the area can be written as

$$A = W \times S$$

Substituting this in the formula,

$$E = \frac{L \times \text{CU} \times \text{LLF}}{W \times S}$$

$$S = \frac{L \times \text{CU} \times \text{LLF}}{E \times W}$$

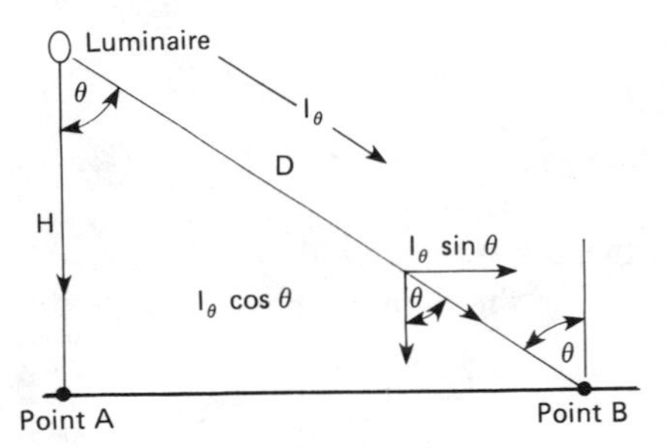

Fig. 9-8. For the same illuminance at points A and B, candlepower emitted at angle θ must be equal to nadir candlepower times the reciprocal of $\cos^3 \theta$.

Assuming that all factors are known, it is obvious from the formula that the spacing between luminaires can be directly proportional to the amount of lumens of the lamp. Theoretically, the use of highest-wattage lamp will space the luminaires farthest apart. In reality, this is not possible, since other considerations, such as uniformity and glare, will have a determining rule over maximum spacings.

In order to get perfect uniformity, it is necessary that the illuminance measured directly underneath a luminaire be the same as that between two poles. This necessitates a substantially higher amount of candlepower emitted sideways rather than directly downwards. Mathematically, the candlepower emitted sideways must increase in proportion to the reciprocal of the $\cos^3 \theta$. This is explained below (see Figure 9-8).

Let us suppose we are to determine the amount of candlepower necessary at an angle θ, to produce the same illuminance at points A and B. By the inverse-square law, illuminance at point A is

$$E_A = \frac{I_0}{H^2}$$

where

I_0 = Candlepower at $\theta = 0$ (nadir)
H = Mounting height
E_A = Illuminance at point A

Now if E_B = Illuminance at point B, then

$$E_B = \frac{I_\theta \cos \theta}{D^2}$$

where

I_θ = Candlepower at θ
D = Distance from luminaire to point B.

Now since cos = H/D, then $D^2 = H^2/\cos^2\theta$.
Substituting these values in the formula above,

$$E_B = \frac{I_\theta \cos\theta}{\dfrac{H^2}{\cos^2\theta}} = \frac{I_\theta \cos^3\theta}{H^2}.$$

For same illuminance at points A and B,

$$E_B = E_A$$

or

$$\frac{I_\theta \cos^3\theta}{H^2} = \frac{I_0}{H^2}$$

or

$$I_\theta = \frac{I_0}{\cos^3\theta}$$

or

$$I_\theta = I_0\,(1/\cos^3\theta).$$

From this relationship, it is obvious that in order to maintain uniformity, candlepower arriving at point B must be equal to the nadir candlepower times a factor equal to reciprocal of $\cos^3\theta$. Note that at an angle $\theta = 0$, the candlepower $I_\theta = I_0$, since $\cos\theta = 1$. But as the angle θ increases, I_θ becomes gradually larger than I_0 and rapidly increases, settling at infinity when it is equal to 90°. Figure 9-9 shows a chart of this multiplication factor ($1/\cos^3\theta$) drawn against various angles of θ. Note that candlepower at 70° needs to be 25 times the nadir candlepower for equal illuminance; in reality it is almost impossible to provide this amount. In order to reduce glare and conserve energy, the intensity at high angles must be redirected to more useful zones, and luminaires must be spaced furthest apart without losing an acceptable minimum uniformity.

As a general rule, maximum light occurs underneath a luminaire, whereas the minimum is in the midpoint of two luminaire spacings. A measure of an acceptable degree of uniformity is expressed by the ratio of average to minimum illuminance. According to the IES, the average-to-minimum ratio should not exceed 3:1 for most applications and should not exceed 6:1 for residential areas:

$$\frac{E_{av}}{E_{min}} < 3.0 \text{ (6.0 for residential areas).}$$

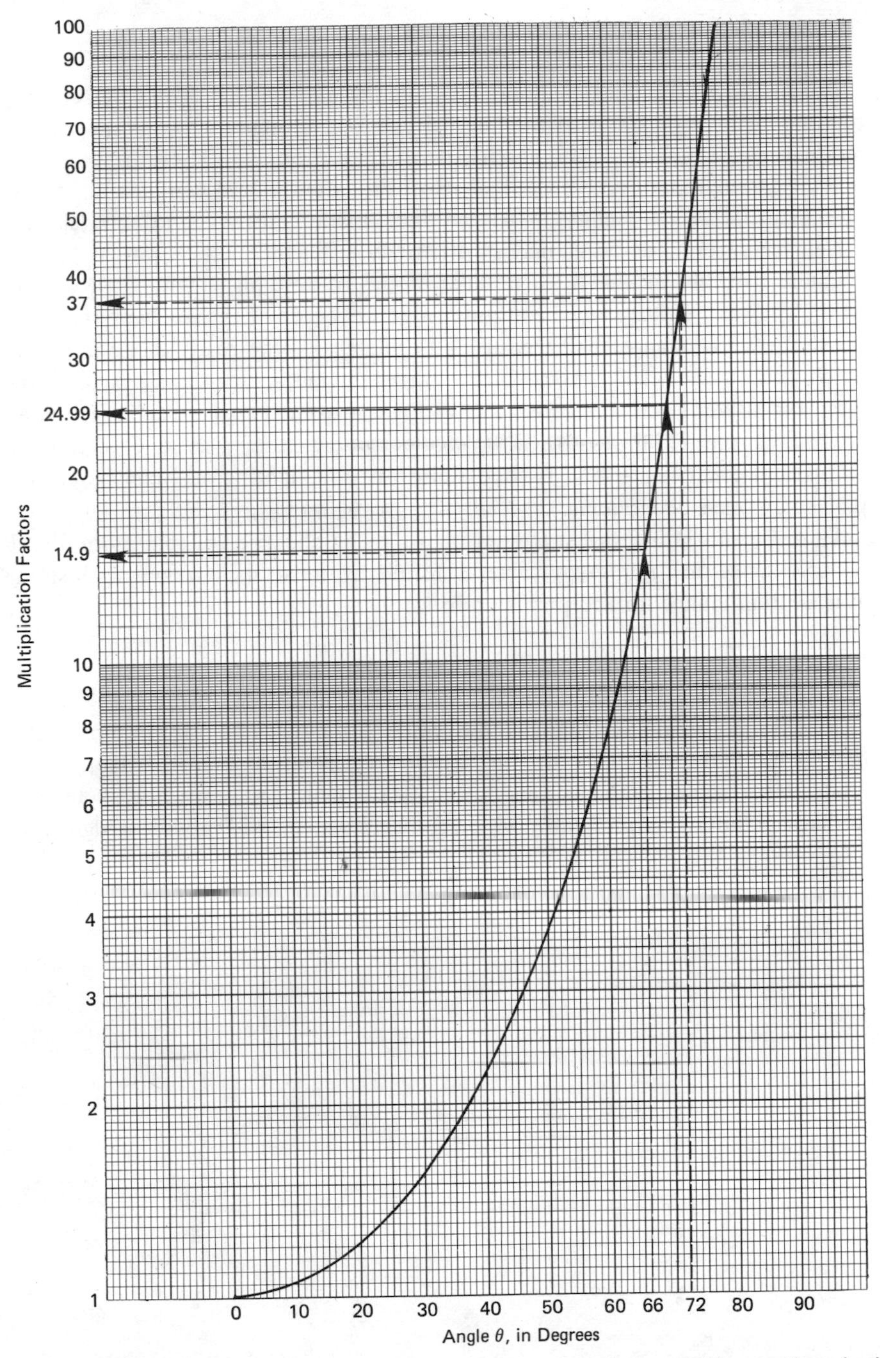

Fig. 9-9. Multiplication factors to be used with nadir candlepower to achieve uniform horizontal illumination. At an angle of 70°, the luminaire must produce about 25 times the nadir candlepower; and at 90°, the amount is infinity.

The simplest way of determining spacings for uniform lighting is by referring to isofootcandle diagrams of the luminaires. Several manufacturers show quick methods of determining such spacings with their products. These are useful for making a preliminary or even a final selection. Care should be taken not to use one manufacturer's technique for another manufacturer's products, since a difference in reflector design may substantially differ the light output.

Figure 9-10 shows the three popular modes of pole placement along a roadway. They are either spaced continuously on one side of the roadway such as is shown in 9-10a, staggered spaced using both sides of the roadway, as in 9-10b, or opposite space, as in 9-10c. Of the three, obviously one-sided spacing is the least expensive, since it involves less wiring and trench cost if service is underground. From a uniformity standpoint, staggered spacing as in Figure 9-10b seems to work best, although good uniformity in lighting still can be accomplished by one-sided spacing, especially for narrow roadways.

A, B, C, and D are the suggested points to be checked for minimum illuminance. Once the minimum value is found, it must be checked against the average illuminance for acceptable uniformity.

Light Loss Factor

Two main factors that attribute to the construction of an LLF for exterior lighting are the LDD and LLD. BF for HID sources is unity; and other factors that relate to surface reflectances are ignored. As a general rule, a well-gasketed and well-sealed luminaire, preferably with a filtering system, will provide a high LLF.

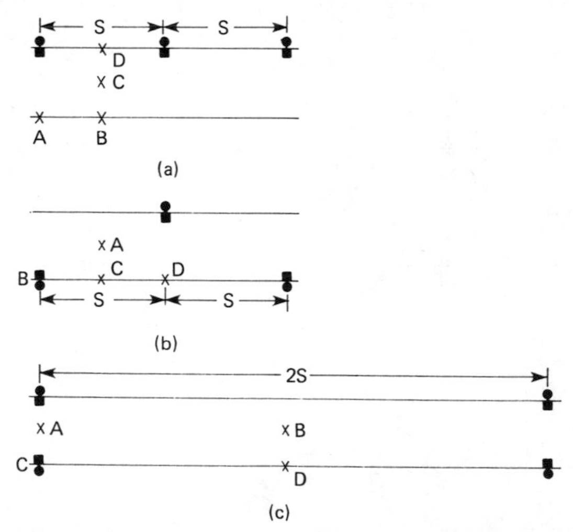

Fig. 9-10. Roadway luminaire spacing arrangements. Cross marks (x) indicate check points for uniformity. (a) One side spacing, (b) Staggered spacing, (c) Opposite spacing.

Selection of Light Sources

Over past years, exterior lighting systems have gone through the complete range of light sources commercially available, from the incandescents through the fluorescents to the HID sources. With the crunch of the energy crisis and sky-rocketing utility costs, however, use of HID sources is imperative. A thorough study in terms of physical and operating characteristics of each of these sources has been shown in Chapters 1, 4, and 5. A brief summary of those applicable for exterior lighting are reviewed here.

Use of incandescents for exterior lighting should be avoided, primarily because they have the lowest efficacy (lumens per watt) and the shortest life. Their initial cost is low, but because of their short life span, the maintenance cost can be high.

Fluorescents offer good efficacy and life. But because of the difficulty in light control because of their physical size and sensitivity to temperature, their out-door use is limited. When enclosed in a gasketed and sealed housing, the built-up temperature provides satisfactory performance during winter, but may cause slight loss of light output during summer.

Use of mercury vapor lamps for exteriors has been in practice for years, primarily because of their extremely long life. Their efficacy, however, is quite low. With the advancement of other sources in the family of HID sources, their popularity is declining. They should be only used when the quality of light is moderately important, along with high longevity.

Metal-halide sources offer good color and efficacy but have relatively lower life when compared with mercury vapor and HPS. These are possibly the most suitable sources for applications where energy saving and color rendering are of equal importance. HPS sources offer the maximum efficacy among all "white-light" sources and has extremely long life. Relatively smaller physical size offers good light control. Color rendering is termed "golden-white" and it is acceptable for most exterior applications. LPS sources have gained special interest in the industry because of their extremely high efficacy. Theoretically, they offer the maximum amount of light for power, when the lamp alone is considered. Their biggest disadvantage is their color rendering, monochromatic yellow, under which all colors except yellow, look gray, brown, or black. This has eliminated their application for most interiors. For exteriors, however, LPS may find some applications where color rendering is not of much importance.

All roadway and area lighting, thus, is basically bonded by metal-halide, HPS, and LPS sources. If color is as important as energy saving, use of metal-halide probably is the only answer. Where color is not important, and energy saving with long life is the main concern, the choice is limited to HPS or LPS. A comparative view of the use of these two sources will be covered in the following examples.

ROADWAY LIGHTING

Design Procedure and Example

Given: Roadway, 30 feet wide
Required illuminance level (E): = 1.4 fc.
MH: 25 feet
Staggered spacing
Investigate the use of HPS and LPS
(color rendering is not important)

For the ease of making some preliminary decisions, luminaire mounting height, required footcandle level, and the choice limitations of the type of light sources are already mentioned in the model example. In reality, the level of illuminance is one of the first information that needs to be known. This can be looked up in the handbook of the IES for the specific application. The type of light source to be used also depends on several factors, as discussed earlier. In this example it is given that color rendering is not important, so the selection would be primarily limited to HPS and LPS because of their energy-saving characteristics.

Choice of lamp wattage and MH has to be made almost simultaneously. As a general rule, MH increases with higher wattage, and it should not be larger than the width of the roadway. A higher MH will spill much of the light beyond the roadway. Local rules and regulations on the restriction of MH constitute another factor of consideration. A selection of 25 feet is in the range, since this is less than the roadway width and high enough to provide uniformity with spacings far apart. Selection of the lamp wattage depends upon spacing, MH, and illuminance level requirements. A tentative selection of a 250-watt HPS lamp has been made because of the low illuminance level requirements. A 150-watt lamp may also be suitable, but this must be investigated. From the standpoint of energy saving, uniformity, and cost, the choice will be using the lowest power-consuming lamp, spaced farthest apart, with highest MH and least glare. From all points considered, a 250-watt HPS mounted on a 25-foot pole is a good preliminary selection.

Next we select the luminaire. This selection will depend upon the type of distribution, high angle control, and spread selection. A cut-off luminaire will provide the least glare, but it will require shorter spacings for uniformity, and this will increase cost and energy consumption. A non-cut-off type, on the other hand, will provide good uniformity with spacings farthest apart, but there will be high glare. From all points considered, the selection is a medium/semi-cut-off, type III. (See Figure 9-1). It is assumed that the luminaire will be mounted on an arm 2 feet long. Note that if the distribution selected were type I or II, the arm would be longer, so a majority of the light would fall on the road. For all other types of distribution, except for type V, it will be better to have the luminaire mounted as close as possible to the pole. The exact location of the pole in relation to the curb is another factor in determining the arm length. If the pole is located further towards the house side, the length of arm will be proportionately longer to allow most of the light to fall on the roadway. A shorter arm equipped with type-III distribution luminaire is a good preliminary selection from the standpoint of cost and because of staggered spacing. If the road is narrow, a long-arm mounted type-I distribution luminaire with one-sided spacing will probably be a better choice. Figure 9-6 shows the preliminary selection arrangement of this example.

Another important item to be checked in luminaire selection is the nadir candlepower and, hence, the nadir illuminance level. From the view of uniformity, it is necessary that the illuminance level between luminaires and directly underneath a luminaire be as close as possible. The expected initial footcandles directly underneath the luminaire can be either calculated or directly read from photometric reports. For this example, the initial footcandle at nadir can be found from the General Information of Figure 9-1, where it is 0.0959. However, this value is per 1000 lamp lumens only and for a luminaire mounted at 30 feet. The footcandle value must be multiplied by an MH and rated lamp-lumens correction factors. The correction factor for a 25-foot mounting height, 1.44, can be read from the table. If the rated lumens of the lamp in use is 27,500, the corrected illuminance is then

$$0.0959 \times 1.44 \times \frac{27{,}500}{1000} = 3.79 \text{ fc.}$$

Note that this value can also be determined by knowing the candlepower at nadir and using the inverse-square law. Given that candlepower at nadir is 86 candela. Now using the inverse-square formula,

$$\text{Corrected illuminance} = \frac{86}{(30)^2} \times \frac{27{,}500}{1000} \times 1.44 = 3.78 \text{ fc.}$$

The illuminance level found directly underneath the luminaire is 3.78/1.4 = 2.7 times what is required. This is all right since the value represents the initial condition only and possibly represents the higher part of the required average illuminance level. With a consideration to LLF, the initial nadir footcandle should not exceed more than about four times the required average illuminance value.

Spacing

As a general rule, for the purpose of energy saving, economy and uniformity, the spacing of luminaires (poles) for a roadway-lighting system should not be lower than three, and larger than five times the mounting height. Any closer spacing increases cost and power consumption, whereas with a larger spacing, it is almost impossible to achieve the 3 : 1 average-to-minimum uniformity. For residential areas, where the acceptable average-to-minimum uniformity is 6 : 1, the spacing should not be lower than eight, and larger then ten times the luminaire mounting height. The spacing can be calculated with the help of lumen formula as follows:

$$S = \frac{L \times \mathrm{CU} \times \mathrm{LLF}}{E \times W}$$

where

S = Spacing in feet
L = Lamp initial lumens
CU = Coefficient of utilization
LLF = Light loss factor
E = Maintained illuminance in fc.
W = Width of road

In determining the value of S, it is necessary to find the values of CU and LLF. The CU value can be determined by reviewing Figure 9-6 and the utilization curve of Figure 9-2. The ratio of TD to MH is to be determined first, and then CU values can be looked up from the curve:

House side: TD/MH = 2/25 = 0.08;

Street side: TD/MH = 28/25 = 1.12.

From the utilization curve in Figure 9-2, the corresponding CU values are 0.12 and 0.42, respectively. The total CU is then, 0.12 + 0.42 = 0.54.

LLF consists of many factors, the most important of which are the LLD and the LDD. As seen in Chapter 1, the LLD of all HPS lamps is approximately 0.73 at the end of their life. Assuming that lamps are replaced only at burn-out, the LLD can be considered to be 0.73. A sealed, gasketed, and properly filtered luminaire produces an LDD of about 0.95. The net light loss is then 0.73 × 0.95 = 0.693. Applying this value in the formula,

$$S = \frac{27{,}500 \times 0.54 \times 0.693}{1.4 \times 30} = 245 \text{ ft.}$$

The spacing is much larger than recommended since,

$$\left(\frac{245}{25}\right) = 9.8 > 5.$$

At this point, there are two alternatives to stay within the recommended spacing limit: (1) increase the MH or (2) use lower-wattage lamps. Increasing the MH beyond 25 feet will give good uniformity, but will lower the illuminance level. This will also waste energy, since much of the light will spill beyond the 30-foot width of roadway. A better solution will probably be the use of a lower-wattage lamp, such as 150 watts.

Figure 9-11 represents the photometric report of a 150-watt HPS, of medium/semi-cut-off, type III classification. Reviewing the report, the amount of foot-candle at nadir is $0.12317 \times (16{,}000/1000) \times 1.44 = 2.83$, which is within the recommended range, since $2.83/1.4 = 2 < 4$. Utilization factors can be found from the utilization curve, corresponding to the previously determined TD/MH ratio data of 0.08 and 1.12 for house side and street side, respectively. These values are 0.12 and 0.38. So the net CU is $0.12 + 0.38 = 0.50$. LLF is 0.693, as found earlier. Applying these values in the spacing formula,

$$S = \frac{16{,}000 \times 0.50 \times 0.693}{1.4 \times 30} = 131 \text{ ft.}$$

This is still slightly beyond the recommended range, since $131/25 = 5.24 > 5$.

A fine tuning can be made by deleting the arm of the pole or mounting the luminaire as close as possible to the curb, so no calculations will be required for the house-side CU.

The new CU can be determined from the utilization curve, with $\text{TD/MH} = 30/25 = 1.2$, which is found to be 0.39. So,

$$S = \frac{16{,}000 \times 0.39 \times 0.693}{1.4 \times 30} = 102 \text{ ft.,}$$

rounded off to 100 feet. This is within the range since, $100/25 = 4$, which is $3 < 4 < 5$.

Figure 9-12 shows the spacing layout of the example. An investigation in uniformity now must be made to complete the design. As discussed earlier, for an acceptable minimum uniformity, the ratio of average to minimum illuminance should not exceed 3:1. This has to be made by a trial-and-error method since no specific formula can directly determine where the minimum lighting will occur. Minimum-lighting location can occur at any place, depending on the distribution characteristics of the luminaire, spacings, arm length, etc. It is best to check on points through experience. These points were shown in Figure 9-9, and are repeated in Figure 9-12.

The various steps involved in determining the average-to-minimum ratios are as follows:

1. For each point of examination, determine the TD/MH ratio and the LD/MH ratio for all luminaires contributing to this point.

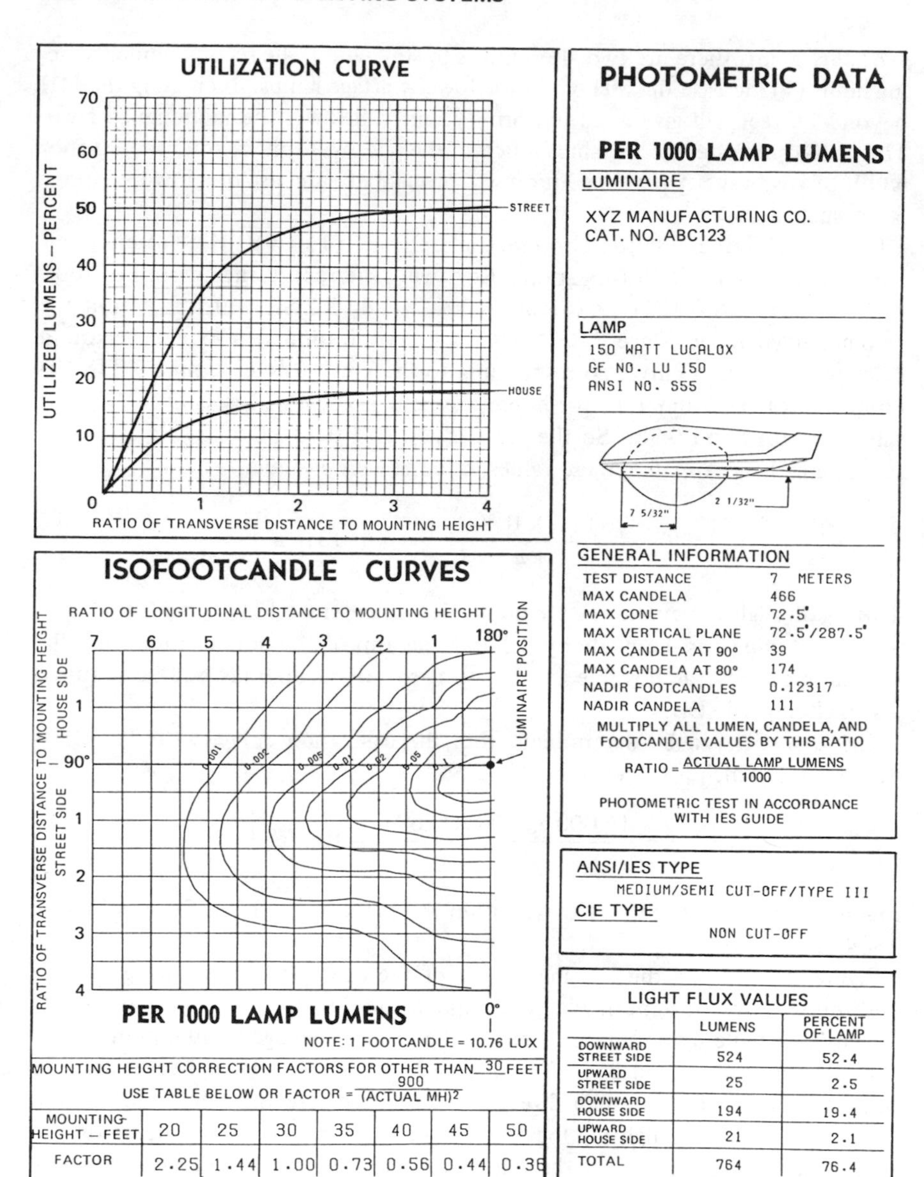

MOUNTING HEIGHT CORRECTION FACTORS FOR OTHER THAN 30 FEET
USE TABLE BELOW OR FACTOR = $\frac{900}{(\text{ACTUAL MH})^2}$

MOUNTING HEIGHT – FEET	20	25	30	35	40	45	50
FACTOR	2.25	1.44	1.00	0.73	0.56	0.44	0.36

LIGHT FLUX VALUES

	LUMENS	PERCENT OF LAMP
DOWNWARD STREET SIDE	524	52.4
UPWARD STREET SIDE	25	2.5
DOWNWARD HOUSE SIDE	194	19.4
UPWARD HOUSE SIDE	21	2.1
TOTAL	764	76.4

Fig. 9-11. Photometric data of a 150-watt, HPS, medium/semi-cut-off, type III luminaire.

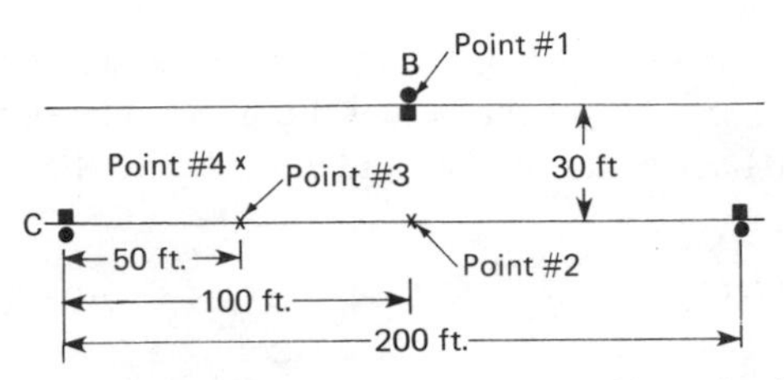

Fig. 9-12. Luminaire spacings and checking points of uniformity.

2. Knowing the two ratios, refer to isofootcandle curves and determine the corresponding footcandle values.
3. The footcandle values as determined directly from the curves represent the initial levels corresponding to the reference conditions of the photometrics only. These values must be multiplied by a lamp factor (LF), MH correction factor, and finally, by a light loss factor (LLF). The table of the calculations is in Table 9-1.

Table 9-1. In Determining Lighting Uniformity, a Table Such as Follows, Will Be Helpful.

	TD/MH	LD/MH	FC
Point 1			
Pole A	30/25 = 1.2	100/25 = 4	0.005
Pole B	0/25 = 0	0/25 = 0	0.123
Pole C	30/25 = 1.2	100/25 = 4	0.005
			Total = 0.133
Point 2			
Pole A	0/25 = 0	100/25 = 4	0.002
Pole B	30/25 = 1.2	0/25 = 0	0.035
Pole C	0/25 = 0	100/25 = 4	0.002
			Total = 0.039
Point 3			
Pole A	0/25 = 0	150/25 = 6	0
Pole B	30/25 = 1.2	50/25 = 2	0.02
Pole C	0/25 = 0	50/25 = 2	0.0175
			Total = 0.0375
Point 4			
Pole A	15/25 = 0.6	150/25 = 6	0
Pole B	15/25 = 0.6	50/25 = 2	0.025
Pole C	15/25 = 0.6	50/25 = 2	0.025
			Total = 0.050

As can be seen from Table 9-1, the minimum value seems to occur at point 3 and is equal to 0.0375. To get the actual illuminance level, this must be multiplied by an LF, MH correction factor, and an LLF. So the illuminance at point 3 is

$$0.0375 \times \frac{16{,}000}{1000} \times 1.44 \times 0.693 = 0.598 \text{ fc.}$$

The average-to-minimum ratio is 1.4/0.598 = 2.34 : 1. This value is within the range of 3 : 1 and does provide the minimum uniformity. If the uniformity were closer to 2 : 1 or less, a shorter MH and farther spacings should be investigated.

Low-Pressure Sodium

The same roadway will now be designed with LPS sources and investigated to determine whether further energy savings can be achieved. The lamp selected is a 135-watt one, since it produces 22,500 lumens, which is close to the 16,000 lumens of the 150-watt HPS used in the previous example. The lamp with next lower wattage is a 90-watt, producing 13,500 lumens. Use of 90-watt luminaires will cause an increase in quantity, and will be investigated, if 135-watt lamps prove to be too large for the purpose.

Figure 9-13 shows the photometric data of a 135-watt LPS luminaire having medium/semi-cut-off, type IV IES/ANSI classification. The candlepower at nadir is 3268 candela. Although an LPS source is not a "point source," the inverse-square law is still valid since the MH is at least five times larger than the length of the source. The illuminance level is, then, $3268/25^2 = 5.2$ fc. This amount is within the recommended range, since $5.2/1.4 = 3.7 < 4$. Next step is to determine the CU and LLF towards finding spacing. CU can be determined from the given utilization curve, after knowing the TD/MH ratio. This ratio, as found earlier, is 1.2. Corresponding to 1.2 in the horizontal axis, the CU value is found to be 0.26.

The LLF value can be constructed after knowing LLD and LDD. The latter for an LPS type is extremely favorable since the rated lumens remain unaltered throughout the 18,000 hours of life and actually increase to about 103% for some lamps. For the purpose of determining LLF, this value is assumed to be 1.0. The LDD factor will vary from luminaire to luminaire, depending on the quality of construction. Assuming that the luminaire used in the example is gasketed and sealed, LDD value will be 0.95, the same as for HPS. The net LLF is, then, 1.0 × 0.95 = 0.95. Using this value in the formula, the spacing is

$$S = \frac{22{,}500 \times 0.26 \times 0.95}{1.4 \times 30} = 132.3 \text{ ft.}$$

This spacing is too large, since $32.3/25 = 5.3 > 5$.

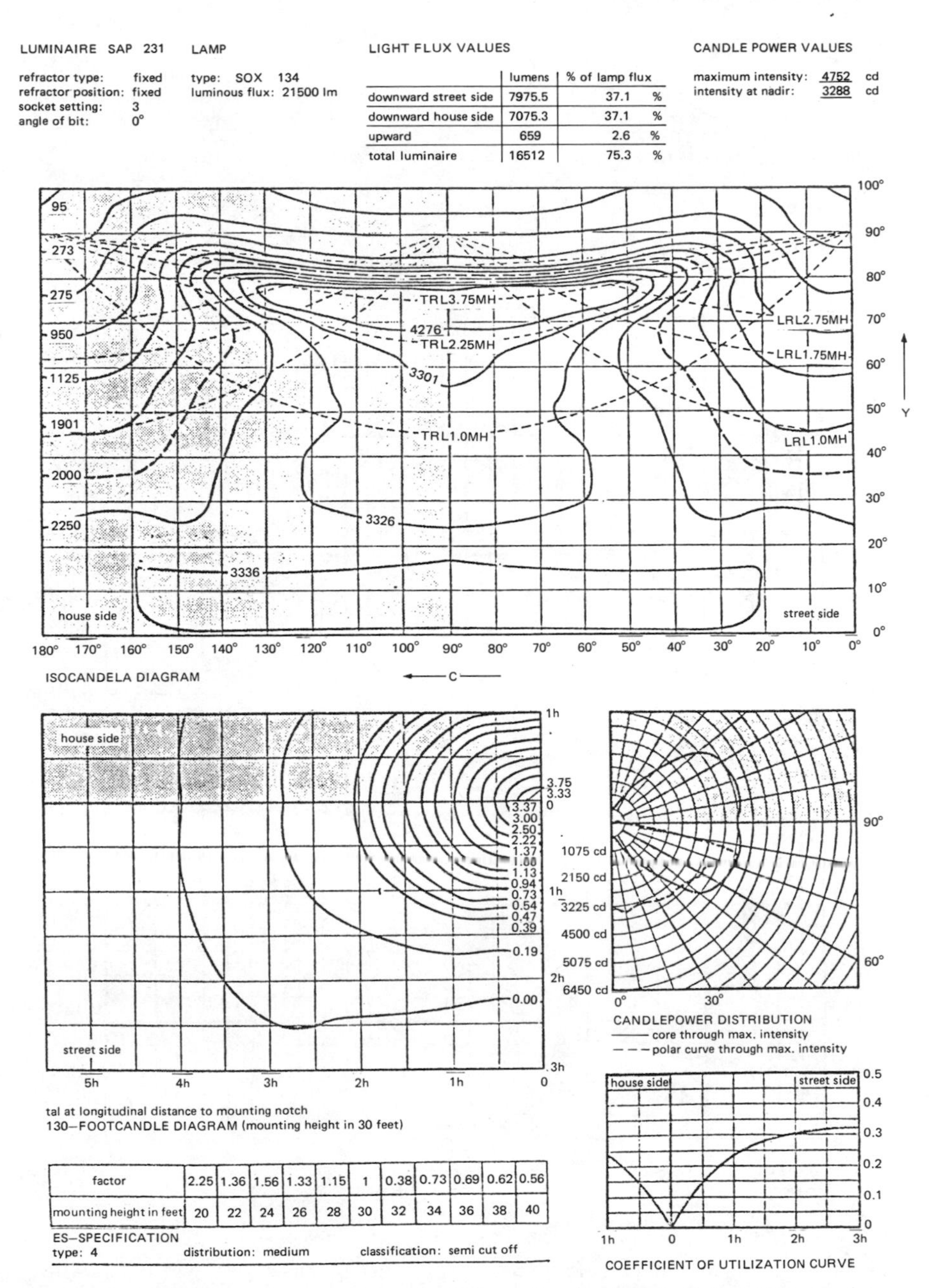

	lumens	% of lamp flux
downward street side	7975.5	37.1 %
downward house side	7075.3	37.1 %
upward	659	2.6 %
total luminaire	16512	75.3 %

factor	2.25	1.36	1.56	1.33	1.15	1	0.38	0.73	0.69	0.62	0.56
mounting height in feet	20	22	24	26	28	30	32	34	36	38	40

Fig. 9-13. Photometric data of a 135-watt, LPS, medium/semi–cut-off, type IV luminaire.

90 WATT LPS. 13500 LUMENS
MED./SEMI-CUTOFF/TYPE IV
NADIR CANDLEPOWER 1461
NADIR FOOTCANDLES 1.63

LIGHT FLUX VALUES

		PERCENTAGE OUTPUTS
DOWNWARD:	STREET SIDE	= 30.6%
	HOUSE SIDE	= 26.0%
	TOTAL	= 56.6%
UPWARD:	STREET SIDE	= 1.1%
	HOUSE SIDE	= 0.5%
	TOTAL	= 1.6%
TOTAL EFFICIENCY		= 58.2%

UTILIZATION AND ISOFOOTCANDLE CURVES:

MOUNTING HEIGHT FOR ISOLUX 30.0 FEET

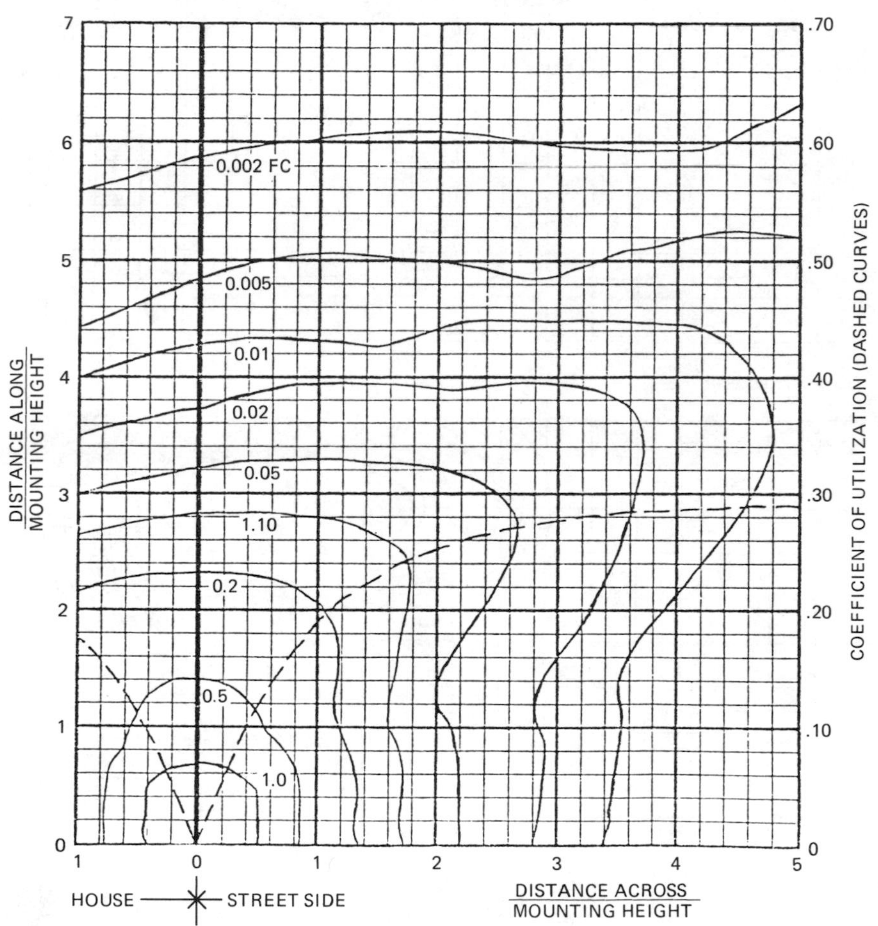

Fig. 9-14. Photometric report of a 90-watt, LPS, medium/semi-cut-off, type IV luminaire.

If luminaires are spaced at 132 feet apart, uniformity will be poor and it will be almost impossible to keep the average-to-minimum ratio within 3 : 1. Increasing the MH will create uniformity, but it will reduce the maintained footcandle level by spilling much light beyond the roadway. A lower-wattage lamp should be investigated at this time. Let us suppose the lamp used is a 90-watt producing 13,500 lumens. The photometric report of this luminaire is shown in Figure 9-14. With the luminaire mounted at 25 feet above ground, and TD/MH = 1.2, the new CU is found to be 0.21. The spacing is then

$$S = \frac{13{,}500 \times 0.21 \times 0.95}{1.4 \times 30} = 64.12$$

rounded to 65 feet. This spacing is not favorable since the ratio of spacing to mounting height is 65/25 = 2.6, which is less than the lower limit of 3. For the purpose of economy and energy saving, a lower pole height should be investigated at this time, so that the luminaires may be spaced as far apart as possible within the recommended limit. Let us suppose the mounting height is 20 feet. With this MH, the new TD/MH ratio is 1.5. From the utilization curve in Figure 9-14, the CU value is found to be 0.23. The new spacing is then

$$S = \frac{13{,}500 \times 0.23 \times 0.95}{1.4 \times 30} = 70.2$$

rounded to 70 feet. This spacing is a better selection since the spacing to mounting height ratio is 70/20 = 3.5, which is larger than 3 and lower than 5. So, a mounting height of 20 feet is our choice.

The uniformity now must be verified with average-to-minimum illuminance ratio, which should not exceed 3 : 1. The points of checking are the same as in the previous example. The calculations are shown in tabular form (Table 9-2). As can be seen from the table, the minimum value seems to occur at point 2 and is equal to 0.22. Note that this reading is good for luminaires mounted on a 30-foot pole and at the initial condition only. No LF correction is necessary since all values are direct readings for a 90 W LPS. For a maintained illuminance level and 20-foot MH this value must be multiplied by the LLF and a MH correc-

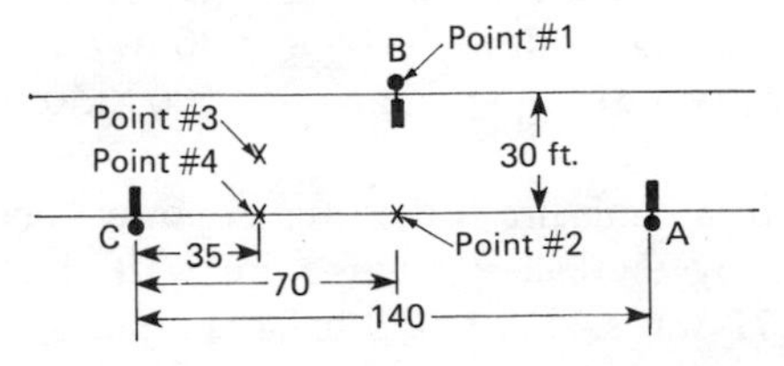

Fig. 9-15. Luminaire spacings and uniformity checkpoints of depicted example.

Table 9-2. Uniformity Checking Table.

	TD/MH	LD/MH	FC
Point 1			
Pole A	30/20 = 1.5	70/20 = 3.5	0.04
Pole B	0/20 = 0	0/20 = 0	1.63
Pole C	30/20 = 1.5	70/20 = 3.5	0.04
			Total = 1.71
Point 2			
Pole A	0/20 = 0	70/20 = 3.5	0.035
Pole B	30/20 = 1.5	0/20 = 0	0.15
Pole C	0/20 = 0	70/20 = 3.5	0.035
			Total = 0.22
Point 3			
Pole A	0/20 = 0	105/20 = 5.25	0.004
Pole B	30/20 = 1.5	35/20 = 1.75	0.15
Pole C	0/20 = 0	35/20 = 1.75	0.35
			Total = 0.504
Point 4			
Pole A	15/20 = 0.75	105/20 = 5.25	0.004
Pole B	15/20 = 0.75	35/20 = 1.75	0.35
Pole C	15/20 = 0.75	35/20 = 1.75	0.35
			Total = 0.704

tion factor. So maintained illuminance at point 2 is 0.22 X 2.25 X 0.95 = 0.47 footcandles. The average-to-minimum ratio is 1.4/0.47 = 2.97 : 1. This value is within the range of 3 : 1 and does provide the minimum uniformity. If the uniformity were closer to 2 : 1 or less, a short MH and farther spacings should be investigated. The uniformity ratio is so close to 3 : 1 that it is doubtful that any other combination will make the system more uniform with further energy savings.

An energy-consumption study must be made between the two examples (HPS and LPS), to determine which results in better energy savings.

HPS, 150 watts: The spacing between poles = 100 ft.
Roadway width = 30 ft.
Area covered per luminaire = 100 X 30 = 3000 sq. ft.

Power consumption of a luminaire is the total of power consumed by lamp plus the power consumed by the ballast. For a 150-watt HPS luminaire it is 150 watts + 21 watts = 171 watts. Most HPS ballasts typically tend to operate the lamp below rated watts initially and near the end of life, but slightly above rated

Table 9-3. Power Consumption of a 90 Watt LPS.

	Power Consumption		
	Lamp	Ballast	Total
Initial	90	35	125
Mean	116	35	151
End of Life	122	35	157

watts at the midpoint of lamp life. The average is thus assumed to be the same as in the figure above. For the purpose of comparison, the power density of this example is, then, 171 watts/3000 square feet = 0.057 watts/square feet.

LPS, 90 watt: Spacing between poles = 70 ft.
Roadway width = 30 ft.
Area covered per luminaire = 70 X 30 = 2100 sq. ft.

As shown in Chapter 1, unlike other sources, the power consumption of an LPS lamp gradually increases with age and peaks at the end of life. With no change in lumen output, the efficacy (lumen per watt) actually decreases with age. This is shown in Table 9-3 for a 90-watt LPS lamp.

Power density
Initial 125/2100 = 0.059 watts/square feet
Mean 151/2100 = 0.071 watts/square feet
End of life 157/2100 = 0.074 watts/square feet

It is obvious from the above that LPS luminaires starting with a slightly higher power consumption in the beginning, ultimately require approximately 29% more power than the HPS system. This, along with other benefits, such as cost savings because of the lower quantity of luminaires and poles, superior color rendering, longer life, and better lighting-control ability, makes the HPS a better choice.

AREA LIGHTING

All luminaires designed for roadway lighting can be used for area lighting. Of the lot, luminaires with type III and type IV are the most suitable, since they tend to produce more light ahead of the luminaires, compared to types I and II. There is no straight formula, however, that can be directly applied to determine the luminaire type, pole height, spacings, and other related aspects involved in area lighting. Irregular area shape, MH restriction, color rendering, and spacing limitations are some of the major determining factors. Each application has its own characteristics that may differ from others, and the solution to the problem is almost a matter of trial and error, on an individual basis.

In order to make a good preliminary selection, the following factors should be observed:

1. Spacing between poles should not be more than 4.5 times the MH.
2. Spacing between the edge of the area to the nearest pole should not be more than 2.25 times the MH.
3. The maximum candlepower should be no less than 8 and no higher than 15 times the nadir candlepower, and it should be emitted at approximately 66° to get an even lighting between two poles.
4. The maximum candlepower should be no less than 10 and no higher than 37 times the nadir candlepower, and it should be emitted at approximately 72° to get an even lighting at the midpoint of four poles in a square array.

Unless it is absolutely necessary, for area lighting, luminaires should not be located at the edge, since this wastes all light emitted at house side. A minimum of two luminaires, 180° apart, should be used per pole, in order to distribute even lighting. Use of types III or IV in both luminaires will tend to produce even lighting with a rectangular array; use of one luminaire of type V distribution will offer a circular pattern, evenly, all around the pole. A maximum of 4.5 S/MH height ratio should be observed, since this will provide uniformity without exceeding the 3 : 1 average-to-minimum illuminance ratio. The distance between the edge of the area and the nearest pole should be no more than 4.5/2 = 2.25 MH for the same reason.

In order to receive an even lighting, the candlepower aimed towards the midpoint between two poles should be no less than 8 and no greater than 15 times the ratio candlepower. The farther apart the poles, the higher the angle at which this maximum should occur. To avoid a disturbing glare, however, the maximum candlepower should not be emitted above 75° and should be limited to within 66–72°, as mentioned earlier. When luminaires are arranged in a square pattern, the midpoint of the square usually is the most difficult place to light. Maximum candlepower of all luminaires should be aimed at approximately 72° to make up for the loss and each should have no less than 10 and no higher than 37 times its nadir candlepower. Note that the illuminance level at this point is actually shared by all four luminaires. In order to produce the same illuminance level at this point as that beneath each luminaire, the required candlepower aimed at this point is actually to be shared by all four luminaires. This is explained below.

Refering to Figure 9-16a, illuminance at point 1 is due to light arriving from

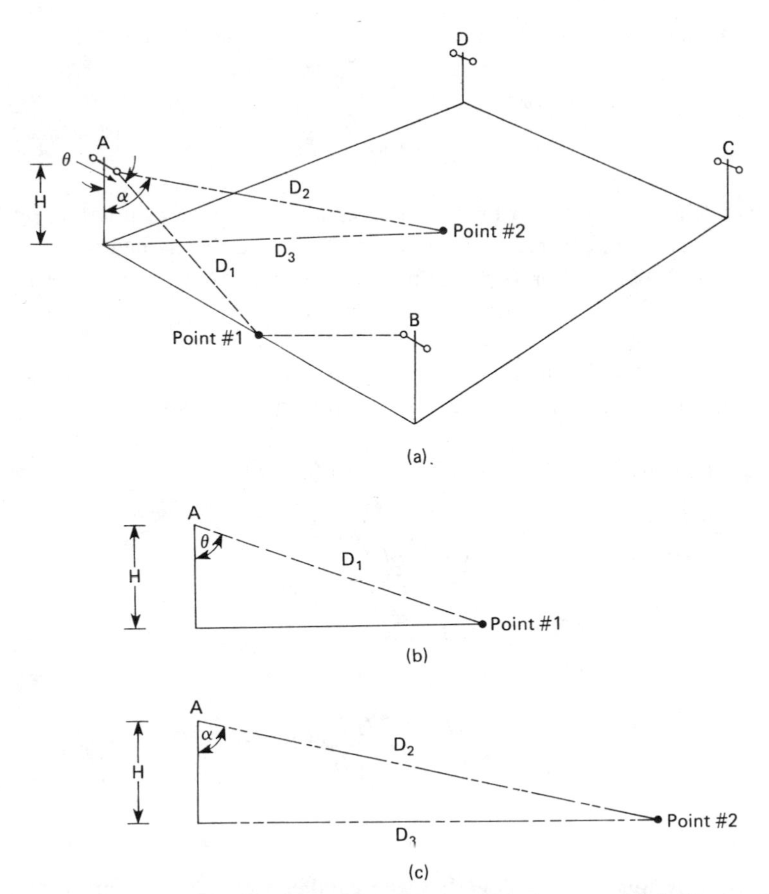

Fig. 9-16. Point 2 followed by point 1 are the most difficult areas to light for an uniform illumination. If the luminaires are positioned properly, illuminance at point 1 will be shared equally by luminaires on poles A and B; and that at point 2 by all luminaires on poles ABCD.

two luminaires, A and B. If this illuminance was to be produced by only one luminare–say, A–it will have to arrive at an angle given by the relation,

$$\cos\theta = \frac{H}{D_1}$$

or

$$\cos\theta = \frac{H}{\sqrt{H^2 + (2.25H)^2}} = 0.406, \text{ or } \theta = 66^\circ.$$

This is shown in Figure 9-16b.

Now applying this value in the formula developed earlier, the candlepower at angle θ is given by:

$$I_\theta = I_0 \, (1/\cos^3 \theta)$$

or

$$I_\theta = I_0 \, (1/0.406^3) = I_0 \, 14.9.^*$$

In other words, in order to produce the same illuminance at point 1, by luminaire A alone, it has to produce approximately 15 times its nadir candlepower at an angle $\theta = 66^\circ$. For two luminaires, however, each has to have 15/2, or approximately 8 times the candlepower at nadir.

A similar argument holds true for point 2, which is the midpoint of four poles. The illuminance at this point is shared by all four, so, the minimum and maximum candlepower and its angle can be determined as follows:

from Figure 9-16a,

$$D_3^2 = (2.25\,H)^2 + (2.25\,H)^2 = 10.125\,H^2$$

from Figure 9-16c,

$$D_2 = \sqrt{H^2 + D_3^2} = \sqrt{H^2 + 10.125\,H^2} = \sqrt{11.125\,H^2},$$

or

$$D_2 = 3.33\,H.$$

Also, from Figure 9-16c,

$$\cos\alpha = \frac{H}{D_2} = \frac{H}{3.33\,H} = 0.300,$$

or

$$\alpha = 72^\circ.$$

Using the formula developed earlier, the candlepower requirement at angle $\alpha = 72^\circ$ is given by

$$I_\alpha = I_0 \, (1/\cos^3 \alpha) = I_0 \, (1/0.3^3) = I_0 \, 37.$$

This shows that to achieve the same illuminance at point 2 as in the nadir, one luminaire has to produce 37 times its nadir candlepower at an angle of 72°. If all four luminaires are contributing evenly to this point, each luminaire needs to produce only 37/4, or approximately 10 times the nadir candlepower at an angle of $\alpha = 72^\circ$.

Each luminaire producing 10 times the nadir candlepower at 72° will provide an even lighting, but this assumption may be too optimistic. Any error in luminaire positioning or the burning out of any of the four lamps will cause a loss in evenness. Each luminaire, providing 37 times their nadir candlepower,

*Note that this multiplication factor can also be obtained directly from Figure 9-9.

on the other hand, will increase the midpoint illuminance fourfold. The best solution is then to have a candlepower somewhere in between the two extremes, such as 24 or 25 times the nadir candlepower at an angle of $\alpha = 72°$.

It is important to remember that all these candlepower factors apply only for luminaires arranged at a 4.5 S/MH ratio. For other ratios, obviously, the factors will vary. To achieve a uniformity of 3:1 average-to-minimum illuminance ratio and with least power consumption, it is recommended that this limit be maintained. A S/MH ratio over 4.5 will reduce the number of poles and luminaires but increase glare and give poor uniformity. A lower ratio, on the other hand, will increase uniformity, but with higher cost and power consumption.

Design Procedure and Sample Problem

The first step involved in area lighting is to determine the required illuminance level. This can be found from the IES handbook. Most area lighting is involved with automobile parking, which in general is of two types: self-parking areas and attended areas. The attended-parking area requires twice the illuminance of self-parking for better visibility and security. Some parking areas are monitored by a television surveillance system, which requires even higher level of illuminance. Color rendition of the luminaires is important if a color TV surveillance system is used. Metal-halide sources are the most suitable for this type of application. Where color is not important, the selection will be limited to HPS and LPS, since these have the most savings potential. A preliminary but realistic selection of the number of poles, MH, the number of luminaires, and luminaire spacings can be made with the guidelines mentioned earlier. To get the effect of even lighting, a minimum of two luminaires per pole (with type III or IV distribution) should be installed, 180° apart.

Sample Problem.

Given: Area to be lighted = 540 feet long × 250 feet wide
Illuminance required = 1.5 fc.
Luminaire MH restriction = 30 feet
Color rendering not important.
Investigate HPS and LPS.

High Pressure Sodium. First, determine the layout and luminaire MH. Note that, as given, the maximum MH is not to exceed 30 feet. With the assumption of using a 30-foot pole, the spacing between poles should not exceed 30 × 4.5 = 135 feet, and that from the edge to the first pole should not exceed 30 × 2.25 = 67.5 feet. With this background, the pole locations can be arranged as shown on Figure 9-17, with two rows of poles along the width and four rows along the

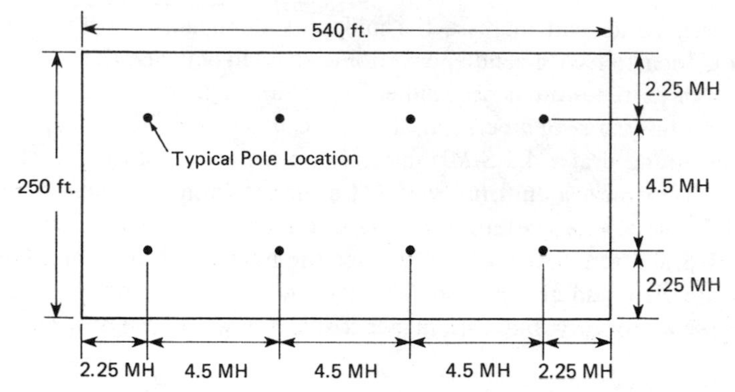

Fig. 9-17. Luminaire spacings of area lighting of depicted example.

length. For the purpose of economy, uniformity, and the least power consumption, it is recommended to stay at 4.5 MH spacings, although parking layout, landscaping, and similar factors may finally determine exact location. With the arrangement shown in Figure 9-17, the exact pole height may be determined as follows:

Lengthwise: $2.25\ \text{MH} + 4.5\ \text{MH} + 4.5\ \text{MH} + 4.5\ \text{MH} + 2.25\ \text{MH} = 550,$

or $18\ \text{MH} = 550$

$$\text{MH} = 30.5\ \text{ft.}$$

This is so close to a 30-foot MH that the layout is almost perfect. However, using exactly 30-foot poles, the spacing between poles should be $4.5 \times 30 = 135$ feet, which will leave the edge spacing equal to $[540 - (135 + 135 + 135)]/2 = 62.5$ feet. This is within the recommended range, since $62.5/30 = 2.08 < 2.25$.

With a spacing of 135 feet between poles, the spacing between the edge and the nearest luminaire, widthwise, can be determined similarly: $(250 - 135)/2 = 57.5$ feet. This is also within the range, since $57.5/30 = 1.9 < 2.25$.

With two luminaires per pole, and with this layout, total number of luminaires is $8 \times 2 = 16$.

In order to determine the lamp wattage, total lumens produced by each luminaire needs to be known. This can be found with the following formula:

$$L = \frac{A \times E}{N \times \text{CU} \times \text{LLF}}$$

where

L = Lamp initial lumens
A = Area

E = Illuminance in footcandles
N = Number of luminaires
CU = Coefficient of utilization
LLF = Light loss factor

All factors are known quantities here except for CU and LLF, which have to be determined. With no specific luminaire selected at this point, a tentative, but realistic CU and LLF have to be used. Assuming that a luminaire similar to that used in a roadway is being used here, the LLF is equal to LLD × LDD = 0.693. The CU value can be determined from its photometric report in Figure 9-2:

Street side CU + house side CU = 0.56 + 0.16 = 0.72.

(Note that in evaluating the CU values, the TD/MH ratios for street side and house side have been taken as 4.5 and 2.25, respectively.)

Using these values in the above formula,

$$L = \frac{(540 \times 240) \times 1.5}{16 \times 0.72 \times 0.693} = 25{,}375 \text{ lumens.}$$

Use of a 250-watt HPS luminaire will be suitable, since it produces 27,500 lumens, which is close to the requirement. Note, too, that the total lumens requirement per pole is actually 25,375 × 2 = 50,750 lumens, since there are two luminaires per pole. With this fact, both luminaires could be replaced by one 400-watt HPS lamp producing 50,000 lumens, which is quite close to the required value. This luminaire must be of type V distribution for even lighting distribution at all sides. Use of 400-watt luminaire will reduce the illuminance to 1.5 × 50,000/ 50,750 = 1.48, which is quite close to 1.5 for all practical purposes. For the purpose of determining exact illuminance, CU values must be used from specific photometric data of the luminaire in use.

Low-Pressure Sodium. For the purpose of an energy-savings comparison, an investigation will be made with LPS luminaires. Using utilization curve of Figure 9-13, the CU value is 0.325 + 0.31 = 0.635. These are based on TD/MH ratios of 4.5 and 2.25 for street side and house side, respectively. The LLF, as found earlier, is 0.95. Using these values in the formula,

$$L = \frac{(540 \times 250) \times 1.5}{16 \times 0.635 \times 0.95} = 20{,}980 \text{ lumens.}$$

Use of a 135-watt LPS luminaire will be quite suitable, since it produces 22,500 lumens, which is close to the required amount.

Energy Consumption Study. A comparative energy consumption study will be made for the three situations determined above: (1) use of 16 luminaires at 250 watts, HPS, (2) use of 8 luminaires at 400 watts, HPS, and (3) use of 16 luminaires at 135 watts, LPS.

1. Power consumption of a 250-watt HPS luminaire:
   Lamp power consumption + ballast power consumption = 250 + 46 = 296 watts.
   Total power is 16 X 296 watts = 4736 watts.
   Power density is 4736/(540 X 250) = 0.035 watts/square feet.
2. Power consumption of a 400-watt HPS luminaire:
   400 + 71 = 471 watts.
   Total power consumed is 8 X 471 watts = 3768 watts.
   Power density is 3768/(540 X 250) = 0.0270 watts/square feet.
3. Power consumption of a 135-watt LPS luminaire:
   Initial: 130 + 43 = 173 watts.
   Mean: 173 + 43 = 216 watts.
   End of life: 178 + 43 = 221 watts.

   Total power consumption:
   Initial: 173 X 16 = 2768 watts.
   Mean: 216 X 16 = 3456 watts.
   End of life: 221 X 16 = 3536 watts.

   Power density:
   Initial: 2768/(540 X 250) = 0.020 watts/square feet.
   Mean: 3456/(540 X 250) = 0.0256 watts/square feet.
   End of life: 3536/(540 X 250) = 0.026 watts/square feet.

The order of choice in obviously 3, 2, and 1. The performance of LPS in area lighting seems always to be better than on roadways, primarily because all of its light output in street side and house side is utilized here. In general, LPS produces equal amounts of light on both sides because of the inherent difficulty in lighting control of large sources. In a roadway lighting, almost half of its light is wasted if the luminaire is mounted directly above the edge of curb.

Use of one type V luminaire as opposed to having two types III or IV at 180° on a pole produces better uniformity, particularly if the layout is in a square pattern, as in the example. As seen in Case 2, as opposed to Case 1, better uniformity along with low initial cost and energy savings can be accomplished if type V luminaires are used. Although a type V distribution is circular, many manufacturers have type V luminaires that produce square light patterns. A proper use of these luminaires will provide better uniformity and waste less light, particularly in a square layout. This is shown in Figure 9-18

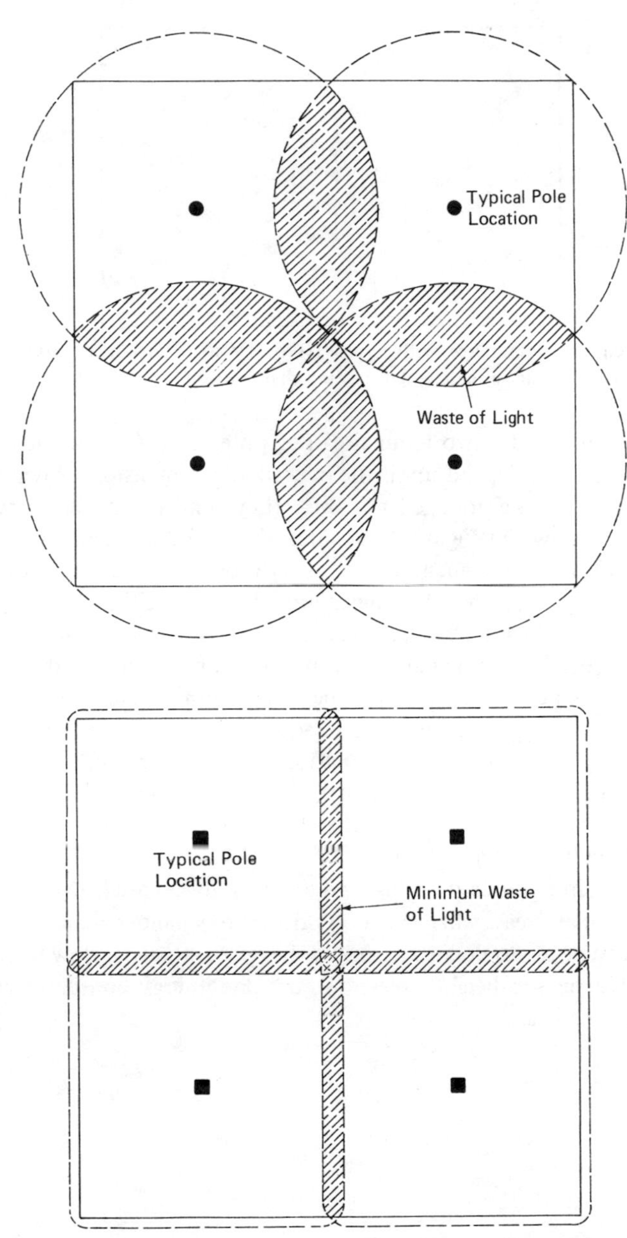

Fig. 9-18. Round and square patterns of a type V lighting distribution. A square pattern may provide better uniformity with less waste.

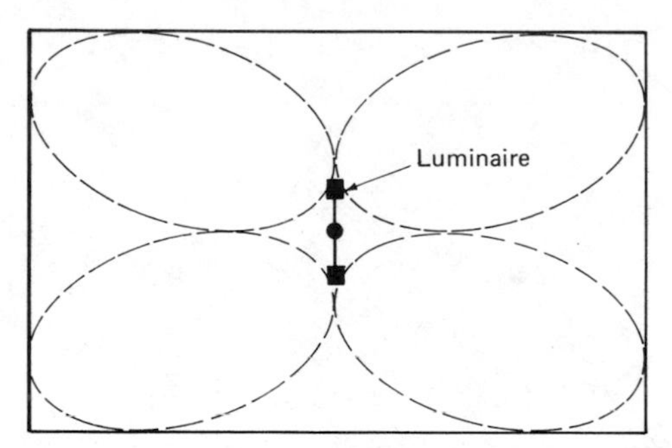

Fig. 9-19. When two luminaires with types III or IV distribution are arranged at 180° apart, they tend to produce a rectangular lighting distribution.

If it is necessary to use two luminaires per pole, as in Case 1, the orientation of the luminaires will depend upon the overall area dimension. When two types III or IV luminaires are arranged at 180°, they tend to produce a rectangular distribution, as shown in Figure 9-19, since they produce maximum light sideways. Based on this phenomenon, it is usually better to have the common axis of the luminaires going across the longitudinal distance of the area.

The examples used earlier represent typical exterior area lighting systems, and discussions were made to involve several common situations that affect the overall layout. A perfect layout with the right luminaires can be made successfully, if the area is large and, preferably, square or rectangular. Many areas are of irregular shape, however. Each situation has to be treated on an individual basis, depending upon its requirements. When the area is irregular, the best method is to divide the total area into several square or rectangular areas and then proceed with the technique mentioned above. If the area is circular, consideration of a large source on a high pole, centrally located, should be considered. For irregular areas, where breaking down to smaller square or rectangular areas is impossible, considerations should be given to using low-wattage luminaires mounted on low-height poles. A cost and energy consumption analysis must be made in all cases.

Appendix

GLOSSARY OF LIGHTING TERMINOLOGY CONVERSION TABLES, AND ILLUMINANCE SELECTION

GLOSSARY OF LIGHTING TERMINOLOGY

This section contains lighting terminology frequently used in this book and in lighting practice. These are presented in alphabetical order and explained so that they are easy to understand. For the complete range of lighting terminology and precise definitions the reader is referred to the IES *Lighting Handbook*, 1981, Reference Volume or C42 Series of *American National Standard Definitions of Electrical Terms*.

Absorption. A general term for the process by which incident flux (light) is converted to another form of energy (ultimately to heat).

Ambient lighting. Lighting throughout an area that produces general lighting.

Baffle. A single translucent or opaque material in a luminaire that shields the source from direct view at some angles or absorbs unwanted light.

Ballast. A device used with gaseous discharge lamps that provides necessary conditions to start and operate the lamps.

Ballast factor. The percentage of light produced by a commercial ballast, compared to a reference or standard ballast.

Bare lamp. A light source without any shielding.

Bulb. A source of light operated with electricity, as distinguished from an assembled unit consisting of a light source in a housing called a *lamp*.

Candela. The metric (SI) unit of luminous intensity. Formerly "candle."

Candlepower. Luminous intensity expressed in candelas.

Candlepower distribution curve. A curve, generally polar, representing the variation of luminous intensity of a luminaire in a plane through the light center.

Cavity ratio. A number indicating cavity proportions calculated from length, width, and height.

Coefficient of utilization. The ratio of lumens (luminous flux) from a luminaire received on the work plane to the lumens produced by the lamps alone.

Coffer. A recessed panel or dome in a ceiling.

Color rendering. An expression for the effect of a light source on the color appearance of objects in comparison with their color appearance under a reference light source.

Contrast rendition factor. The ratio of visual task contrast with a given lighting environment to the contrast of sphere illumination.

Diffused lighting. Lighting provided on the work plane that does not arrive from a particular direction.

Diffuser. A device to redirect and scatter the light from a source in all directions.

Direct glare. Glare resulting from high brightness or insufficiently shielded light sources in the field of view.

Direct-indirect lighting. An expression of general diffused lighting that predominantly occurs above and under the luminaire.

Direct lighting. Lighting by luminaires distributing the maximum light in the general direction of the surface to be illuminated.

Disability glare. Glare resulting in reduced visual performance and visibility.

Discomfort glare. Glare producing discomfort that does not necessarily interfere with visual performance.

Efficacy. The ratio of lumens (luminous flux) to power consumed.

Efficiency. The ratio of lumens produced by a luminaire to that produced by its lamp alone.

Equivalent sphere illumination. The level of sphere illumination that would produce task visibility equivalent to that produced by a specific lighting environment.

Fluorescent lamp. A low-pressure mercury electric-discharge lamp in which a fluorescing coating (phosphor) transforms some of the ultraviolet energy to light.

Flush mounted luminaire. A luminaire that is mounted above the ceiling with the opening level with the surface.

Footcandle. The unit of illuminance in the classical system.

Footlambert. The unit of brightness in the classical system.

Gaseous discharge. The emission of light from gas atoms excited by an electric current.

High-intensity discharge lamp. An electric discharge lamp in which the light producing arc is stabilized by wall temperature. HID.

High mast lighting. Illumination of a large area by means of a group of luminaires mounted on the top of a high mast.

High-pressure sodium lamp. An HID lamp whose output is mainly dependent upon sodium vapor.

Illuminance. The density of the luminous flux incident on a surface; it is the quotient of the luminous flux by the area of the surface when the latter is uniformly illuminated. This unit was formerly known as *illumination*.

Illumination. The act of illuminating or state of being illuminated.

Incandescent. A lamp in which light is produced by a filament heated to intense brightness by an electric current.

Indirect lighting. Lighting by luminaires producing the majority light upwards.

Interflection. The repeated reflection of light by the room surfaces before it reaches the work plane or a specific surface of the room.

Inverse-square law. This law states that illuminance at a point on a surface varies directly with the intensity of a point source, and inversely as the square of the distance between the source and the point.

Isocandela line. A line plotted on any appropriate set of coordinates to show directions in space, around a source of light, in which the intensity is same.

Isofootcandle (isoflux) line. A line plotted on any appropriate set of coordinates to show all the points on a surface where the illuminance is the same.

Lamp. A general term for a man-made source of light that operates with electricity.

Light loss factor. A factor used in calculating illuminance mainly taking into account the light losses because of dirt accumulation, light depreciation, maintenance intervals, and temperature and voltage variations. This term formerly was known as a "maintenance factor." LLF.

Lighting effectiveness factor. The ratio of equivalent sphere illumination to illumination measured or calculated by classical method. LEF.

Low-pressure sodium. A discharge lamp in which light is produced by radiation from sodium vapor operating at a low pressure. LPS.

Lumen. Unit of luminous flux.

Luminaire. A complete lighting unit consisting of lamp(s), ballast(s), reflector(s), and the housing.

Luminous ceiling. A ceiling-area lighting system comprising a continuous surface of transmitting material of a diffusing character with light sources mounted above it.

Lux. Unit of illuminance in metric (SI) system. One lux is one lumen per square meter.

Maintenance factor. A factor formerly used for LLF. See Light loss factor.

Mercury vapor lamp. An HID lamp, whose light is mainly produced by radiation from mercury vapor.

Metal halide lamp. An HID lamp, whose light is mainly produced by radiation from metal halides.

Photoelectric cell. A device that responds electrically in a measurable manner to incident radiant energy.

Point-method. A procedure of lighting design for determining the illuminance at various locations with the help of luminaire photometric data.

Polarization. The process by which the horizontal vibrations of light waves are retained, and only the vertical vibrations are allowed to travel.

Reflection. A general term for the process by which the incident flux leaves a surface without a change in frequency.

Refractor. A device used to redirect the luminous flux from a source mainly by the process of refraction.

Starter. A device used with ballasts for the purpose of starting the lamp.

Task lighting. A special lighting arrangement that illuminates for visual tasks.

Task-ambient lighting. A combination of task lighting and ambient lighting.

Troffer. A recessed luminaire with the opening flush with the ceiling.

Veiling reflection. Regular reflection superimposed upon diffuse reflections from an object that obscure the details and reduce contrast.

Visual comfort probability. The rating of a lighting system expressed as the percentage of people who will be expected to find it acceptable in terms of discomfort glare. VCP.

Work plane. The plane at which work is usually done. In general, it is about 30 inches above the floor.

CONVERSION TABLES

Because of their continuous popularity, standard U.S. systems are used in this book. The following conversion tables to metric (SI) systems are provided to help during the transition of one system to another. In using the tables, locate the known quantity in the standard, or metric, system in the middle column and read the converted quantity in one of the two remaining columns. For instance, to determine the equivalent metric quantity of 100 fc, locate 100 in the middle column of the illuminance conversion table and read 1076 under lx. "lx" is an abbreviation of lux, which is the metric (SI) equivalent of the footcandle in standard U.S. systems. One fc is equal to 10.76 lx.

Illuminance and Length Conversion Table.

Locate known quantity in the middle column. Read the converted figure on the side column.

Illuminance			Length		
lx	lx/fc	fc	m	m/ft	ft
10.76	1	0.09	0.30	1	3.3
21.5	2	0.19	0.61	2	6.6
32.3	3	0.28	0.91	3	9.8
43.0	4	0.37	1.22	4	13.1
53.8	5	0.47	1.52	5	16.4
64.6	6	0.56	1.82	6	19.7
75.3	7	0.65	2.13	7	23.0
86.1	8	0.74	2.44	8	26.2
96.8	9	0.84	2.74	9	29.5
1076	100	9.3	30.5	100	328
1184	110	10.2	33.5	110	361
1291	120	11.1	36.6	120	394
1399	130	12.1	39.6	130	427

Illuminance and Length Conversion Table (*Continued*)

Locate known quantity in the middle column. Read the converted figure on the side column.

Illuminance			Length		
lx	lx/fc	fc	m	m/ft	ft
1506	140	13.0	42.7	140	459
1614	150	13.9	45.7	150	492
1722	160	14.9	48.8	160	525
1829	170	15.8	51.8	170	558
1937	180	16.7	54.9	180	591
2044	190	17.7	57.9	190	623
2152	200	18.6	61.0	200	656
2260	210	19.5	65.0	210	689
2367	220	20.4	67.1	220	722
2475	230	21.4	70.1	230	755
2582	240	22.3	73.2	240	787
2690	250	23.2	76.2	250	820
2798	260	24.2	79.2	260	853
2905	270	25.1	82.3	270	886
3013	280	26.0	85.3	280	919
3120	290	26.9	88.4	290	951
3228	300	27.9	91.4	300	984
3336	310	28.8	94.5	310	1017
3443	320	29.7	97.5	320	1050
3551	330	30.7	100.6	330	1083
3658	340	31.6	103.6	340	1116
3766	350	32.5	106.7	350	1148
3874	360	33.4	109.7	360	1181
3981	370	34.4	112.8	370	1214
4089	380	35.3	115.8	380	1247
4196	390	36.2	118.9	390	1280
4304	400	37.2	121.9	400	1312
4412	410	38.1	125.0	410	1345
4519	420	39.0	128.0	420	1378
4627	430	39.9	131.1	430	1411
4734	440	40.9	134.1	440	1444
4842	450	41.8	137.2	450	1476
4950	460	42.7	140.2	460	1509
5057	470	43.7	143.3	470	1542
5165	480	44.6	146.3	480	1575
5272	490	45.5	149.4	490	1608

1 fc = 10.76 lx
1 ft = 0.3048 m
1 in = 2.54 cm

Temperature Conversion Table.

Locate known temperature in °C/°F column. Read converted temperature in °C or °F column.

°C	°C/°F	°F	°C	°C/°F	°F	°C	°C/°F	°F
-45.4	-50	-58	15.5	60	140	76.5	170	338
-42.7	-45	-49	18.3	65	149	79.3	175	347
-40	-40	-40	21.1	70	158	82.1	180	356
-37.2	-35	-31	23.9	75	167	85	185	365
-34.4	-30	-22	26.6	80	176	87.6	190	374
-32.2	-25	-13	29.4	85	185	90.4	195	383
-29.4	-20	-4	32.2	90	194	93.2	200	392
-26.6	-15	5	36	95	203	96	205	401
-23.8	-10	14	37.8	100	212	98.8	210	410
-20.5	-5	23	40.5	105	221	101.6	215	419
-17.8	0	32	43.4	110	230	104.4	220	428
-15	5	41	46.1	115	239	107.2	225	437
-12.2	10	50	48.9	120	248	110	230	446
-9.4	15	59	51.6	125	257	112.8	235	455
-6.7	20	68	54.4	130	266	115.6	240	464
-3.9	25	77	57.1	136	275	118.2	245	473
-1.1	30	86	60	140	284	120.9	250	482
1.7	35	95	62.7	145	293	123.7	255	491
4.4	40	140	66.5	150	302	126.5	260	500
7.2	45	113	68.3	155	311	129.3	265	509
10	50	122	71	160	320	132.2	270	518
12.8	55	131	73.8	165	329	135	275	527

°F = (9/5 X °C) + 32.
°C = 5/9 (°F − 32).

IES Recommended Levels of Illuminance.

The following tables give a partial list of minimum, maintained levels of illuminance as recommended by the IES *Lighting Handbook*, 5th edition.

Table 1. General Lighting.

Area	fc*	dlx*
Art galleries		
General	30	32
On paintings (supplementary)	30	32
On statuary and other displays	100	110
Auditoriums		
Assembly only	15	16
Exhibitions	30	32
Social activities	5	5.4
Banks		
Lobby		
General	50	54
Writing areas	70†	75†
Tellers' stations	150†	160†
Posting and keypunch	150†	160†
Barber shops and beauty parlors	100	110
Breweries		
Brewhouse	30	32
Boiling and keg washing	30	32
Filling (bottles, cans, kegs)	50	54
Churches and synagogues		
Altar, ark, reredos	100	110
Choir and chancel	30	32
Classrooms	30†	32†
Pulpit, rostrum (supplementary)	50	54
Main worship area		
Light and medium interior finish	15	16
For churches with special zeal	30	32
Courtooms		
Seating area	30	32
Court activity area	70†	75†
Garages, automobile and truck		
Service garages		
Repairs	100	110
Active traffic area	20	22
Parking garages		
Entrance	50	54
Traffic lanes	10	11
Storage	5	5.4
Library		
Reading areas		
Reading printed material	30†	32†
Study and note taking	70†	75†
Conference areas	30†	32†
Seminar rooms	70†	75†
Book stack (30 in above floor)		
Active stacks	30	32
Inactive stacks	5	5.4
Book repair and binding	70	75
Cataloging	70†	75†
Card files	100†	110†
Carrels, individual study areas		
Circulation desks	70†	75†
Rare-book rooms, archives		
Storage areas	30	32
Reading areas	100†	110†
Map, picture, and print rooms		
Storage areas	30	32
Use areas	100†	110†
Audio-visual areas		
Preparation rooms	70	75
Viewing rooms (variable)	70	75
TV room	70	75
Audio listening areas		
General	30	32
For note taking	70†	75†
Record inspection table	100	110

IES Recommended Levels of Illumination.

The following tables give a partial list of minimum, maintained levels of illumination as recommended by the IES *Lighting Handbook*, 5th edition.

Table 1. *(Continued)*

Area	fc*	dlx*
Microform areas		
Files	70†	75†
Viewing areas	30	32
Locker rooms	20	22
Meat packing		
Slaughtering	30	32
Cleaning, cutting, cooking, grinding, canning, packing	100	110
Offices		
Drafting rooms		
Detailed drafting, designing, and cartography	200†	220†
Rough-layout drafting	150†	160†
Accounting offices		
Auditing, tabulating, bookkeeping, business-machine operation, computer operation	150†	160†
General offices		
Reading poor reproductions, business-machine operation, computer operation	150†	160†
Reading handwritting in hard pencil or poor paper, reading fair reproductions, active filing, mail sorting	100†	110†
Reading handwriting in ink or medium pencil on good-quality paper, intermittent filing	70†	75†
Private offices		
Reading poor reproductions, business-machine operation	150†	150†
Reading handwriting in hard pencil or on poor paper, reading fair reproduction	100†	110†
Reading handwriting in ink or medium pencil or good-quality paper	70†	75†
Reading high-contrast or well-printed materials	30†	32†
Conferring and interviewing	30	32
Conference rooms		
Critical seeing tasks	100†	110†
Conferring	30	32
Note taking during projection	30†	32†
Corridors	20	22
Post offices		
Lobby, on tables	30	32
Sorting, mailing, etc.	100	110
Schools		
Tasks		
Reading printed material	30†	32†
Reading pencil writing	70†	75†
Spirit duplicated material		
Good	30†	32†
Poor	100†	110†
Drafting, benchwork	100	110
Lip reading, chalkboards, sewing	150	160
Classrooms		
Art rooms	70	75
Drafting rooms	100†	110†
Home economics rooms		
Sewing	150	160
Cooking	50	54
Ironing	50	54
Sink activities	70	75
Note-taking areas	70†	75†
Laboratories	100	110
Lecture rooms		
Audience area	70†	75†
Demonstration area	150	160
Music rooms		
Simple scores	30†	32†
Advanced scores	70†	75†

IES Recommended Levels of Illumination.

The following tables give a partial list of minimum, maintained levels of illumination as recommended by the IES *Lighting Handbook*, 5th edition.

Table 1. *(Continued)*

Area	fc*	dlx*
Shops	100	110
Sight-saving rooms	150†	160†
Study halls	70†	75†
Corridors and stairways	20	22
Dormitories		
General	10	11
Reading books, magazines, newspapers	30†	32†
Study desk	70†	75†
Service stations		
Service bays	30	32
Sales room	50	54
Shelving displays	100	110
Rest rooms	15	16
Storage	5	5.4
Theaters and motion picture houses		
Auditorium		
During intermission	5	5.4
During picture	0.1	0.1
Foyer	5	5.4
Lobby	20	22
Badminton		
Tournament	30	32
Club	20	22
Recreation	10	11
Basketball		
College and professional	50	54
College intramural and high school	30	32
Recreation (outdoor)	10	11
Bowling		
Tournament		
Approaches	10	11
Lanes	20	22
Pins	50	54
Recreational		
Approaches	10	11
Lanes	10	11
Pins	30	32
Gymnasium		
Exhibitions, matches	50	54
General exercising and recreation	30	32
Assemblies	10	11
Dances	5	5.4
Locker and shower rooms	20	22
Handball		
Tournament	50	5.4
Club		
Indoor, four-wall or squash	30	32
Outdoor, two-court	20	22
Recreational		
Indoor, four-wall or squash	20	22
Outdoor, two-court	10	11
Tennis, lawn (indoor)		
Tournament	50	54
Club	20	22
Recreational	20	22
Tennis, lawn (outdoor)		
Tournament	30	32
Club	20	22
Recreational	10	11
Tennis, table		
Tournament	50	54
Club	30	32
Recreational	20	22
Volley ball		
Tournament	20	22
Recreational	10	11

*Minimum on the task at any time for young adults with normal and better than 20/30 corrected vision. Dekalux is an SI unit equal to 1.076 fc. 1 dekalux = 10 lux.
†Equivalent sphere illumination.

Index